INTERNATIONAL CENTRE FOR MECHANICAL SCIENCES

COURSES AND LECTURES - No. 288

UNILATERAL PROBLEMS IN STRUCTURAL ANALYSIS

PROCEEDINGS OF THE SECOND MEETING
ON UNILATERAL PROBLEMS IN
STRUCTURAL ANALYSIS
RAVELLO, SEPTEMBER 22-24, 1983

EDITED BY

G. DEL PIERO
UNIVERSITA' DI UDINE

F. MACERI
II UNIVERSITA' DI ROMA

SPRINGER-VERLAG WIEN GMBH

Le spese di stampa di questo volume sono in parte coperte da contributi
del Consiglio Nazionale delle Ricerche.

This volume contains 83 illustrations.

Originally published by Springer-Verlag Wien New York in 1985

ISBN 978-3-211-81859-6 ISBN 978-3-7091-2632-5 (eBook)
DOI 10.1007/978-3-7091-2632-5

PREFACE

In Nature, unilateral constraints are more frequent than bilateral; one could say that the first ones are the rule, and the second ones the exception. For this reason, unilateral problems were considered early in Mechanics. Unfortunately, their mathematical formulation involves inequalities, and this causes serious mathematical difficulties. For this reason, it became customary to develop the mechanical theories under the conventional assumption that all constraints were bilateral.

Only in recent years the development of branches of Mathematics such as Linear Programming, Convex Analysis and Variational Inequalities determined a renewed interest in Unilateral Problems. Moreover, the new solution techniques looked very promising for solving some problems in Structural Engineering, and this supplied the theory with an important field of application.

At present, Unilateral Problems appear a particularly favourable meeting point for mathematicians, mechanicians and engineers. In Italy, many specialists from these areas are now working on this subject, under a research project supported by the National Ministry of Education.

The desire of recording the progress made in the different branches of the subject, that is, proper statement of problems, characterization of solution, numerical computation techniques, induced us to organize a first workshop on Unilateral Problems. It was held in the Prescuding Valley, near Udine, in May 1982. The character of the seminar was informal, with oral communications and large time left to free discussions.

All participants agreed with the proposal of a second meeting, with enlarged participation and with publication of the presented communications. The second meeting was held in Ravello in September 1983, and was attended by about thirty-five participants, coming from seven Countries.

The present Volume collects the communications presented at the Meeting in Ravello. They deal with unilateral problems coming from various branches of Mechanics: Contact, Friction, Fracture and Fluid Mechanics. Looking at these papers, we get convinced that much of the theory and a good number of solution techniques have been well established, so that many unilateral problems arising from natural phenomena can now be properly analyzed. On the other hand, the increasing variety of the applications, and the fact that some questions are still in progress, seem to suggest that much attention should be devoted to these topics also in the future.

We wish to acknowledge gratefully the support, in resources and personnel, provided by the International Centre for Mechanical Sciences (CISM), by the University of Udine and by the Second University of Rome. The sponsorship of the Italian Association for Theoretical and Applied Mechanics (AIMETA) is also acknowledged.

Finally, we wish to express our deep appreciation to the city of Ravello and to its Major, Mr. Salvatore Sorrentino, whose warm hospitality in a magnificent environment greatly contributed to the pleasant and fruitful development of the meeting.

Gianpietro Del Piero, Franco Maceri

LIST OF PARTICIPANTS

Donato ABBRUZZESE, Istituto di Tecnica delle Costruzioni, Facoltà di Ingegneria, Piazzale Tecchio, 80125 Napoli, Italy.

Luigi ASCIONE, Dipartimento di Strutture, Università della Calabria, 87036 COSENZA, Italy.

Alessandro BARATTA, Istituto di Costruzioni, Facoltà di Architettura, Via Monteoliveto 3, 80134 Napoli, Italy.

Stefano BENNATI, Istituto di Scienza delle Costruzioni, Università di Pisa, Via Diotisalvi 2, Pisa, Italy.

Luigi BIOLZI, Istituto di Meccanica Teorica ed Applicata, Università di Udine, Viale Ungheria 43, 33100 Udine, Italy.

Franco BREZZI, Dipartimento di Meccanica Strutturale, Istituto di Analisi Numerica del C.N.R., 27100 Pavia, Italy.

Elio CABIB, Istituto di Meccanica Teorica ed Applicata, Università di Udine, Viale Ungheria 43, 33100 Udine, Italy.

Mario COMO, Istituto di Tecnica delle Costruzioni, Facoltà di Ingegneria, Piazzale Tecchio, 80125 Napoli, Italy.

Edorado COSENZA, Istituto di Tecnica delle Costruzioni, Facoltà di Ingegneria, Piazzale Tecchio, 80125 Napoli, Italy.

Alain CURNIER, Département de Mécanique, Ecole Polytechnique Fédérale de Lausanne, Ecublens ME, CH - 1015 Lausanne, Switzerland.

Gianpietro DEL PIERO, Istituto di Meccanica Teorica ed Applicata, Università di Udine, Viale Ungheria 43, 33100 Udine, Italy.

Marino DE LUCA, Istituto di Tecnologia, Università di Reggio Calabria, Via Amendola 8/b, 89100 Reggio Calabria, Italy.

Michel FREMOND, Laboratoire Central des Ponts et Chaussées, 58 Boulevard Lefebvre, 757322 Paris, France.

Ahmed FRIAA, Ecole Nationale d'Ingenieurs de Tunis, BP 37 Le Bélvédère, Tunis, Tunisia.

Antonio GRIMALDI, II Università di Roma, Via Orazio Raimondo - La Romanina, 00173 Roma, Italy.

J.J. KALKER, Department of Mathematics and Informatics, Delft University of Technology, Julianalaan 132, 2628 BL Delft, Holland.

Marzio LEMBO, Facoltà di Ingegneria, II Università di Roma, Via Orazio Raimondo - La Romanina, 00173 Roma, Italy.

Angelo LEONARDI, Facoltà di Ingegneria, II Università di Roma, Via Orazio Raimondo - La Romanina, 00173 Roma, Italy.

Aldo MACERI, Istituto di Scienza e Tecnica delle Costruzioni, Facoltà di Architettura, Viale A. Gramsci, 00100 Roma, Italy.

Franco MACERI, Facoltà di Ingegneria, II Università di Roma, Via Orazio Raimondo - La Romanina, 00173 Roma, Italy.

Luisa Donatella MARINI, Istituto di Analisi Numerica del C.N.R., Corso C. Alberto 5, 27100 Pavia, Italy.

Jean-Jacques MOREAU, Institut de Mathématiques, Université des Sciences et Techniques du Languedoc, Place Eugéne Bataillon, 34060 Montpellier, France.

Panagiotis D. PANAGIOTOPOULOS, School of Technology, Aristotelian University, Thessaloniki, Greece.

Michel POTIER-FERRY, Mécanique Théorique, Université Pierre et Marie Curie, 4 Place Jussieu, 75230 Paris, France.

J.N. REDDY, Department of Engineering Science and Mechanics, Virginia Polytechnic Institute and State University, Blacksburg, Virginia 24061, U.S.A.

Michel RAOUS, Laboratoire de Mécanique et d'Acoustique CNRS, 31, ch. Joseph-Aiguier, BP71 1327 Marseille, France.

Giovanni ROMANO, Istituto di Scienza delle Costruzioni, Facoltà di Ingegneria, Piazzale Tecchio, 80125 Napoli, Italy.

Manfredi ROMANO, Istituto di Scienza delle Costruzioni, Facoltà di Ingegneria, Viale Andrea Doria 6, 95125 Catania, Italy.

Pierre SUQUET, Mécanique des Milieux Continus, Université Montpellier II, Place E Bataillon, 34060 Montpellier, France.

Raffaele TOSCANO, Istituto di Matematica, Facoltà di Ingegneria, Via Claudio 21, 80125 Napoli, Italy.

CONTENTS

ON THE DELAMINATION PROBLEM OF TWO-LAYER PLATES

L. Ascione, D. Bruno
Dipartimento di Strutture
University of Calabria

SUMMARY: In this paper we analyze the delamination problem of a two-layer plate by means of a unilateral contact approach. The mathematical formulation of the problem is discussed and a finite element approximation is presented. Two numerical examples concerning one-dimensional and two-dimensional problems are examined. Some comparisons with analytical results are also given, which show the effectiveness of the unilateral approach.

SOMMARIO: In questo lavoro si esamina, mediante un approccio di tipo contatto unilaterale, il problema di delaminazione di pannelli compositi a due strati. Si discute la formulazione matematica del problema, di cui si presenta una approssimazione mediante elementi finiti.

I risultati numerici ottenuti riguardano un problema di delaminazione monodimensionale ed un altro bidimensionale.

Il confronto di questi risultati con soluzioni analitiche disponibili mostra l'efficienza del modello proposto.

1. INTRODUCTION

During the last years the unilateral costraint problems have been an active subject of research. From a theoretical point of view, many interesting results have been obtained by analyzing these problems in the context of the variational inequalities [1-2-3]. In particular, applications relative to the static unbonded contact of plates or beams resting on an elastic foundation can be found in [4-5-6-7].

A different class of problems relies upon the hypothesis of contact with finite bonding strenght. Such a hypothesis can be usefully utilized to model lack of adhesion [8] or delamination problems between two elastic plates. These problems are very interesting in the analysis of the composite materials, which are sensitive to the delamination phenomena.

The main purpose of the present paper is a numerical investigation on this subject, by examining, via finite elements, some one-dimensional and two-dimensional problems.

2. FORMULATION OF THE CONTACT PROBLEM WITH FINITE BONDING STRENGHT

Let us consider the equilibrium problem of a plate (Fig. 1) resting on an elastic foundation with finite bonding strength.

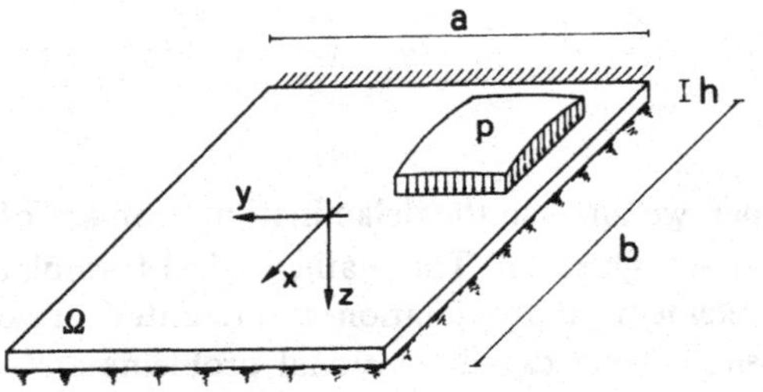

Fig. 1

We suppose that the contact is frictionless and that the boundary conditions are able to avoid rigid displacements.

The hypothesis of finite bonding strenght can be formulated by assuming that the spring reaction r is a function of the displacement w (Fig. 2) such as:

$$r(w) = \begin{cases} Kw & \text{if } w \leqslant w_0 , \\ 0 & \text{if } w > w_0 . \end{cases} \tag{2.1}$$

where K is a positive constant.

Consequently, the strain energy J of the spring is not convex (Fig. 2b):

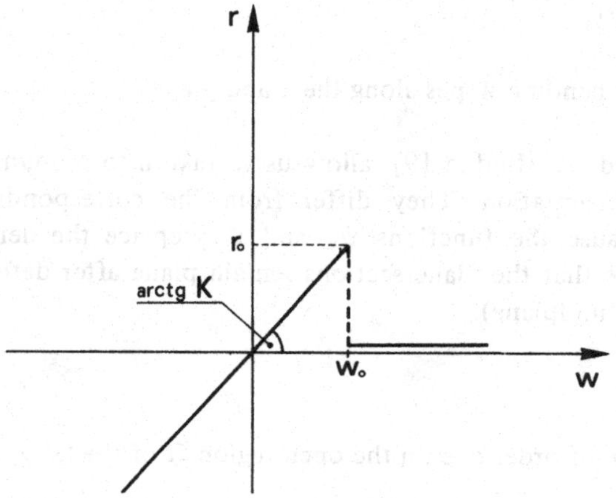

Fig. 2a - Spring response.

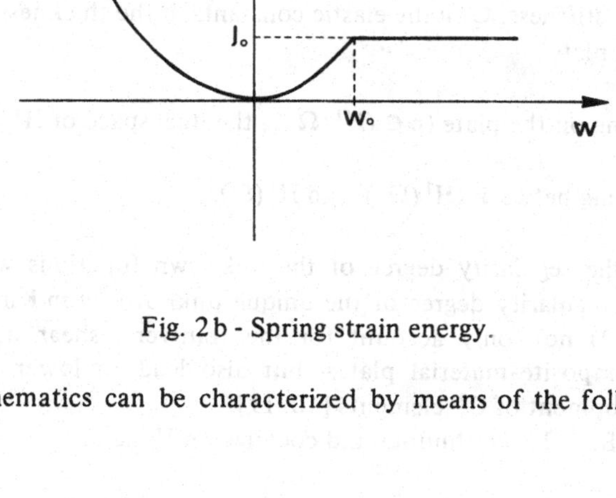

Fig. 2b - Spring strain energy.

The plate kinematics can be characterized by means of the following displacement field components:

$$u_x(x,y,z) = -z\,\psi_x(x,y), \tag{2.2a}$$

$$u_y(x,y,z) = -z\,\psi_y(x,y), \tag{2.2b}$$

$$u_z(x,y,z) = w(x,y), \tag{2.2c}$$

where ψ_x and ψ_y are the bending slopes along the x and y axes.

Eqs. (2.2), proposed by Mindlin [9], allow us to take into account the shear stress effects on the plate deformation. They differ from the corresponding equations of Kirchhoff's theory, because the functions ψ_x and ψ_y replace the derivatives $w_{,x}$ and $w_{,y}$ of w (i.e. we assume that the plane sections remain plane after deformation but not necessarily normal to the midplane).

Set:

$H^1(\Omega)$: the Sobolev space of order one on the open region Ω of the (x, y) plane,

$V = H^1(\Omega) \times H^1(\Omega) \times H^1(\Omega)$: the space of the admissible dispacements (w, ψ_x, ψ_y),

$B[(w, \psi_x, \psi_y), (u, \phi_x, \phi_y)]$: the bilinear form on V associated with the strain energy of the plate

$$B[(w,\psi_x,\psi_y),(u,\phi_x,\phi_y)] =$$
$$D\iint_\Omega [\psi_{x,x}\phi_{x,x} + \psi_{y,y}\phi_{y,y} + \nu(\psi_{x,x}\phi_{y,y} + \psi_{y,y}\phi_{x,x}) + \frac{1-\nu}{2}(\psi_{x,y} + \psi_{y,x})\cdot$$
$$\cdot(\phi_{x,y} + \phi_{y,x})]\,d\Omega + \chi G h \iint_\Omega [(w_{,x} - \psi_x)(u_{,x} - \phi_x) +$$
$$+ (w_{,y} - \psi_y)(u_{,y} - \phi_y)]\,d\Omega, \tag{2.3}$$

where D is the flexural stiffness, G, ν the elastic constants, h the thickness and χ the shear correction factor of the plate,

p: the vertical load acting on the plate ($p \in (H^1(\Omega))'$, the dual space of $H^1(\Omega)$),

$<.,.>$: the duality pairing between $(H^1(\Omega))'$ and $H^1(\Omega)$.

We observe that the regularity degree of the unknown functions w, ψ_x, ψ_y in eq. (2.2) is lower than the regularity degree of the unique unknown w in Kirchhoff's theory. Consequently, eqs. (2.2) not only account for the transverse shear strains, which are quite significant in composite-material plates, but also lead to lower order equations that facilitate the development of C°-elements [10-11].

The bilinear form $B[\cdot,\cdot]$ is continuous and coercive on V, i.e.:

$$B[(w, \psi_x, \psi_y), (u, \phi_x, \phi_y)] \leq c \, \|(w, \psi_x, \psi_y)\|_V \, \|(u, \phi_x, \phi_y)\|_V \,, \tag{2.4a}$$

$$c \geq 0, \; \forall (w, \psi_x, \psi_y), (u, \phi_x, \phi_y) \in V,$$

$$B[(w, \psi_x, \psi_y), (w, \psi_x, \psi_y)] \geq c' \, \|(w, \psi_x, \psi_y)\|_V^2 \,, \tag{2.4b}$$

$$c' \geq 0, \; \forall (w, \psi_x, \psi_y) \in V,$$

where $\| \cdot \|_V$ is the norm on the space V.

After these preliminaries, the equilibrium problem in Fig. 1 can be put in the form (virtual work equation):

"Find $(w, \psi_x, \psi_y) \in V$ such that:

$$B[(w, \psi_x, \psi_y), (u, \phi_x, \phi_y)] - \langle p - r, u \rangle = 0 \tag{2.5}$$

$$\forall (u, \phi_x, \phi_y) \in V."$$

We observe that the reaction $r \in (H^1(\Omega))'$ cannot be characterized as differential of the spring strain energy J, because this functional is not differentiable. Consequently, eq. (2.5) cannot be viewed as the stationary condition in the minimum problem:

$$\text{Min } \mathcal{E}(w, \psi_x, \psi_y), \qquad (w, \psi_x, \psi_y) \in V, \tag{2.6}$$

involving the potential energy functional $\mathcal{E}$ of the elastic system of Fig. 1 (plate and elastic foundation):

$$\mathcal{E}(w, \psi_x, \psi_y) = \frac{1}{2} B[(w, \psi_x, \psi_y), (w, \psi_x, \psi_y)] + \frac{1}{2} \langle r(w), w \rangle - \langle p, w \rangle \tag{2.7}$$

As well known, the variational problem (2.6) is basic in discussing existence and uniqueness problems in Elasticity.

In order to overcome the difficulty relative to the not differentiability of $\mathcal{E}$ a regularization of the original problem can be introduced. More precisely, set:

$$r_\epsilon(w) = \begin{cases} Kw & \text{if } w \leq w_0 , \\ Kw_0 - \dfrac{1}{\epsilon}(w - w_0) & \text{if } w_0 < w \leq w_\epsilon , \\ 0 & \text{if } w > w_\epsilon , \end{cases} \tag{2.8}$$

where $w_\epsilon = (\epsilon k + 1) w_0$, and ϵ is a positive constant (Fig. 3).

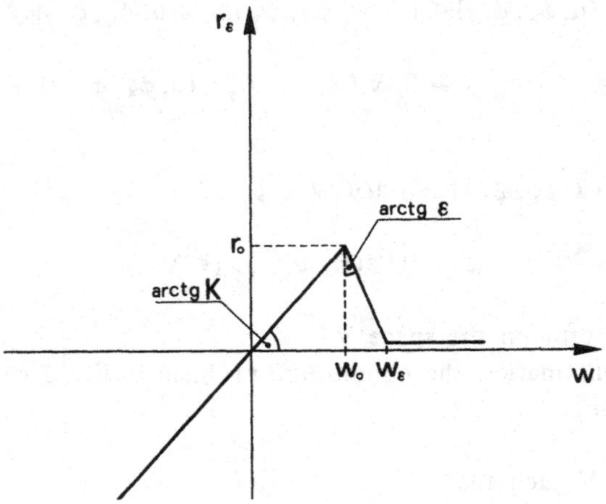

Fig. 3 a

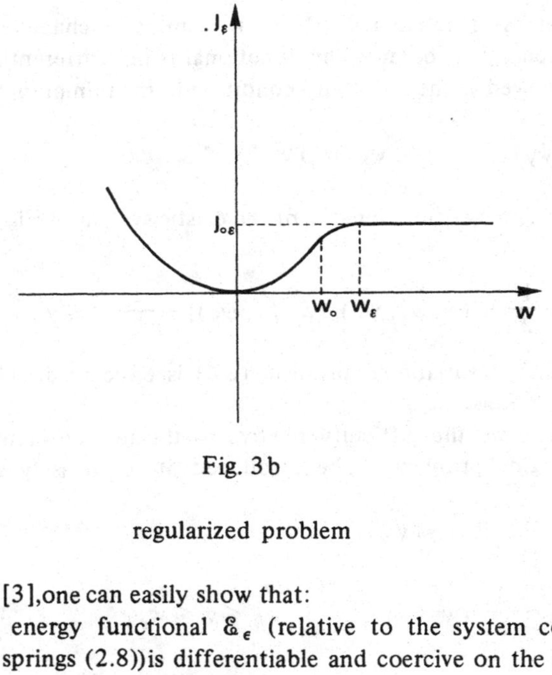

Fig. 3 b

regularized problem

By standard arguments [3],one can easily show that:

i) The potential energy functional $\mathcal{E}_\epsilon$ (relative to the system composed by the plate and the springs (2.8))is differentiable and coercive on the space V; further on, it is weakly lower semi-continuous.

ii) Eq. (2.5), written in terms of the regularized reaction r_ϵ, can be characterized as the stationary condition of the minimum problem:

$$\text{Min } \mathcal{E}_\epsilon (w, \psi_x, \psi_y), \qquad (w, \psi_x, \psi_y) \in V \tag{2.9}$$

iii) There exists an element $(\bar{w}, \bar{\psi}_x, \bar{\psi}_y)_\epsilon \in V$ which minimizes $\mathcal{E}_\epsilon$ on V.
iv) There exists a subsequence of $\{(\bar{w}, \bar{\psi}_x, \bar{\psi}_y)_\epsilon\}$ which converges weakly, as $\epsilon \to o$, to an element $(\bar{w}, \bar{\psi}_x, \bar{\psi}_y) \in V$ satisfying eq. (2.5).

The coerciveness and the weakly lower semi-continuity properties of the functional $\mathcal{E}_\epsilon$ ensure the existence of a minimum point of this functional on the space V [12]. The uniqueness cannot be proven in general, because the functional $\mathcal{E}_\epsilon$ is not convex, due to the presence of the term J_ϵ corresponding to the springs strain energy (Fig. 3 b).

3. FINITE ELEMENT APPROXIMATION

In this section we present a finite element model of problem (2.5) and a simple iterative scheme that can be utilized to get a solution of the discrete problem. The plate is discretized by means of four-node isoparametric elements and the unknown functions are interpolated as:

$$w(x, y) = \sum_{i=1}^{N_G} w_i f_i(x, y) , \tag{3.1a}$$

$$\psi_x (x, y) = \sum_{i=1}^{N_G} \psi_{xi} f_i(x, y) , \tag{3.1b}$$

$$\psi_y (x, y) = \sum_{i=1}^{N_G} \psi_{yi} f_i (x, y), \tag{3.1c}$$

where w_i, ψ_{xi}, ψ_{yi} are the values of the unknown functions in the N_G global nodes of the mesh and $f_i (x, y)$ are the global interpolants [11].

By substituting eqs. (3.1) into eq. (2.5), it is easy to get the following equations set for the discrete problem:

$$\underset{\sim}{K}^{(w,w)} \underset{\sim}{w} + \underset{\sim}{K}^{(w.x)} \underset{\sim}{\psi}_x + \underset{\sim}{K}^{(w,y)} \underset{\sim}{\psi}_y = \underset{\sim}{q} , \tag{3.2a}$$

$$\underset{\sim}{K}^{(x,w)} \underset{\sim}{w} + \underset{\sim}{K}^{(x,x)} \underset{\sim}{\psi}_x + \underset{\sim}{K}^{(x,y)} \underset{\sim}{\psi}_y = \underset{\sim}{0} , \tag{3.2b}$$

$$\underset{\sim}{K}^{(y,w)} \underset{\sim}{w} + \underset{\sim}{K}^{(y,x)} \underset{\sim}{\psi}_x + \underset{\sim}{K}^{(y,y)} \underset{\sim}{\psi}_y = \underset{\sim}{0} , \tag{3.2c}$$

where $\underset{\sim}{w}$, $\underset{\sim}{\psi}_x$ and $\underset{\sim}{\psi}_y$ are the vectors of the global nodal values w_i, ψ_{xi}, ψ_{yi} $(i = 1, 2, \ldots N_G)$ and the matrices $\underset{\sim}{K}^{(\cdot,\cdot)}$ and $\underset{\sim}{q}$ have components $(i, j = 1, 2, \ldots N_G)$:

$$K_{ij}^{(w,w)} = \bar{K}_{ij}^{(w,w)} + \tilde{K}_{ij}^{(w,w)} , \tag{3.3}$$

$$\bar{K}_{ij}^{(w,w)} = \chi Gh \left[\iint_\Omega f_{i,x} f_{j,x} \, d\Omega + \iint_\Omega f_{i,y} f_{j,y} \, d\Omega \right], \tag{3.4a}$$

$$\tilde{K}_{ij}^{(w,w)} = \sum_{e=1}^{N_e} \sum_{\gamma_e=1}^{G_e} K P_{\gamma_e} f_i(x_{\gamma_e}, y_{\gamma_e}) \, f_j(x_{\gamma_e}, y_{\gamma_e}), \tag{3.4b}$$

$$K_{ij}^{(w,x)} = K_{ji}^{(x,w)} = -\chi Gh \iint_\Omega f_{i,x} f_j \, d\Omega, \tag{3.4c}$$

$$K_{ij}^{(w,y)} = K_{ji}^{(y,w)} = \chi Gh \iint_\Omega f_{i,y} f_j \, d\Omega, \tag{3.4d}$$

$$K_{ij}^{(x,y)} = K_{ij}^{(y,x)} = D \left[\nu \iint_\Omega f_{i,x} f_{j,y} \, d\Omega + \frac{1-\nu}{2} \iint_\Omega f_{i,y} f_{j,x} \, d\Omega \right], \tag{3.4e}$$

$$K_{ij}^{(x,x)} = D \left[\iint_\Omega f_{i,x} f_{j,x} \, d\Omega + \frac{1-\nu}{2} \iint_\Omega f_{i,y} f_{j,x} \, d\Omega \right] + \chi Gh \iint_\Omega f_i f_j \, d\Omega, \tag{3.4f}$$

$$K_{ij}^{(y,y)} = D \left[\iint_\Omega f_{i,y} f_{j,y} \, d\Omega + \frac{1-\nu}{2} \iint_\Omega f_{i,x} f_{j,y} \, d\Omega \right] + \chi Gh \iint_\Omega f_i f_j \, d\Omega, \tag{3.4g}$$

$$q_i = \iint_\Omega p \, f_i \, d\Omega. \tag{3.4h}$$

In eq. (3.4b) the sum is extended over all elements N_e and over all Gauss points G_e of each element e. In the same equation the coefficient P_{γ_e} are defined as:

$$P_{\gamma_e} = \begin{cases} W_{\gamma_e} \text{ (Gaussian weight relative to the point } (x_{\gamma_e}, y_{\gamma_e})) \text{ if} \\ \sum_{i=1}^{N_G} w_i f_i(x_{\gamma_e}, y_{\gamma_e}) < w_0, \\ 0 \quad \text{if } \sum_{i=1}^{N_G} w_i f_i(x_{\gamma_e}, y_{\gamma_e}) \geq w_0. \end{cases} \tag{3.5}$$

Eqs. (3.2) can be solved by means of the following iterative procedure:

$$\underset{\sim}{K}^{(w,w)} (\underset{\sim}{w}^{(i-1)}) \, \underset{\sim}{w}^{(i)} + \underset{\sim}{K}^{(w,x)} \, \underset{\sim}{\psi}_x^{(i)} + \underset{\sim}{K}^{(w,y)} \, \underset{\sim}{\psi}_y^{(i)} = \underset{\sim}{q}, \tag{3.6a}$$

$$\underset{\sim}{K}^{(x,w)} \, \underset{\sim}{w}^{(i)} + \underset{\sim}{K}^{(x,x)} \, \underset{\sim}{\psi}_x^{(i)} + \underset{\sim}{K}^{(x,y)} \, \underset{\sim}{\psi}_y^{(i)} = \underset{\sim}{0}, \tag{3.6b}$$

$$\underset{\sim}{K}^{(y,w)} \, \underset{\sim}{w}^{(i)} + \underset{\sim}{K}^{(y,x)} \, \underset{\sim}{\psi}^{(i)} + \underset{\sim}{K}^{(y,y)} \, \underset{\sim}{\psi}_y^{(i)} = \underset{\sim}{0}, \tag{3.6c}$$

i.e. at the i-th step we solve the system (3.6) by evaluating the matrix $\tilde{K}^{(w,w)}$ from eq. (3.4b) for $\underset{\sim}{w} = \underset{\sim}{w}^{(i-1)}$. The first step corresponds to the bonded contact solution.

4. NUMERICAL RESULTS

The numerical applications, presented here, concern the delamination problem of a two-layer plate. In particular, two examples are examined, corresponding to the cases of a narrow plate and of a square plate.

4.1. One-dimensional example

Firstly, let us consider the case of two narrow plates of lenght a and width b ($a/b \gg 1$), welded to each other. Further on, let us assume that a portion of the weld, corresponding to the segment $[0, l_0]$, is defective (i.e. not welded at all). Our goal is to investigate the load-deflection relationship, when the plate is loaded as shown in Fig. 4. The problem can be solved either by fracture mechanics approach [13] (mode I of fracture) or by the unilateral approach.

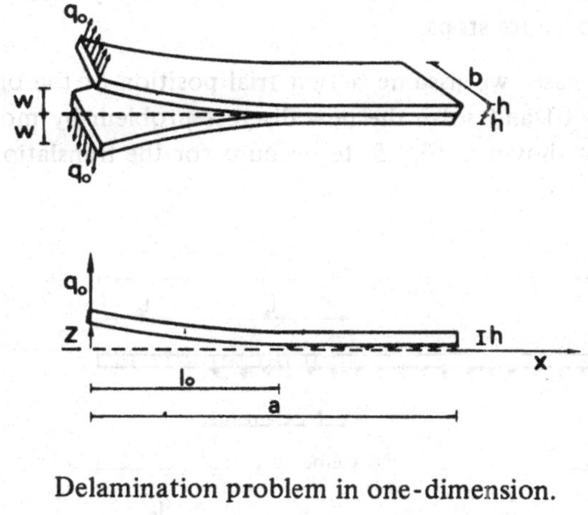

Delamination problem in one-dimension.

Fig. 4

In the latter one the constraint between the two plates is modeled by means of elastic springs (Fig. 4) of the type presented in Sec. 2.

If K denotes the spring stiffness, the limit deflection w_0 in the constitutive law (2.1) is defined by means of following equation:

$$\frac{1}{2} K w_0^2 = \gamma \,, \tag{4.1}$$

where γ is the surface energy per unit area (of opening) depending on the nature of the material.

It is easy to prove [14] that, when $k \to \infty$, the unilateral solution yields the fracture mechanics solution.

One of the more delicate computational aspect in this approach is related to the approximation of the beam wave lenghts, which become progressively smaller in the limit as $k \to \infty$. The greatest accuracy is evidently required in the neighbourhood of the opening front, which is unknown.

In order to deal with dimensionless quantities, we put:

$$L = \sqrt{2D/\gamma}\ , \qquad Q = q_0/\gamma, \qquad W = \frac{3}{2}\, w/L, \qquad \xi = l/L \qquad (4.2)$$

The problem can be solved by the following iterative scheme:

i) We start from the initial position ξ_0 of the opening front and discretize the beam as shown in Fig. 5. The finite element mesh refinement is limited to a beam segment of length l_1 (approximately equal to 5λ, where $\lambda = (4D/k)^{1/4}$).

ii) We solve the discrete contact problem by using the algorithm of Sec. 3.

iii) We check if delamination occurred. If $w \leqslant w_0$ everywhere (no delamination occurred) the procedure stops.

iv) In the opposite case, we assume a first trial position of the opening front $\overline{\xi}_0^{(1)} = \xi_0 + \Delta\xi$ $(\Delta\xi > 0)$ and solve the new discrete problem by modifying the finite element mesh, as shown in Fig. 5, to account for the translation $\Delta\xi$ of the opening front.

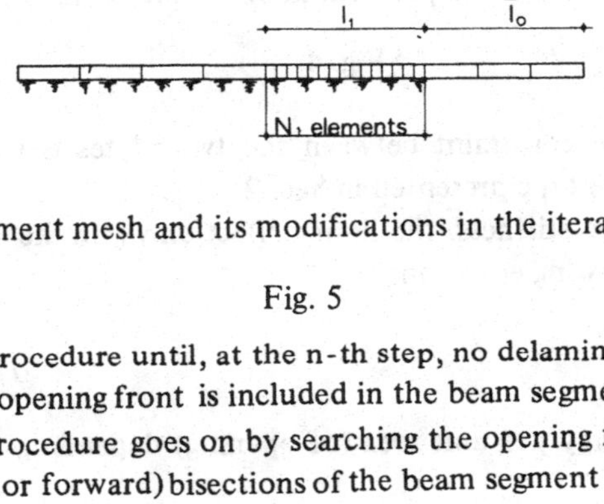

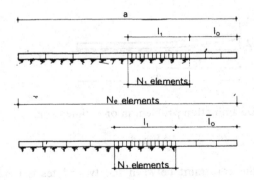

(Finite element mesh and its modifications in the iterative scheme).

Fig. 5

v) We repeat the procedure until, at the n-th step, no delamination occurs. The actual position of the opening front is included in the beam segment $[\overline{\xi}_0^{(n-1)}, \overline{\xi}_0^{(n)}]$.

vi) The iterative procedure goes on by searching the opening front position by successive (backward or forward) bisections of the beam segment $[\overline{\xi}_0^{(n-1)}, \overline{\xi}_0^{(n)}]$.

At each step the finite element mesh in modified as previously shown. The procedure stops when the distance between two successive trial positions of the opening front is smaller than an assigned tolerance.

Tab. 1 shows some numerical results obtained by the previous procedure. To test the influence of the mesh size, two beam discretizations have been examined, corresponding to $N_e = 10$ ($N_1 = 15$) and $N_e = 40$ ($N_1 = 30$) elements respectively. A comparison of these results with the theoretical ones, obtained via fracture mechanics approach, is also given. We observe a good convergence of the finite element approximation as well as a good agreement between theoretical and numerical results.

Finally, plots of Q versus W and ξ versus W are shown in Fig. 6.

	Analytical solution (fracture analysis)		Finite element solution			
			$N_e = 20$ ($N_1 = 15$)		$N_e = 40$ ($N_1 = 30$)	
W	ξ	Q	ξ	Q	ξ	Q
0.5	0.783	1.04	0.783	0.987	0.783	0.984
1	1	1	0.981	1.087	0.987	0.989
5	2.236	0.447	2.218	0.420	2.223	0.419
10	3.160	0.316	3.122	0.304	3.153	0.307

Tab 1 ($a/\lambda = 2 \times 10^3$, $a/h = 10^5$, $a/b = 10$, $a/L = 5.22$, $\nu = 0.3$, $\xi_0 = 0.783$)

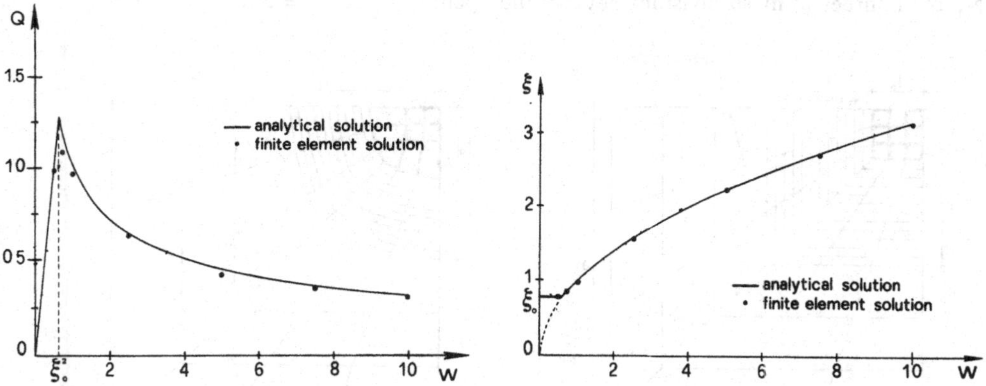

(W dimensionless beam deflection; Q dimensionless load;
ξ dimensionless lenght of the delaminated beam segment).

Fig. 6

4.2. Two-dimensional example

The unilateral contact approach is more convenient to analyze the two-dimensional delamination. Indeed, the fracture mechanics approach is more complicated when the delamination is two-dimensional. Therefore, the second example, that we present, concerns the delamination of a two-layer square plate of side a (Fig. 7).

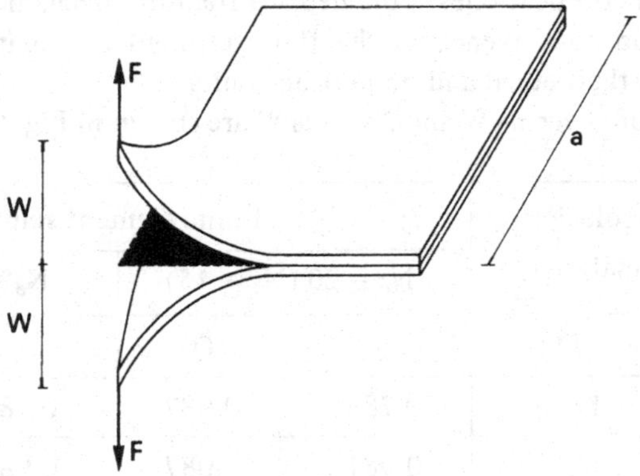

Fig. 7 - Delamination of a two-layer square plate.

The initially defective portion of the weld is a square of side l_0, as shown in Fig. 8. In the same figure the finite element mesh that we utilize to get the numerical results is depicted. We denote by N_r the number of involved vector radii ($N_r = 7$ in Fig. 8) and by N_1 the number of mesh divisions beyond the opening front ($l_1 \cong 5\,\lambda$).

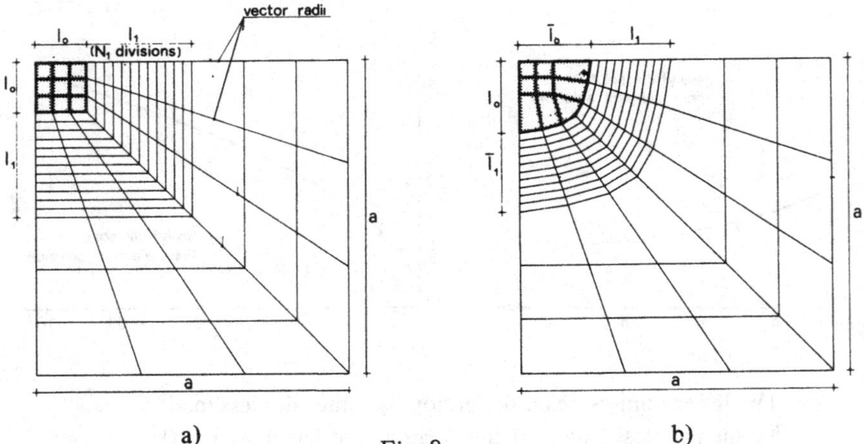

a) Fig. 8 b)

Finite element mesh. Modification of the finite element mesh in the iterative scheme.

We adopt an iterative scheme similar to the previously described one by assigning successive trial positions of the opening front and modyifing the finite element mesh as shown in Fig. 8b.

Fig. 9 gives some numerical results relative to the load-deflection relation and to the shape of the opening front ($N_r = 9$, $N_1 = 10$): the quantities W, L, λ, γ are defined as in Sec. 4.1 and $Q = F/\gamma L$. Finally, a comparison between numerical results corresponding to different rations a/λ is given in Tab. 2, while the influence of the mesh is shown in Tab. 3.

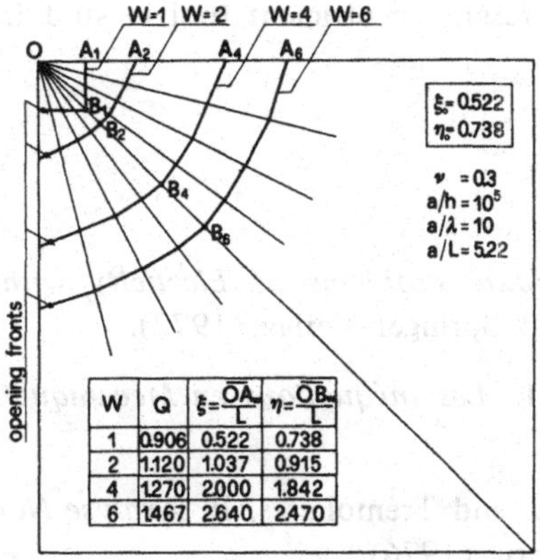

Fig. 9
(Evolution of the opening front).

N_r	Q	ξ	η
7	1.175	1.023	0.974
9	1.120	1.037	0.915

Tab. 2 ($a/\lambda = 10$, $W = 2$)

a/λ	Q	ξ	η
10	1.120	1.037	0.915
20	1.379	1.310	1.325
40	2.090	1.391	1.650
100	3.462	1.330	2.181

Tab. 3 ($N_r = 9$, $W = 2$)

5. CONCLUSIONS

A unilateral contact approach for the delamination problem of a two-layer plate has been analyzed.

Two numerical examples have been examined via finite elements; the first one corresponding to a one-dimensional problem, the second one to a two-dimensional problem. In the first case some analytical results are available in the context of the fracture mechanics: a good agreement between these results and the present ones is found.

In two-dimensional cases the fracture mechanics approach is difficult to apply, and the proposed analysis can represent an efficient tool in studying such kind of problems.

REFERENCES

1. Fichera, G. *Boundary - Value Problems of Elasticity with Unilateral Constraints,* Enc. of Physics, Vol. VIa/2, Springer-Verlag, (1972).

2. Duvaut, G. and Lions, J.L. *Les Inéquations en Mécanique et en Physique,* Dunod, Paris, (1972).

3. Glowinski, R., Lions, J.L. and Trémolieres, R. *Analyse Numérique des Inéquations Variationnelles,* Dunod, Paris, (1976).

4. Toscano, R. and Maceri, A., On the problem of the elastic plate on one-side foundation, *Meccanica,* Vol. 15, No. 2, 95-106, (1980).

5. Ascione, L., Grimaldi, A. and Maceri, F., Modeling and analysis of beams on tensionless foundations, *Int. J. of Modelling and Simulation,* Vol. 3, No. 2, (1983).

6. Ascione, L. and Grimaldi, A., Unilateral contact between a plate and an elastic foundation, *Meccanica,* to appear, (1983).

7. Panagiotopoulos, P.D. and Talaslidis, D., A linear analysis approach to the solution of certain classes of variational inequality problems in structural analysis, *Int. J. Solids Structures,* Vol. 16, 991-1005, (1980).

8. Frémond, A., Adhésion de solides élastiques, Lecture delivered at the *First Symposium on Unilateral Problems in Mechanics,* CISM - Udine, Italy, May 1982.

9. Mindlin, R.D. Influence of rotatory inertia and shear on flexural motions of isotropic elastic plates, *J. Appl. Mech.*, vol. 18, 31-88, (1951).

10. Reddy, J.N., A penalty plate-bending element for the analysis of laminated anisotropic composite plates, *Int. J. Numer. Meth. Eng.*, Vol. 15, pp. 1187-1206, (1980).

11. Reddy, J.N., *An Introduction to the Finite Element Method,* Mc Graw-Hill, New York, (1983).

12. Oden, J.T. and Reddy, J.N., *Variational Methods in Theoretical Mechanics*, Springer-Verlag, (1976).

13. Bunidge, R. and Keller, J.B., Peeling, slipping, and cracking – some one dimensional free - boundary problems in mechanics, *SIAM Review*, Vol. 20, No. 1, (1978).

14. Grimaldi, A. and Reddy, J.N., On delamination in plates: a unilateral-contact approach, *This Meeting*, (1983).

ERROR ESTIMATES IN THE APPROXIMATION OF A FREE BOUNDARY

F. Brezzi
Dipartimento di Meccanica Strutturale
and Istituto di Analisi Numerica del C.N.R.
Pavia

Introduction

Free boundary problems appear in a great variety of physical problems. See for instance [1,2,3] for the mathematical setting of many examples. We shall consider here a model case in which the free boundary is the line (or, more generally, the (n-1) dimensional surface) that splits the domain (say, D) where the solution (say, u) is sought, into $D = D^+ \cup D^0$ so that $u>0$ in D^+ and $u \equiv 0$ in D^0. Many free boundary problems can be reduced to this case. If the free boundary is smooth and u is also smooth separately in each subdomain D^0 and D^+, then the <u>global regularity</u> of u will depend on the behaviour of u, in D^+, near the free boundary: for instance, if u behaves like $(d(x))^{k+\alpha}$ (k integer and nonnegative, $0<\alpha\leq 1$, d(x) = dis-

Research supported by M.P.I. 40%.

tance of x from the free boundary) then $u \in C^{k+\alpha}(D)$. Assume now that we have got, by means of some discretization process, a sequence of discrete solutions $\{u_h\}_h$ which converges to u in the L^∞ norm. We set

$$e(h) := \|u-u_h\|_{L^\infty} \to 0 \text{ for } h \to 0 \tag{1.1}$$

As a general matter, when u is (globally) more regular one expects better estimates for e(h). We try here to estimate the errors between continuous and discrete free boundaries F, F_h in terms of powers of e(h):

$$\text{dist}(F,F_h) \leq (e(h))^s. \tag{1.2}$$

Surprisingly enough, under reasonable assumptions on u_h, we see that the exponent s in (1.2) behaves as

$$s = (k+\alpha)^{-1} \tag{1.3}$$

for $u \in C^{k,\alpha}(D)$, that is: for a more regular u we have a worse s, and hence a worse estimate in (1.2). Assuming, as it is reasonable, that e(h) in (1.1) is like $h^{k+\alpha}$, we have from (1.2) and (1.3) that the error in the free boundaries goes like h, regardless to the global regularity of u (to tell the truth: in the concrete examples we have some $|\log h|$ factor to make the estimate worse!). The present paper, that is a sort of philosophy out of [4], makes the above statements precise. For the sake of simplicity, we do not consider a definite problem or class of problems, but we assume that we are given a solution u of an unspecified problem and a sequence $\{u_h\}_h$ of discrete solutions having suitable properties. An example of problems where our assumptions hold is quickly sketched at the end, referring to [4] for more details.

Main results

Let D be an open domain in $\mathbb{R}^n$ and let u(x) be continuous on $\bar{D}$ and nonnegative. We set

$$D^+ = \{x | x \in D,\ u(x) > 0\},$$
$$D^0 = \{x | x \in \bar{D},\ u(x) = 0\},$$
$$F = D^0 \cap (\partial D^+),$$
$$d(x) = \text{dist}(x,F).$$

Definition 1 - We say that u has the growth property of order r(r real $\geq$ 0) if $u(x) \simeq (d(x))^r$ near F, that is

$$\exists c_1,c_2,c_3>0 \text{ s.t. } c_1(d(x))^r<u(x)<c_2(d(x))^r \quad \forall x \in D^+ \text{ with } d(x)\leq c_3. \tag{2.1}$$

We shall note, for $\varepsilon > 0$ and for any closed subset $A \subset \bar{D}$:

$$B_\varepsilon(A):=\{x | x \in \bar{D},\ \text{dist}(x,A)\leq\varepsilon\} \tag{2.2}$$

Proposition 1 - If u has the growth property of order r, then $\exists \varepsilon_1>0$, $c_4>0$ such that $\forall\ \varepsilon$, $0<\varepsilon<\varepsilon_1$,

$$\{x | x \in \bar{D},\ 0<u(x) <c_4\varepsilon^r\} \subset B_\varepsilon(F). \tag{2.3}$$

Proof - From (2.1) we have

$$c_1d^r < u < c_2d^r \qquad \text{in } D^+ \cap B_{c_3}(F). \tag{2.4}$$

Let m be the minimum value of u in $\overline{D^+ \setminus B_{c_3}(F)}$; clearly m>0 (we are away from F!). Let $\varepsilon_1>0$ be such that $c_2\varepsilon_1^r<m$. Now, for $0<\varepsilon<\varepsilon_1$ let $\bar{x}$ be a point in D^+ where $0<u(\bar{x})<c_1\varepsilon^r$; we have $u(x)<c_1\varepsilon^r<c_1\varepsilon_1^r<c_2\varepsilon_1^r<m$. Since m is the minimum of u in $D^+ \setminus B_{c_3}(F)$, this implies that $\bar{x} \in B_{c_3}(F) \cap D^+$ and hence (2.4) gives $c_1d^r < u$ at $\bar{x}$. Therefore

$c_1(d(\bar{x}))^r<u(\bar{x})<c_1\varepsilon^r$ which implies $d(\bar{x})<\varepsilon$, that is $\bar{x}\in B_\varepsilon(F)$. Hence (2.3) holds with $c_4=c_1$.

Assume now that we are given a sequence $\{u_h\}_{h>0}$ of nonnegative functions in $C^0(\bar{D})$ and define, for any $h>0$ the sets D_h^+, D_h^0 and F_h as above. It would be nice, now, to assume that u_h has a growth property similar to (2.1). A first difficulty is given by the fact that, in the applications, u_h will often be, say, piecewise linear, so that (2.1) could only hold with $r=1$. It seems then natural to assume that, for any $h>0$, we are given special points in $\bar{D}$ that we call <u>nodes</u>. We suppose that for each node P there is another node Q at a distance $|P-Q|\le h$. Finally we assume that u_h is characterized by its values at the nodes. A second, major difficulty, is due to the fact that we do not want to assume that F_h is "smooth" (whatever the meaning you are willing to give it), and a property like (2.1) says little for a nasty $\{F_h\}$. For instance, if F_h has oscillations like $\sin(x/h)$ a result analogous to (2.3) will never be true with constants independent of h. It will then be more useful to assume a different property, which is somehow related to (2.1) but not equivalent.

<u>Definition 2</u> - <u>We say that the sequence</u> $\{u_h\}$ <u>has uniformly the ball property of order</u> r (r real >0) <u>at</u> F_h <u>if there exist</u> $c_5>0$ <u>and, for all</u> $h>0$, $\rho_0(h)$, $\rho_1(h)$ ($\rho_0<\rho_1$) <u>such that</u>

$$\begin{cases} \forall \text{ node } P\in F_h \quad \forall \rho\in(\rho_0(h),\rho_1(h)) \quad \exists \text{ node } Q \text{ such that} \\ |P-Q|\le\rho \text{ and } u_h(Q)\ge c_5\,\rho^r. \end{cases} \tag{2.5}$$

Note that (2.5) still says, somehow, that u_h leaves its free boundary with a speed which is at least as d^r.

We are now able to prove an error estimate for the distance between the free boundaries.

<u>Theorem 1</u> - <u>Assume that</u> u <u>has the growth property of order</u> r <u>and that</u>

$\{u_h\}$ has uniformly the ball property of order r at F_h. Let $e(h)$ be given by (1.1) and assume that $\exists c_6>0$ such that $\rho_0(h)\leq(e(h)/c_6)^{1/r}$ and that $c_5(\rho_1(h))^r > e(h)$ for h small enough. Then there exists $h_0>0$ and $c_0>0$ such that

$$\forall\, h \leq h_0 \quad \forall\, \varepsilon \geq c_0\,(e(h))^{1/r} \qquad F_h \subseteq B_\varepsilon(F). \tag{2.6}$$

Proof. Assume $P \in F_h$ to be a node, and let us prove that

$$\varepsilon \geq c_0 e^{1/r} \Rightarrow P \in B_\varepsilon(F) \text{ (that is: } \mathrm{dist}(P,F)\leq\varepsilon) \tag{2.7}$$

Suppose first that $P \in D^+$; then $u(P)>0$ and since $u_h(P)=0$ we have $u(P)\leq e(h)$. For h small enough, $e(h)\leq c_4\varepsilon_1^r$ (c_4 and ε_1 given in Prop. 1). Hence for $(e/c_4)^{1/r}<\varepsilon<\varepsilon_1$ we have $0<u(P)<e<c_4\varepsilon^r$ which implies (2.7) by Proposition 1. Suppose now that $P \in D^0$ and set $\rho = d(P) = \mathrm{dist}(P,F)$. If $\rho<\rho_0(h)$ then (2.7) holds for $\varepsilon\geq\rho_0(h)$. If $\rho>\rho_1(h)$, then (2.5) implies the existence of Q such that $|P-Q|\leq\rho_1(h)$ and $u_h(Q)\geq c_5(\rho_1(h)^r)$. Clearly $Q \in D^0$, because $\mathrm{dist}(P,Q)\leq\rho_1(h)<\rho=\mathrm{dist}(P,F)$. Hence $u_h(Q)\geq c_5(\rho_1(h))^r>e(h)$ is contradictory for h small enough. If finally $\rho \in (\rho_0(h),\ \rho_1(h))$, then (2.5) implies the existence of Q such that $|P-Q|\leq\rho$ and $u_h(Q)\geq c_5\rho^r$. Clearly $Q \in D^0$ (because $|P-Q|\leq\mathrm{dist}(P,F)$) so that $u_h(Q) = u_h(Q)-u(Q)\leq e(h)$; hence $e\geq c_5\rho^r$, that is, $\mathrm{dist}(P,F)=\rho\leq(e/c_5)^{1/r}$. Combining all the cases one sees that (2.6) holds with $c_0 = [\max(c_4,\ c_6,\ c_5)]^{-1/r}$.

Remark - It has to be pointed out that (2.6) does not imply that $F_h\neq\emptyset$. When both D^+ and D^0 have positive measure, one can ensure the existence of a non empty free boundary by requiring (2.5) also for the points P where $u_h(P)>0$ or, at least, where $0<u_h\leq e(h)$. Actually, if $D^+\neq\emptyset$ then $D_h^+\neq\emptyset$ for h small enough (because $e(h) \to 0$). On the other hand assume that D^0 has positive measure and (by contradiction) that $D_h^0=\emptyset$ (for a sequence of h's $\to 0$). In that case $0<u_h\leq e(h)$ in D^0; as soon as D^0 contains a ball of radius $\rho \in (\rho_0(h),\ \rho_1(h))$ and

$c_5\rho^r > e(h)$ (this must happen for h small enough, due to the assumptions of theorem 1), (2.5) ensures the existence of a point Q in D^0 where $u_h(Q) > e(h)$ which is contradictory.

<u>Example</u> - Assume that Ω is a smooth bounded convex domain in $\mathbb{R}^2$, f a smooth function in Ω with $f \leq \alpha < 0$, and g a smooth function on $\partial\Omega$ with $g \geq 0$. Set

$$K = \{v \mid v \in H^1(\Omega),\ v=g \text{ on } \partial\Omega,\ v \geq 0 \text{ a.e. in } \Omega\},$$

$$J(v) = \int_\Omega (\tfrac{1}{2}|\nabla v|^2 - fv)dx,$$

and let u be the unique solution of

$$u \in K, \quad J(u) \leq J(v) \quad \forall\, v \in K.$$

Then u has the growth property of order 2 near the free boundary. Let now $\{\mathcal{T}_h\}$ be a regular sequence of triangulations of convex polygons Ω_h inscribed in Ω and let V_h be the space of continuous functions on Ω, piecewise linear on Ω_h and constant, in each piece of $\Omega \setminus \Omega_h$, along the direction normal to $\partial\Omega_h$. Set

$$K_h = \{v \mid v \in V_h,\ v=g \text{ at nodes on } \partial\Omega,\ v \geq 0 \text{ in } \Omega\}.$$

Let u_h be the unique solution of

$$u_h \in K_h\ , \quad J(u_h) \leq J(v) \quad \forall v \in K_h.$$

If all the angles are less than or equal to $\pi/2$ then $\{u_h\}$ has uniformly the ball property of order 2, with $\rho_0(h) = 2h$ and $\rho_1(h)$ independent of h.

Usual error estimates [5, 6] give

$$e(h) = \|u-u_h\|_{L^\infty} \leq c\, h^2|\log h|\ \|D^2u\|_{L^\infty}\ ,$$

so that the error in the free boundaries goes like $h|\log h|^{1/2}$. For more details on this example see [4].

References

1. Baiocchi, C. & A.C.S. Capelo, *Variational and quasi-variational Inequalities*, Wiley, New York, 1984

2. Friedman, A., *Variational principles and free boundary problems*, Wiley, New York, 1982

3. Magenes, E. (Editor), *Free boundary problems* (Proc. Sem. 1979 Pavia) Istituto Nazionale di Alta Matematica, Roma, 1980

4. Brezzi, F. & L.A. Caffarelli, Convergence of the free boundary for finite element approximations , R.A.I.R.O. Anal. Numér. 17, 385, 1983

5. Baiocchi, C., Estimations d'erreur dans L∞ pour les inéquations à obstacle, in *Mathematical aspects of F.E.M.*, Lecture Notes in Math. N. 606, Magenes, E. & I. Galligani Eds., Springer, Berlin, 1977

6. Nitsche, J., L∞ convergence of finite element approximations, in *Mathematical aspects of F.E.M.*, Lecture Notes in Math. N. 606, Magenes, E. & I. Galligani Eds., Springer, Berlin, 1977

A UNILATERAL MODEL FOR THE LIMIT ANALYSIS OF MASONRY WALLS

M. Como
Istituto di Tecnica delle Costruzioni
University of Naples

A. Grimaldi
Dipartimento di Ingegneria Civile Edile
II University of Rome

SOMMARIO

In questo lavoro si analizza il collasso di pareti murarie sotto spinta, utilizzando per il materiale muratura un legame costitutivo di tipo elastico- non resistente a trazione. La definizione della condizione di collasso è ottenuta dalle condizioni di esistenza della soluzione del problema dell'equilibrio elastico.

ABSTRACT

In this paper the collapse of masonry walls is examined. For the masonry material a constitutive elastic model with zero tensile strength is assumed. The existence of solutions of the elastic equilibrium problem is analyzed and a definition of the collapse condition is provided. The corresponding kinematical and statical theorems, for the evaluation of the collapse load, are established.

1. INTRODUCTION

The aim of this paper is to establish a method for the evaluation of the lateral strength of masonry walls. This problem is of worthy interest in the analysis of the behaviour of masonry buildings under earthquake loadings.

To develop this analysis a crucial starting point is the choice of the constitutive equation of the masonry material. It is well known that the masonry material exhibits very low tensile strength. It can be useful, therefore, to assume as constitutive model the elastic material with zero tensile strength. In fact this unilateral model has been studied in several papers, both of theoretical and applied character [1,7].

After recalling the constitutive equation of the assumed model, this paper analyzes the problem of the elastic equilibrium of the masonry solid and provides a definition of the collapse condition. The corresponding kinematical and statical theorems are then established. They are very similar to the well known theorems of standard limit analysis of elastic-plastic structures.

The lateral strength of the single masonry panel is then evaluated. For more complex masonry walls an elastic-plastic behaviour of the architraves is assumed and bounds of the corresponding lateral collapse loads are obtained.

A first development of the approach presented in this paper has been given in [1] where the physical and technical aspects of the problem have been mainly examined.

2. THE CONSTITUTIVE MODEL OF THE ELASTIC MASONRY MATERIAL WITH NO TENSILE STRENGTH

The constitutive model is based on the following assumptions:

– tensile stresses are not allowed. Consequently, the principal values of the stress tensor σ are not positive. This condition will be referred to as

$$\sigma \leqslant 0 \tag{1}$$

– The deformation tensor is given by the sum of an elastic and an inelastic part, the last due to material cracking

$$\epsilon = \epsilon_e + \epsilon_c , \tag{2}$$

where

$$\epsilon_e = C^{-1} \sigma \tag{3}$$

with C the elastic tensor of the material.

– The cracking strains ϵ_c produce only dilatations of the material at any direction. This statement implies that the principal values of the cracking strain tensor ϵ_c are always non negative

$$\epsilon_c \geqslant 0. \tag{4}$$

– The normality condition

$$\sigma \cdot \epsilon_c = 0 \tag{5}$$

is also assumed. This assumption implies coaxiality between tensors σ and ϵ_c.

Hence we reduce to the following set of constitutive equations

$$\begin{array}{ll} \sigma = C(\epsilon - \epsilon_c) & \sigma \leqslant 0 \\ \epsilon_c \geqslant 0 & \sigma \cdot \epsilon_c = 0. \end{array} \tag{6}$$

In the one dimensional case the assumed constitutive model yields the following stress-strain law

$$\sigma = \theta\, E\epsilon \tag{7}$$

where

$$\theta = \begin{cases} 0 \text{ for } \epsilon \geqslant 0, \\ 1 \text{ for } \epsilon < 0. \end{cases} \tag{8}$$

The assumed model therefore implies that the stress tensor σ is a single-valued function $\sigma(\epsilon)$ of the strain tensor ϵ and defines a non linear elastic behaviour of the material (fig. 1).

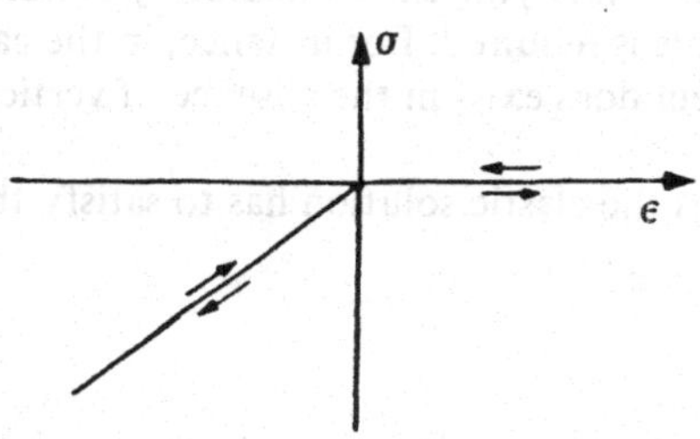

Fig. 1

This model is of course very different from the elastic-plastic scheme with zero yield stress (fig. 2) for which dilatations are not reversible, because of the elastic unloading.

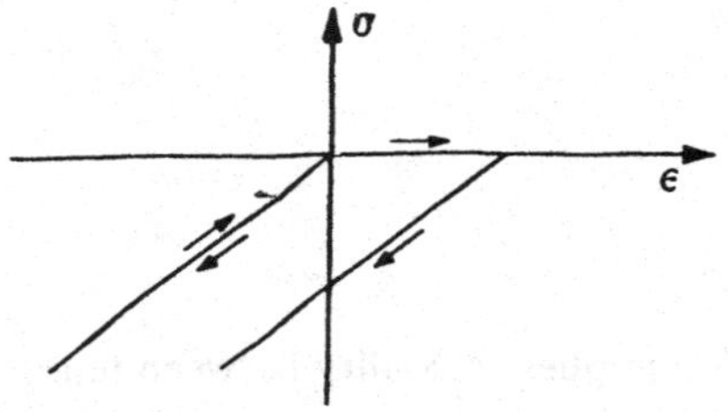

Fig. 2

3. EXISTENCE OF SOLUTIONS TO THE EQUILIBRIUM PROBLEM FOR THE MASONRY SOLID

Let us consider a masonry solid loaded by vertical loads g and horizontal loads λ q, increasing with a load parameter λ (fig. 3).

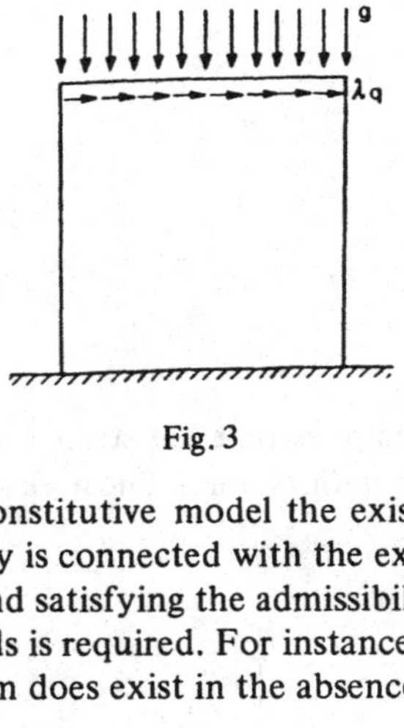

Fig. 3

Because of the assumed constitutive model the existence of solutions of the elastic equilibrium of the masonry body is connected with the existence of stress fields σ in equilibrium with the loads g + λ q and satisfying the admissibility condition (1). Consequently, a restriction on the external loads is required. For instance, in the case of fig 3, no solution of the elastic equilibrium problem does exist in the absence of vertical loads.

– In the absence of body forces the elastic solution has to satisfy the following equations:

– equilibrium:

$$\text{Div}\ \sigma = 0 \tag{9}$$

– constitutive equations

$$\epsilon = C^{-1}\,\sigma + \epsilon_c \qquad \sigma \leqslant 0$$
$$\epsilon_c \geqslant 0 \qquad \sigma \cdot \epsilon_c = 0 \tag{10}$$

– strain compatibility

$$\epsilon = D\,u \tag{11}$$

where D is the differential operator that maps displacements into strains.

On the boundary either displacements u must satisfy the constraints or the stresses σ must be in equilibrium with the applied loads.

– The research of solution can be performed by using the variational approaches corresponding to the minimum principles of the total potential energy or of the complementary energy, as shown by G. Romano and M. Romano [2,3]. The total potential energy functional of the loaded masonry solid is defined by the differentiable and convex functional

$$E(u, \epsilon_c) = \frac{1}{2}\langle C(Du - \epsilon_c), Du - \epsilon_c\rangle - \langle g + \lambda\, q, u\rangle. \tag{12}$$

The first term represents the strain energy produced in the body by the elastic strains while the second one is the potential energy of the surface external loads. The functional $E(u, \epsilon_c)$ is defined, of course, in a suitable functional space of the variables u and ϵ_c. The solution of the problem, i.e. the solution of eqs. (9, 10, 11), is equivalent to finding the minimum of the functional $E(u, \epsilon_c)$ in the space of the displacements u and of the cracking strains $\epsilon_c \geqslant 0$.

– The existence of solutions, as previously pointed out, is strictly connected to satisfying a compatibility condition for loads. Such a condition can be stated as

$$\langle g + \lambda q, u\rangle \leqslant 0 \qquad u \in M, \tag{13}$$

where

$$M = \{u : Du \geqslant 0\}. \tag{13'}$$

The set M defines strain fields which produce only dilatation of the material; i.e. displacement fields with zero strain energy. They will be defined as "mechanisms".

It is easy to prove that *condition (13) is necessary for the existence of the solution.*

This statement comes out if we take into account that, for $u \in M$ and $\epsilon_c = Du$, it is

$$E(u, \epsilon_c) = -\langle g + \lambda q, u\rangle. \tag{14}$$

Hence, when

$$\langle g+\lambda q, u\rangle > 0 \qquad u \in M, \tag{15}$$

with

$$\bar{u} = \alpha u \qquad \alpha > 0 \tag{16}$$

we get

$$\lim_{\alpha \to \infty} E(\bar{u}, \epsilon_c) = -\infty \tag{17}$$

which means that the functional $E(u, \epsilon_c)$ does not admit a minimum.

No solution of the elastic equilibrium exists if, at least for one mechanism displacement u, condition (15) holds.

This result is also evident by observing that, in term of stresses, the solution σ has to be found in the set S of the admissible stress field $\sigma \leqslant 0$ which are in equilibrium with the applied loads $g + \lambda q$. The statically admissible set S is therefore defined as the set of the stress field σ which satisfies the relations

$$\langle \sigma, \delta \epsilon\rangle = \langle g+\lambda q, \delta u\rangle \tag{18}$$

$$\sigma \leqslant 0 \tag{18'}$$

for any admissible displacement field δu.

Hence, if a solution σ exists, then from condition (18) for $\delta u \in M$ we have $\delta \epsilon \geqslant 0$.

Thus, taking in account (18′), we get

$$\langle g+\lambda q, \delta u\rangle \leqslant 0 \qquad \forall \delta u \in M. \tag{19}$$

A detailed analysis of the relation between the inequality (19) and the existence and uniqueness of the solution has been recently developed by G. Romano and M. Romano [2,3].

The research of the elastic solution can be also worked out by using the minimum principle of the complementary energy

$$E^*(\sigma) = \frac{1}{2}\langle C^{-1}\sigma, \sigma\rangle. \tag{20}$$

Likewise to the linear elastic case it is possible to show that the research of the solution is equivalent to the evaluation of the statically admissible stress field σ which minimizes the functional $E^*(\sigma)$. Therefore another necessary condition for the existence is required: *the set S of the statically admissible stress fields cannot be empty*. The necessary condition of the existence of the solution of the elastic equilibrium of the loaded masonry solid is the existence of at least one admissible stress field σ in equilibrium with the applied loads. Because of the differentiability and strict convexity of $E^*(\sigma)$ this condition is also *sufficient* for the existence and uniqueness of the stress field [2,3]. However uniqueness of the stress does not imply, as a rule, uniqueness of the corresponding displacements and strains. With reference, for instance, to the example of fig. 4, the uniform compression stresses $\sigma_y = -p$ represent the unique stress solution of the problem.

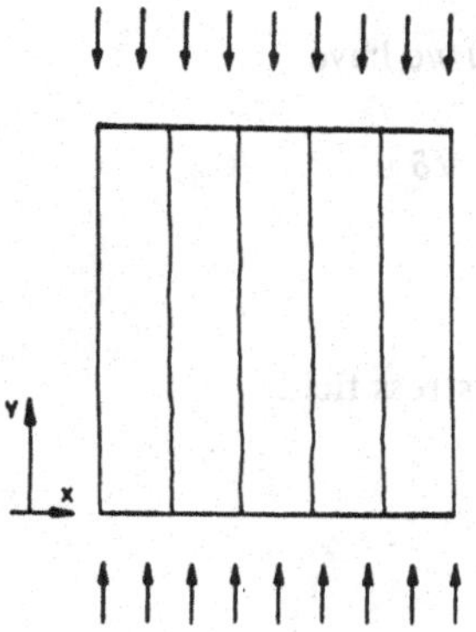

Fig. 4

We can associate, in fact, cracking strains of arbitrary intensity to this stress field.

4. COLLAPSE CONDITIONS OF THE MASONRY SOLID UNDER VERTICAL AND HORIZONTAL LOADS

We will now examine the existence problem along the loading process

$$\lambda q, \qquad \lambda \geqslant 0 \tag{21}$$

of horizontal forces q.

At $\lambda = 0$, i.e. for masonry solid only loaded by dead forces g, we assume that the corresponding stress solution does exist and is unique. Consequently we have that

$$\langle g,u \rangle \leqslant 0 \qquad u \in M \tag{22}$$

When increasing the load multiplier λ, this existence condition is verified up to a value λ_c beyond which the problem will non admit solution any more. In fact it is easy to recognize that *if solution does'nt exist, for instance at $\lambda = \lambda_1$, solutions cannot exist for any $\lambda \geqslant \lambda_1$.*

To prove this statement it is equivalent to show that if solutions exist at $\lambda = \lambda'$, we have solutions for any λ such that $0 \leqslant \lambda \leqslant \lambda'$. In fact the stress field σ' solution at $\lambda = \lambda'$ satisfies the conditions:

$$\langle \sigma', \delta \epsilon \rangle = \langle g, \delta u \rangle + \lambda, \langle q, \delta u \rangle \qquad \forall \delta u$$

$$\sigma' \leqslant 0 \tag{23}$$

Similarly for the solution σ_g at $\lambda = 0$ we have

$$\langle \sigma_g, \delta \epsilon \rangle = \langle g, \delta u \rangle \qquad \forall \delta u$$

$$\sigma_g \leqslant 0 \tag{23'}$$

Let us consider, for $0 \leqslant \lambda \leqslant \lambda'$, the stress field

$$\sigma(\lambda) = \sigma_g \left(1 - \frac{\lambda}{\lambda'}\right) + \frac{\lambda}{\lambda'} \sigma'$$

From eqs. (23) and (23′) we get

$$\langle \sigma(\lambda), \delta \epsilon \rangle = \langle g. \delta u \rangle + \lambda \langle q, \delta u \rangle \qquad \forall \delta u \tag{24}$$

and

$$\langle \sigma(\lambda), \delta \epsilon \rangle \leqslant 0 \qquad \forall \delta u \in M$$

that is

$$\sigma(\lambda) \leqslant 0 \tag{24'}$$

Therefore conditions (24) and (24″) show that, for any λ in the interval $0 \leqslant \lambda \leqslant \lambda'$, the set S of the statically admissible stress fields is non empty , and this is a sufficient condition for the existence of a solution.

Therefore it is of great importance the value λ_c of λ, defined as *collapse multiplier* of tge distribution of horizontal loads λq, that marks out the transition from existence to non existence conditions. Hence, the collapse multiplier λ_c defines the distribution of loading

$$g + \lambda q \tag{25}$$

where, for $\lambda < \lambda_c$, equilibrium configurations exist while for $\lambda > \lambda_c$ elastic equilibrium cannot be satisfied. The evaluation of λ_c is therefore centered in verifying existence conditions.

Finding the collapse multiplier λ_c for a masonry structure thus represents a problem of great importance because it provides its lateral ultimate strength. It can be performed according to two different approaches: the *kinematical* and the *statical* approaches that reproduce the well known procedures of the standard limit analysis structures.

– *The kinematical approach*

We define as kinematical multiplier

$$\lambda^+(u) \tag{26}$$

corresponding to the mechanism u, the value

$$\lambda^+(u) = -\frac{\langle g, u \rangle}{\langle q, u \rangle}, \qquad u \in M \tag{27}$$

that marks the transition, from negative to positive values, of the work of external loads $g + \lambda q$ along the mechanism u.

For $\lambda > \lambda^+(u)$ the necessary condition for the existence is violated along the mechanism u and

$$\lambda^+(u) \geqslant \lambda_c \tag{28}$$

The solution of the elastic equilibrium does exist only when condition (13) holds for any mechanism. The collapse multiplier thus will be given by

$$\lambda_c = \inf_{u \in M} \lambda^+(u) \tag{29}$$

– *The statical approach*

The stress solution σ exists if the set S of the statically admissible stress fields is not empty. Thus we define ***statically admissible multiplier*** the value $\lambda^-(\sigma)$ of λ for which an admissible stress field σ, in equilibrium with the load $g + \lambda^- q$. does exist. As previously shown, if solutions exist for $\lambda = \lambda^-$, solutions will exist also for $\lambda < \lambda^-$. The collapse load multiplier λ_c can be therefore defined as

$$\lambda_c = \sup_{\sigma \in S} \lambda^-(\sigma) \tag{31}$$

5. THE COLLAPSE THRUST FOR THE MASONRY PANEL

The problem of the evaluation of the collapse thrust of the masonry panel, loaded by a given distribution of vertical dead loads g, has relevant interest from a theoretical and a technical point of view.

This problem has been tackled by many Authors in numerous technical and experimental researches [1, 8, 9, 10, 11, 12, 13].

In the framework of the assumed constitutive model of the masonry material, it is possible to give the value of the collapse load of the panel.

Let us consider the simple masonry panel of fig. 3, loaded at the top section by the force distributions

$$g(y) \qquad \lambda\, q(y) \qquad 0 \leqslant y \leqslant b \tag{32}$$

The loads g (y) represent a fixed vertical compression of the panel while λ q (y) represent shearing stresses dependent upon loading parameter λ.

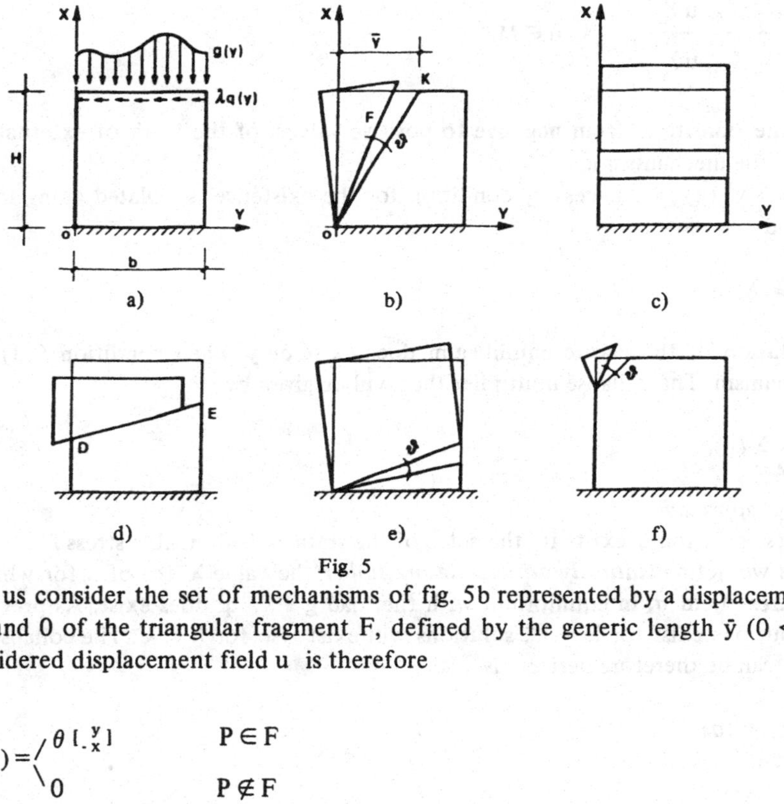

Fig. 5

Let us consider the set of mechanisms of fig. 5b represented by a displacement-rotation around 0 of the triangular fragment F, defined by the generic length $\bar{y}$ ($0 < \bar{y} < b$). The considered displacement field u is therefore

$$u(P) = \begin{cases} \theta \begin{bmatrix} y \\ -x \end{bmatrix} & P \in F \\ 0 & P \notin F \end{cases} \tag{33}$$

The corresponding cracking deformations, orthogonal to the fracture line OK, are distributed along the line OK.

The works of the applied external loads (32) for the mechanism u are

$$\langle g(y), u \rangle = -\theta \int_0^{\bar{y}} g(y)\, y\, dy \qquad \langle q(y), u \rangle = H\, \theta \int_0^{\bar{y}} q(y)\, dy \tag{34}$$

and the corresponding kinematical multipliers will be

$$\lambda^+(u) = \frac{\int_0^{\bar{y}} g(y)\, y\, dy}{\int_0^{\bar{y}} q(y)\, dy} \tag{35}$$

Very useful is the evaluation of the least lower bound of the multiplier (35) for $\bar{y}$ varying between 0 and b. Other significant mechanisms of the panel don't seem in fact possible. For the mechanism of fig. 5c the horizontal forces don't make work; the sliding displacement of fig. 5d does'not define a mechanism since shearing strains occur between the principal directions of stress defined by the sliding line DE and its normal; the mechanisms of fig. 5e and 5f give kinematical multipliers not lower than (35). The quantity

$$\inf_{0<\bar{y}<b} \lambda^+(u(\bar{y})) \tag{36}$$

can therefore represent a close upper bound of λ_c.

Let us assume now, as an example, the following distributions of forces (32)

$$g(y) = g = \text{const} \qquad \lambda q(u) = \lambda \tau_1 \left(\frac{y}{b}\right)^\alpha \qquad \alpha \geqslant 0 \tag{37}$$

for $0 \leqslant y \leqslant b$.

We then obtain

$$\inf_{\bar{y}} \lambda^+(u(\bar{y})) = \begin{cases} \lambda^+(u(b)) = (G/T)(b/2H) & \alpha \geqslant 1 \\ \lambda^+(u(0)) = 0 & \alpha < 1 \end{cases} \tag{38}$$

where G and T respectively represent the resultants of the force distributions g and q (y).

For shearing force distributions with $\alpha \geqslant 1$ quantity (36) gives an upper bound of the collapse multiplier while, for $\alpha < 1$, quantity (36) is zero and gives the collapse multiplier of the panel. For the load distributions (37) characterized by $\alpha < 1$, the set S of the admissible stress fields is empty. In this case, in fact, the work done by dead loads g is of negligible order of magnitude with respect to the work done by shearing forces.

Let us examine now another particular distribution of loads (32), which allows to exactly evaluate the collapse load. Let us consider the stress fields corresponding to the Michell elastic solution of a wedge loaded by a concentrated force R at its apex [13]. The force R is defined by both components R_a and R_n which are respectively directed along the axis of the wedge (fig. 6).

The stress field is radial and in polar coordinates is represented by

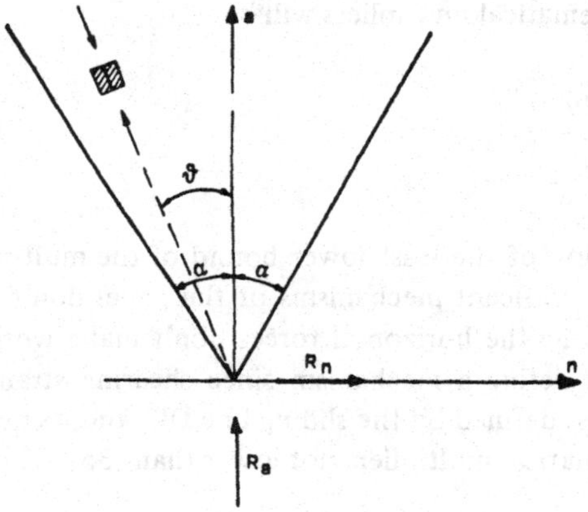

Fig. 6

$$\sigma_{r\theta} = \sigma_\theta = 0 \tag{39}$$

$$\sigma_r = -\frac{R_a \cos\theta}{r(\alpha + 1/2 \sin 2\alpha)} + \frac{R_n \sin\theta}{r(\alpha - 1/2 \sin 2\alpha)}$$

This solution of the elastic equilibrium of the wedge can be used for the evaluation of an admissible stress field in the masonry panel. We can assume in fact that in the portion

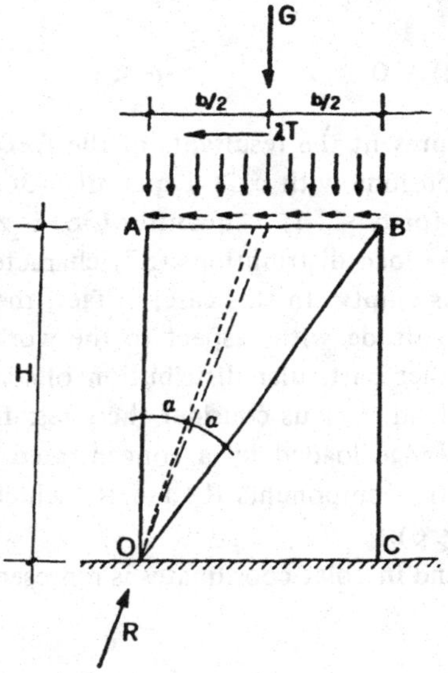

Fig. 7

0 AB of the panel a radial stress state (39) is acting while in the remaining part BC0 stresses are zero (fig. 7). The stress state acting on the top section of the panel is represented by vertical and horizontal resultants G and λ T. They pass through the center of the top section if

$$\lambda T = \frac{G}{2} \operatorname{tg} 2\alpha \tag{39}$$

where

$$\operatorname{tg} 2\alpha = \frac{b}{H} \tag{40}$$

The resultant R of G and λT passes through point 0. The corresponding components R_a and R_n of R in terms of G and λT are

$$R_a = G \cos\alpha + \lambda T \sin\alpha \qquad R_n = - G \sin\alpha + \lambda T \cos\alpha \tag{41}$$

The corresponding stress components emerging from the top section of the panel are then

$$\begin{aligned} \sigma_x (H, y) &= \sigma_r \cos^2 (\alpha - \theta) \\ \tau_{xy} (H, y) &= \sigma_r \sin (\alpha - \theta) \cos (\alpha - \theta) \end{aligned} \tag{42}$$

for $r = H/\cos(\alpha - \theta)$. Thus the load distribution (32) applied to the top section and in equilibrium with the emerging stresses (42) are

$$g(y) = - \sigma_x (H, y) \qquad \lambda q(y) = - \tau_{xy} (H, y) \tag{43}$$

The radial stress field (39) acting in the masonry panel loaded at the top section by the load distributions (43), at least for $H \geqslant b$, is *admissible*. The radial stresses σ_r produce always compression. Hence, the multiplier $\lambda^- (\sigma)$ of the thrust T is statically admissible. We have therefore

$$\lambda^- = \frac{G}{T} \frac{b}{2H} \tag{44}$$

For all the overturning mechanism of fig. 5b the application of virtual work equation always gives for the corresponding kinematical multiplier λ^+ (u) the value

$$\lambda^+ = \frac{G}{T} \frac{b}{2H} \tag{45}$$

Hence it represents the collapse multiplier of the masonry panel under the particular load distribution (43).

The previous solution can be used to evaluate the collapse lateral strength of masonry panels in common experimental loading condition. Collapse tests of masonry panels are usually carried out by applying vertical and horizontal load at the top of the panel by means of a rigid equipment, as shown in fig. 8:

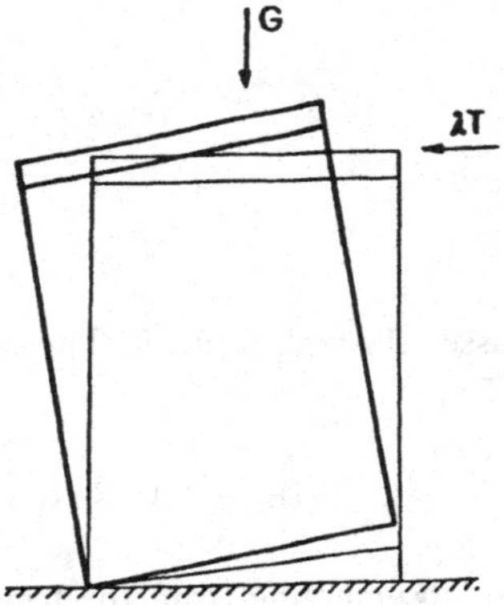

Fig. 8

Hence, the restraint condition of the top section of the panel requires equal horizontal and linearly varying vertical displacements. A collapse mechanism of the panel with such constraint condition is represented by the overturning rotation of fig. 8 which gives the kinematical multiplier (45). The statical theorem can be also applied, by taking as statically admissible stress field the radial stress distribution (39): correspondingly the statical multiplier is still given by (45). At the top section of the panel, in fact, the emerging loads from the assumed admissible stress distribution have to satisfy only equilibrium. Hence the value (45) represents the collapse multiplier of the horizontal thrust T applied to the panel by means of the rigid equipment of fig.8.

The value (45) is able to give a good evaluation of the limit lateral strength of masonry panels. A comparison with a series of experimental data has been performed by the Authors in [1], where it has been also taken in account the finite crushing strength of the masonry by means of a simple correction of equation (45). In this case, in fact eq. (45) can be replaced by

$$\lambda_c = \frac{G}{T} \frac{b}{2H} \left(1 - \frac{\sigma}{\sigma_k}\right) \tag{45'}$$

where σ is the panel compression due to the dead load σ and σ_k the crushing stress of the masonry material [1].

It is worthwhile to remark now that also an approximate approach can be used to evaluate the collapse multiplier of the panel. According to the technical beam theory we can consider the generalized forces M and N. – the bending moment and the axial force – and the corresponding generalized displacements θ and Δ – the rotation and the axial extension of the plane section of the panel –. The collapse condition hence becomes (fig. 9):

$$M = \pm N \frac{b}{2} \tag{46}$$

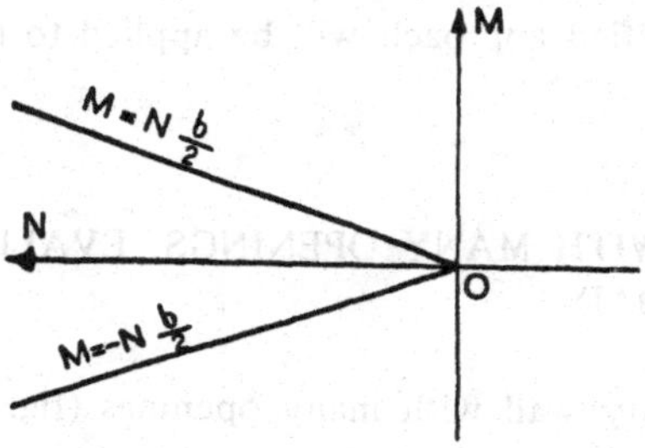

Fig.9

The cracking deformation on the other hand is defined by (fig. 10)

$$\theta_c \;, \qquad \Delta_c = \frac{b}{2}\,\theta_c \tag{47}$$

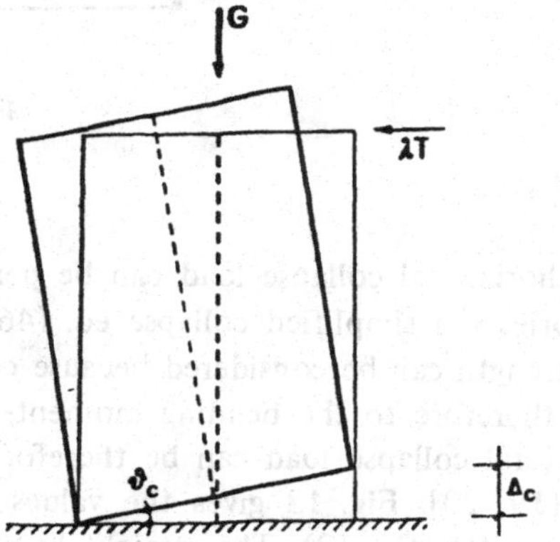

Fig. 10

Consequently the internal collapse work, in connection with the normality condition (6a) is

$$M\,\theta_c - N\,\Delta_c = M\frac{b}{2}\theta_c - N\frac{b}{2}\theta_c = 0 \tag{48}$$

The collapse multiplier λ_c still gives the value (45). We get in fact,

$$\lambda_c\, T\, H\,\theta - G\frac{b}{2}\theta = 0 \tag{49}$$

Applying the beam model to the panel gives therefore the collapse thrust previously obtained by means of the bidimensional scheme of unilateral elastic material. The beam scheme of the panel on the other hand corresponds to the model sistematically applied by Heyman [15, 16, 17] and by Kooharian [18] to the limit analysis of masonry arches. In the next section this simplified approach will be applied to the limit analysis of masonry walls with openings.

6. THE MASONRY WALL WITH MANY OPENINGS. EVALUATION OF THE HORIZONTAL COLLAPSE LOAD

Let us consider a masonry wall with many openings (fig. 11): it can represent a common wall of a masonry building.

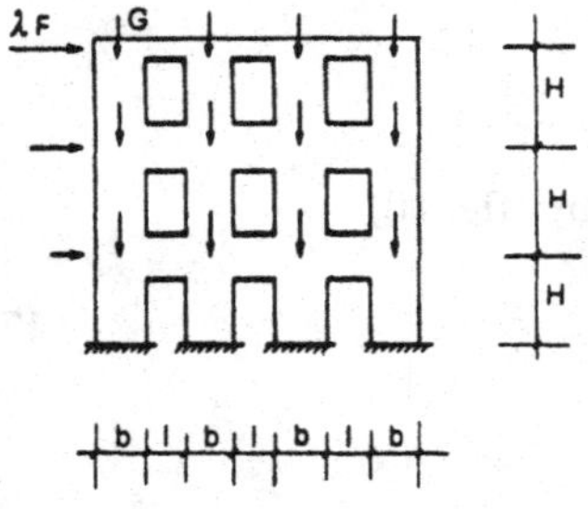

Fig. 11

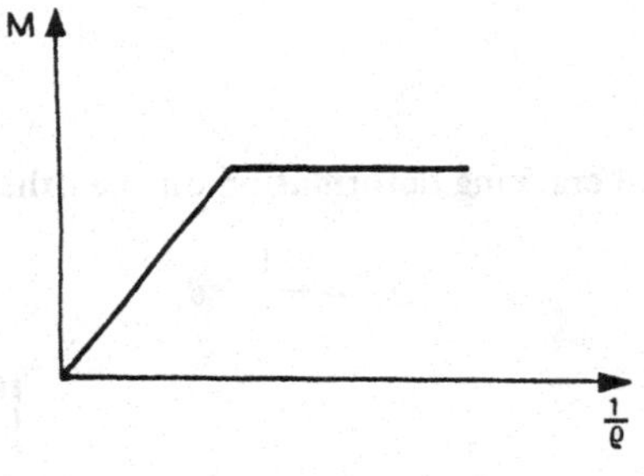

Fig. 12

An evaluation of the horizontal collapse load can be performed by using for the single walls between the stories the simplified collapse eq. (46). For the architraves, on the other hand, a bending strength can be considered because of the presence of tie rods or tendons. With reference therefore to the bending moment-curvature diagram of fig. 12, the evaluation of the lateral collapse load can be therefore performed according to the classical limit analysis [19, 20]. Fig. 13 gives the values of the collapse horizontal thrust for the masonry portal (fig. 13). The straight line 1 - 2; 2 - 3; 3 - 4 of fig. 13 correspond to the various collapse mechanisms of fig. 14 a, b, c that are, at the same time, statically admissible [1].

The parameter Z, i.e. the abscissa in Fig. 14, is

$$Z = \frac{T_0\, l}{G\, b} \tag{50}$$

and represents the ratio between the strengths of architrave and pilaster. The force T_0 of eq. (50) gives in fact the bending strength M_0 of the architrave

$$T_0 = \frac{2\, M_0}{l} \tag{51}$$

where l is the corresponding span length.

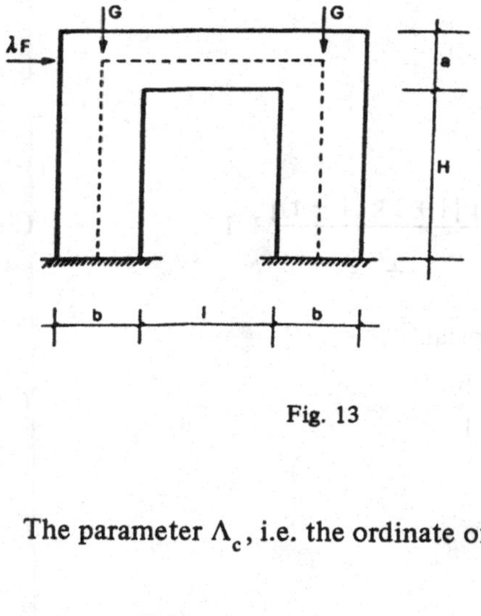

Fig. 13

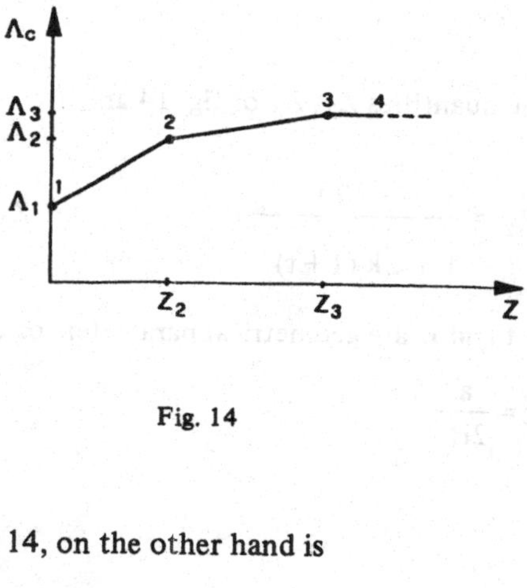

Fig. 14

The parameter Λ_c, i.e. the ordinate of fig. 14, on the other hand is

$$\Lambda_c = \lambda_c \frac{F\, H}{G\, b} \tag{52}$$

and the limit values Λ_1, Λ_2, Λ_3, corresponding to the various collapse mechanisms of fig. 15, respectively are

$$\Lambda_1 = \frac{1}{1+t}\,[1 + Z\,(1+k)] \qquad \Lambda_2 = \frac{(1+\frac{1}{k})(\frac{3}{2}+t+\frac{Z}{2})+t}{(1+t)(1+\frac{1}{k}+t)} \tag{53}$$

$$\Lambda_3 = 2$$

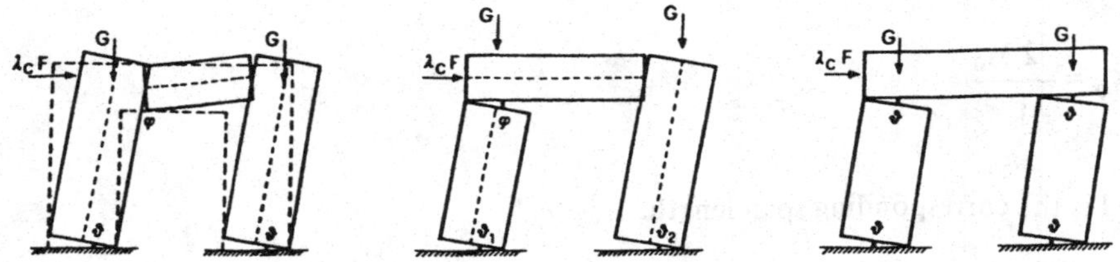

Fig. 15

and the quantities Z_2, Z_3 of fig. 14 are

$$Z_2 = \frac{1+2t}{1+2k(1+t)} \qquad Z_3 = \frac{2t\,[1+2k(1+t)]}{1+k} + 1 \tag{54}$$

where t and K are geometrical parameters of the portal

$$t = \frac{a}{2H} \qquad k = \frac{b}{l}. \tag{55}$$

6. CONCLUSIONS

The unilateral model of elastic material with zero tensile strength has been used to evaluate the ultimate load of masonry walls.

It is shown that this model provides a simple formulation of the collapse problem.

A necessary condition for the existence of solutions of the elastic equilibrium problem is used to define the collapse load and to obtain the classical kinematical and statical theorems.

The method is applied to the evaluation of the lateral strength of a masonry panel subjected to dead load and increasing horizontal forces. For this case a simple and exact expression of the collapse load is obtained. An example of a masonry wall with openings is also examined, showing the possibility to apply this approach to more complex structural problems.

REFERENCES

1. M. Como, A. Grimaldi, "Analisi limite di pareti murarie sotto spinta", Università di Napoli, Atti Ist. Tecnica delle Costruzioni n. 546, nov. 1983.

2. G. Romano, M. Romano, "Equilibrium and compatibility under internal and external convex constraints" Univ. di Catania, Fac. di Ingegneria, Ist. Sc. Costruzioni n. 5/83.

3. G. Romano, M. Romano "Elastostatics of Structures with Unilateral Conditions on Strains and Displacements", Unilateral Problems in Structural Analysis, Ravello, sept. 22-24, 1983.

4. A. Baratta, R. Toscano, "Stati tensionali in pannelli di materiale non resistente a trazione", AIMETA, VI Congr. Naz.le Genova, 7-8-9 Ottobre 1982.

5. S. Di Pasquale "Questioni di meccanica dei solidi non reagenti a trazione", AIMETA, VI Congr. Naz.le Genova, 7-8-9 Ottobre 1982.

6. P. Villaggio "Stress Diffusion in Masonry Walls", Journ. Struct. Mech. 9 (4), 1981.

7. G. Romano, M. Romano, "Sulla soluzione di problemi strutturali in presenza di legami costitutivi unilaterali" Atti Acc. Naz.le Lincei, Serie VIII, Vol. LXVII, 1979.

8. D. Benedetti, M.L. Casella "Shear strength of masonry piers", 7th World Conf. on Earthquake Eng., Instambul, 1980.

9. Y. Yokel, S.G. Fattal, "Failures hypothesis for masonry walls", ASCE, Struct. Div., March 1976.

10. C.K. Murthy, A.W. Hendry "Model experiments in load bearing brickwork" Building Science, Vol. 1, Pergamon Press, London, 1966.

11. D. Benedetti, A. Castellini, L. Formis "Verifica sperimentale di un metodo del calcolo della resistenza alle azioni sismiche", L'Industria delle Costruzioni, Ottobre 1979.

12. V. Turnsek, P. Sheppard "The shear and flexural strength of masonry walls" Int. Research Conf. on Earthquake Eng, Skopije, 1980.

13. P. Jossa, N. Salzano, "Il criterio della biella compressa per la verifica sismica dei pannelli murari", L'Industria delle Costruzioni, Dicembre 1982.

14. J.H. Michell Proc. London Math. Soc., vol 34, p. 134, 1902.

15. J. Heyman "The stone skeleton", Int. J. of Solids and Structures, 2, 1966.

16. J. Heyman "The safety of masonry arches", Int. J. Mech. Sci. 2, 1969.

17. J. Heyman "The masonry arch" Cambridge Press, 1982.

18. A. Kooharian "Limit analysis of voussoir and concrete arches", Journ. Amer. Concrete Inst. 24, 1952.

19. W. Prager "An introduction to Plasticity" Addison - Wesley Publ. Comp., Reading Mass. USA, 1959.

20. C. Massonnet, M. Save "Calcolo plastico a rottura delle Costruzioni", CLUP, Milano, 1980.

A THEORY OF FRICTION*

A. Curnier
Laboratoire de Mécanique Appliquée
Ecole Polytechnique Fédérale de Lausanne

Abstract : A rather general theory of friction, inspired from the classical theory of plasticity is proposed. It includes the contact impenetrability condition as a by-product.

*) This communication is based on a paper by the same Author, to appear in the International Journal of Solids and Structures, 1984.

I. INTRODUCTION

I.1. Objective

In spite of the resemblance which exists between plastic and frictional phenomena, the degrees of development reached by the theories of plasticity and friction are rather disparate.

The paper summarized here aims at reducing the gap accumulated between the two disciplines by proposing a rather general theory of friction (yet limited to small amounts of slip) inspired from the theory of plasticity (at small strains) along a line originally explored by Fredriksson[1] and further pursued by Michalowski and Mroz[2].

By construction this theory is susceptible to include not only the influence of the normal load on the friction force but also other factors such as wear and adhesion.

I.2. Context

A variational formulation of the contact between two deformable bodies constitutes a good basis for the development of a constitutive law of friction including an impenetrability condition.

To this end consider two bodies, one called the striker and the other the target, bound to contact one another within a surface A characterized by the unit outward normal N to the target.

Basically, to the standard weak statements of equilibrium of the two deformable bodies, needs to be added a contact term of the form :

$$\int_A F \cdot \dot{D}\, dA \;=\; \int_A (F_N \cdot \dot{D}_N + F_T \cdot \dot{D}_T)\, dA \qquad (> 0) \tag{1}$$

In the above, the vector D represents the **distance** of contact separating each point S on the striker from the corresponding point T on the target, i.e. $D = S - T$. The vector F is the **force** of contact per unit area (also called stress vector) acting on the target at the point $S = T$ whenever contact occurs, i.e. when $\dot{D}_N = D_N = 0$. Because the amounts of slip and deformation are assumed to remain small, the contact integral (1) may be indifferently defined over the reference contact surface A with unit outward normal N or over the deformed contact surface a with normal n .

The dissipation $F_T \cdot \dot{D}_T$ due to friction vanishes in the two limiting cases of perfect stick $(\dot{D}_T = 0)$ and perfect slip $(F_T = 0)$. In between these extremes, a **law of friction** relating the force of friction to the amount of slip (and other internal variables to be introduced later) of the form $F_T = \hat{F}_T[D_T, \dot{D}_T, \ldots]$ is necessary. It is the purpose of the paper to propose such a law embedding for completeness an impenetrability condition of the form $F_N = \hat{F}_N[D_N, \dot{D}_N, \ldots]$.

II. RATE INDEPENDENT THEORY OF FRICTION WITH IMPENETRABILITY

Relying on basic experimental observation, the theory of friction proposed here is independent of the slip rate, uses a standard slip rule for all internal variables and presents high resistance to penetration. By analogy with plasticity, it rests upon four basic principles.

II.1 Decomposition of the contact distance into adherence and slip
(cf. decomposition of the strain into elastic and plastic parts)

The theory is based upon the decomposition of the distance of contact at a point D into the sum of two parts : one reversible, rather unusual, called **adherence** and denoted D^A and the other irreversible, more familiar, called **slip** and denoted D^S , which can be resolved into normal and tangential components as before

$$\begin{aligned} D &= D^A + D^S && \text{(a)} \\ &= (D^A_N + D^S_N) + (D^A_T + D^S_T) && \text{(b)} \qquad (2) \\ &= D_N + D_T && \text{(c)} \end{aligned}$$

This double decomposition supposes that : a) the contact distance $D = S-T$ is measured with respect to the same origin T throughout the motion, b) the direction of the outward normal N remains nearly constant throughout the sliding process. D^A can be attributed to the elastic deformations of the asperities covering the surfaces in contact whereas D^S may be imputed partly to their plastic deformation and mainly to the rupture of the junctions occuring at their tips.

The decompositions (2) of the contact distance may be compared to the strain decompositions into elastic and plastic parts on one hand and into bulk and deviatoric components on the other hand used in plasticity

$$E = E^E + E^P = (\bar{E}^E + \bar{E}^P) + (E'^E + E'^P) = \bar{E} + E'$$

with an obvious notation.

To complete this kinematic description of a frictional contact, it remains to introduce the **cumulated** slip D^c , defined as

$$D^c = \int_0^t dD^c \qquad \text{where} \qquad dD^c = \sqrt{dD^s_T \cdot dD^s_T} \tag{3}$$

It is the direct analog of the equivalent plastic strain E^c .

The directional slip D^S and the cumulated slip D^c are the two internal variables, (one vector and one scalar), used to take into account the permanent memory characteristic of the phenomenon of friction.

II.2 Laws of adherence, tear and wear
(cf. elastic, kinematic- and isotropic-softening laws)

To the kinematic variables of adherence D^A , directional slip D^S and cumulated slip D^c are associated, by energetic duality, three dynamic quantities which are called the force of **friction** F , the force of **"tear"** F^S and the force of **"wear"** F^c .

The forces of tear and wear are introduced to characterize two different forms of a single and same phenomenon, the running-in of contact surfaces in relative sliding motion. The first form occurs when the sliding motion of the two bodies is monotone and oriented along some preferential direction resulting into an **anisotropic tear** of the contact surfaces which requires a vector entity for its modelization : the tear force.

The second form occurs when the two bodies rub against one another in alternate arbitrary directions, resulting into an **isotropic wear** of the surfaces sufficiently well described by a scalar quantity : the wear force.

In order to define these forces, three constitutive laws are needed. For the sake of simplicity three linear laws are proposed here but any invertible relationship would be perfectly legitimate :

$$
\begin{aligned}
\mathbf{F} &= \mathbf{P}\ \mathbf{D}^A && \text{(a)} \\
\mathbf{F}^S &= \mathbf{Q}\ \mathbf{D}^S && \text{(b)} \qquad (4) \\
F^C &= Q\ D^C && \text{(c)}
\end{aligned}
$$

where $\mathbf{P}$ is a "penalty matrix" representing the elasticity of the asperities, $\mathbf{Q}$ is a "rugosity matrix" which characterizes the directional tear of asperity tips and junctions and Q a "rugosity modulus" playing a similar role for the cumulated wear.

These three constitutive laws of adherence, tear and wear are the analogs of the elastic, kinematic and isotropic softening laws of plasticity :

$$\mathbf{S} = \mathbf{E}\ \mathbf{E}^E , \quad \mathbf{S}^P = \mathbf{H}\ \mathbf{E}^P , \quad S^C = H\ E^C$$

where $\mathbf{S}$ is the stress, $\mathbf{S}^P$ the plastic softening stress, E^C and S^C the equivalent plastic strain and stress, $\mathbf{E}$, $\mathbf{H}$ and H the various associated moduli.

II.3 Slip criterion
(cf. yield criterion)

To activate the kinematic decomposition of the distance of contact into adherence and slip, a **slip criterion** is interrogated to decide which one of the two modes occurs. In general the adherence-slip limit depends on all the state variables which, for the purely mechanical theory under consideration, implies :

$$Y(D^A, D^S, D^c) < 0 \quad \text{contact and adherence} \tag{5}$$
$$= 0 \quad \text{gap or slip}$$

It is emphasized that since both the normal and tangential components of the contact distance enter the definition of adherence, the slip criterion includes the contact criterion as well i.e. the **impenetrability** condition.

Upon substitution of the invertible constitutive laws (4) into the kinematic criterion, a dynamic **friction criterion** is obtained which is equivalent to the slip criterion but more appropriate to state eventual associated slip rules :

$$Y(F, F^S, F^c) \leq 0 \tag{6}$$

The friction-slip criterion is the homologue of the plastic-yield criterion used in plasticity, indifferently written in strain or stress space as

$$Y(E^E, E^P, E^c) \leq 0 \quad \text{or} \quad Y(S, S^P, S^c) \leq 0$$

The normal impenetrability condition is comparable to the incompressibility of certain materials.

As a classical example, it is instructive to mention the criterion of perfect friction (Fig.1)

$$Y(F) = \begin{cases} F_N \leq 0 & \text{contact} \\ |F_T| + f\,F_N - C \leq 0 & \text{slip} \end{cases} \tag{7}$$

where $|F_T| = \sqrt{F_{T_1}^2 + F_{T_2}^2}$ denotes the euclidian norm of F_T , f the **coefficient of friction** and C a constant characterizing adhesion.

In geometric terms the criterion assumes the shape of a truncated cone in the normal-tangential axes attached to the contact point. It is the analogue of the Drucker-Prager criterion in plasticity and constitutes the determining ingredient of Coulomb's law of friction. The replacement of the euclidean norm by an elliptic norm transforms the isotropic criterion into an **anisotropic** one, accounting for an oriented rugosity. Deviations from the law of perfect friction may be accounted for by acting on the **shape** of the slip criterion and including running in or **tear** and **wear** mechanisms similar to plastic softening.

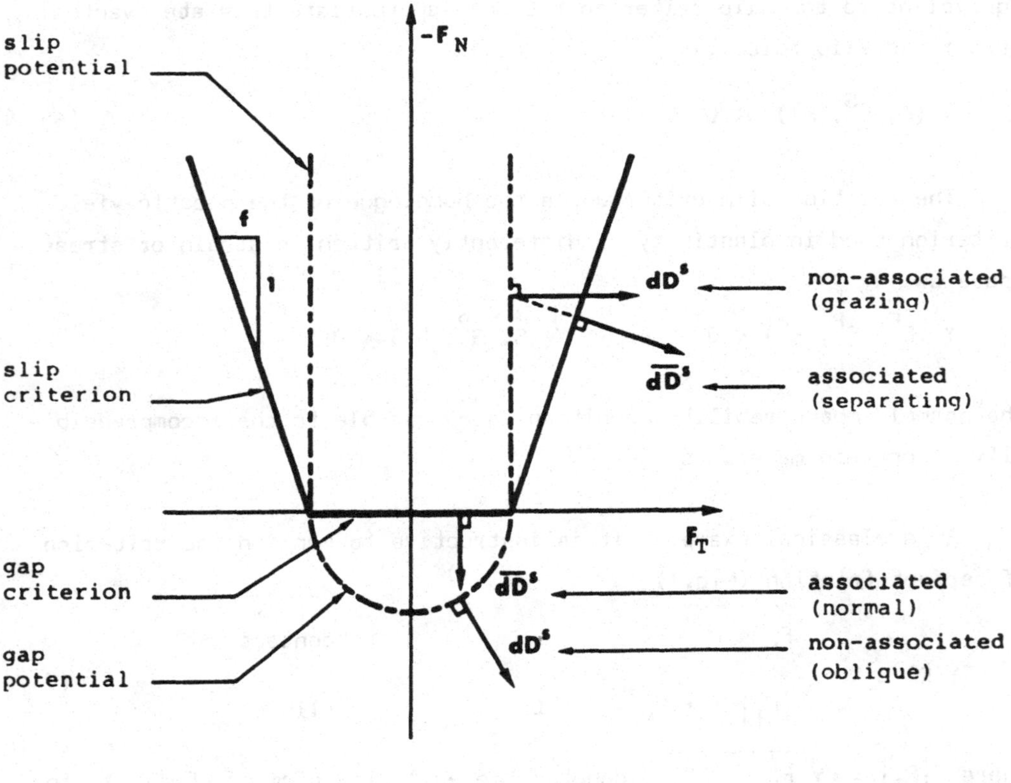

Fig. 1. Associated and non-associated slip rules

II.4 Slip rules
(cf. flow rules)

The slip direction (including take off) is governed by slip rules deriving from a convex potential $Z(F,F^S,F^c)$:

$$
\begin{aligned}
dD^S &= \lambda \frac{dZ}{dF} \\
-\, dD^S &= \lambda \frac{dZ}{dF^S} \qquad (8) \\
-\, dD^c &= \lambda \frac{dZ}{dF^c}
\end{aligned}
$$

where λ is a positive constant expressing the collinearity of the slip increment with the outward normal to the potential Z .

The slip rules are equivalent to the flow rules in plasticity which for a standard generalized material take the form

$$dE^P = \lambda \frac{dZ}{dS} \qquad -\, dE^P = \lambda \frac{dZ}{dS^P} \qquad -\, dE^c = \lambda \frac{dZ}{dS^c}$$

If the slip potential Z is replaced by the slip criterion Y the slip rule (8) becomes **associated** to the criterion. Although in plasticity the flow rules associated to the standard criteria prove realistic for relatively large classes of materials, the slip rules associated to the usual friction criteria like (7) are **not** acceptable (Fig.1). Indeed the incompressibility of plastic deformations implied by a flow rule associated with the Von Mises criterion finds wide applications. On the contrary, the normal component separating the two bodies

$$dD_N^S = \lambda \frac{dY}{dF_N} = \lambda f \qquad (> 0) \qquad (9)$$

produced by a slip rule associated to the classical criterion of perfect friction bears no support. Consequently a **cylindrical** slip potential with a hemispherical cap seems to be the only admissible surface which guarantees that slip begins in the tangent plane common to the two points in contact.

III. CONCLUSION

The basic ingredients of this small slip theory of standard generalised friction independent of the slip rate are recalled to consist of a kinematic decomposition (2), constitutive laws to generate the conjugate dynamics (4), a transition criterion (6) and an additional equation of evolution (8).

The result is a generalisation of Coulomb's law accounting for the influence on the macroscopic coefficient of friction of :

- the normal **load** (hereby resulting into a nonlinear dependence of the force of friction on the normal load);
- the initial **rugosity** of the surfaces in contact (whether it is isotropic or anisotropic);
- the subsequent **wear** of these surfaces (whether it is monotone or alternate).

The combination of the normal impenetrability condition and the tangential law of friction into an integrated law of frictional contact is another feature of this theory which deserves a mention.

REFERENCES

1. Fredriksson B., Finite element solution of surface nonlinearities in structural mechanics with special emphasis to contact and fracture mechanics problems, Computers and Structures, 6, 281, 1976.

2. Michalowski R., Mróz Z., Associated and non-associated sliding rules in contact friction problems, Archives of Mechanics, 30, 259, 1978.

CONTACT UNILATERAL AVEC ADHERENCE

M. Frémond
Laboratoire Central des Ponts et Chaussées
Service de Mathématiques
Paris

1. INTRODUCTION

Considérons du papier adhésif collé sur une table. Il faut exercer un effort vertical notable pour le décoller : différence importante avec les problèmes de contact unilatéral classique où le décollement se produit dès que l'on surmonte la pesanteur. L'effort supplémentaire sert à vaincre ce que l'on appelle les effets de l'adhérence. Nous nous attachons ici à donner une description de l'adhérence [1], [2] basée sur les méthodes de la mécanique des milieux continus [3], [4]. La physique de l'adhérence qui fait l'objet de nombreux travaux est exclue de nos préoccupations [5].

L'idée essentielle que nous mettons en oeuvre est l'insuffisance des champs de déplacement habituels pour décrire le contact : en effet lorsque deux solides sont en contact, ils peuvent s'ignorer et ne pas agir l'un sur l'autre (ils n'adhèrent pas), ils peuvent aussi interagir l'un sur l'autre (il y a adhérence), dans les deux cas les déplacements sont les mêmes pour des situations mécaniques différentes.

On introduit une nouvelle variable cinématique qui décrit l'état du contact : l'intensité d'adhésion β (paragraphe 2). La théorie est construite en utilisant le principe des puissances virtuelles (paragraphes 3, 4 et 5) qui donne les équations d'équilibre pour les efforts intérieurs associés aux variables cinématiques. Le second principe de la thermodynamique que nous rappelons brièvement donne la structure des lois de comportement (paragraphes 6,7,8,9 et 10). Nous examinons par la suite le comportement élastique, c'est-à-dire celui pour lequel la dissipation mécanique est nulle. On démontre ensuite un théorème d'existence pour les problèmes d'équilibre. Ce faisant on retrouve les formulations énergétiques globales qui sont en général le point de départ de la théorie de l'adhérence [5] (paragraphes 11.,12 et 13). On termine par deux exemples simples. Le premier non dissipatif permet de surmonter un paradoxe de la théorie classique de l'adhésion : il est impossible de décoller un point collé sur une surface ! Le second dissipatif décrit le processus de ce décollement.

2. DESCRIPTION CINEMATIQUE

Soit une structure occupant un domaine $\overline{\Omega}\subset\mathbb{R}^n$ $(1\leqslant n\leqslant 3)$ (figure 1), fixée sur une partie Γ_0 de sa frontière et reposant sur un socle rigide sur une autre partie Γ_1. Le contact sur Γ_1 est unilatéral avec adhérence.

La description cinématique de la structure est assurée par les champs de déplacement $\vec{u}$ et de vitesse $\frac{d\vec{u}}{dt}$ habituels fonctions de $x\in\Omega$ et du temps t. Ces variables sont insuffisantes pour décrire I'état du contact sur Γ_1 puisque lorsqu'il y a contact en un point, ($\vec{u}=\vec{u}_s$ déplacement du socle), on doit préciser s'il y a adhérence ou non, ou encore si la liaison est rompue ou non avec le socle rigide. On définit donc une nouvelle variable cinématique β : l'intensité d'adhérence. Si $\beta=0$, il n'y a pas adhérence, $\beta=1$ il y a adhérence totale et si $0<\beta<1$ il y a adhérence partielle. L'intensité d'adhésion est une fonction $\beta(x,t)$ du point et du temps. Elle

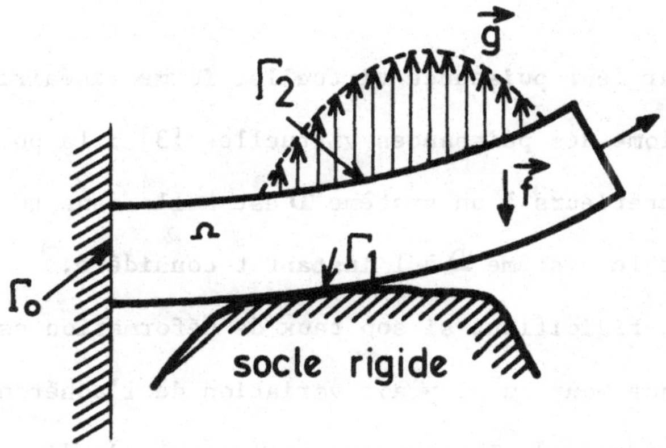

Figure 1. La structure Ω est soumise aux forces surfacique g et volumique f, fixée sur le support Γ_0 et adhère au socle rigide sur Γ_1.

peut être considérée comme une description macroscopique des liaisons microscopiques entre Ω et Γ_1. Une diminution de β correspond à la rupture de liaisons microscopiques. On peut encore considérer β comme la proportion surfacique de liaisons non rompues entre Ω et Γ_1.

Le déplacement $\vec{u}$, le déplacement $\vec{u}_s$ du socle et l'intensité d'adhésion sont soumis aux liaisons internes sur Γ_1

(1) $0 \leq \beta \leq 1$,

(2) $\beta(\vec{u}-\vec{u}_s) = \vec{0}$.

La deuxième exprime que lorsqu'il y a décollement ($\vec{u} \neq \vec{u}_s$), il n'y a plus d'interaction entre la structure et son support ($\beta = 0$).

On définit enfin les espaces vectoriels des champs de déplacements ou de vitesses virtuelles $\mathcal{V}$ de la structure définis sur $\bar{\Omega}$ et $\mathcal{V}_s$ du socle définis sur Γ_1 ainsi que l'espace vectoriel $\mathcal{B}$ des champs d'intensité d'adhésion ou de vitesse virtuelle d'intensité d'adhésion définis sur Γ_1. Ces champs auront la régularité convenable pour assurer la cohérence des calculs. Les déplacements et intensités d'adhésion réels sont notés $\vec{u}, \vec{u}_s, \beta$.

Un élément quelconque de $\mathcal{B}\times\mathcal{V}\times\mathcal{V}_s$ est noté $\gamma,\vec{v},\vec{v}_s$.

3. EFFORTS INTERIEURS

Ils sont définis par leur puissance virtuelle, forme linéaire sur $\mathcal{B}\times\mathcal{V}\times\mathcal{V}_s$ vérifiant l'axiome des puissances virtuelles [3] : la puissance virtuelle des efforts intérieurs à un système $\mathcal{D}$ est nulle dans tout mouvement virtuel rigidifiant le système $\mathcal{D}$ à l'instant t considéré.

Un élément de $\mathcal{V}$ est rigidifiant si son taux de déformation est nul. Il est largement admis que pour qu'il y ait variation de l'adhérence, il faut qu'il y ait déformation [5]. Il est donc raisonnable de dire qu'un élément $(\gamma,\vec{v},\vec{v}_s)\in\mathcal{B}\times\mathcal{V}\times\mathcal{V}_s$ est ridigifiant si $\vec{v}$ est une vitesse de corps rigide (un distributeur), $\vec{v}=\vec{v}_s$ et la vitesse γ est nulle sur Γ_1.

On retient alors comme puissance virtuelle des efforts intérieurs $\hat{\mathcal{T}}_i$

$$\forall\mathcal{D}\subset\Omega,\quad \hat{\mathcal{T}}_i = -\int_{\mathcal{D}} \sigma_{ij}\hat{D}_{ij}d\Omega - \int_{\partial\mathcal{D}\cap\Gamma_1} \{F\gamma + \vec{Q}(\vec{v}-\vec{v}_s)\}d\Gamma$$

avec $\hat{D}_{ij}=\frac{1}{2}(v_{i,j}+v_{j,i})$. L'effort intérieur F est une densité surfacique d'énergie.

La puissance réelle des efforts intérieurs est

$$\forall\mathcal{D}\subset\Omega,\quad \mathcal{T}_i = -\int_{\mathcal{D}} \sigma_{ij}D_{ij}d\Omega - \int_{\partial\mathcal{D}\cap\Gamma_1} \{F\dot{\beta} + \vec{Q}(\dot{\vec{u}}-\dot{\vec{u}}_s)\}d\Gamma$$

avec $D_{ij}=\frac{1}{2}(\dot{u}_{i,j}+\dot{u}_{j,i})$.

La théorie de l'adhérence présentée ici est un exemple d'utilisation du principe des puissances virtuelles dans une situation inusuelle. La richesse de la description cinématique n'est pas obtenue comme dans la théorie du second gradient [6] en augmentant la finesse de la description des taux de déformation mais en introduisant une variable cinématique nouvelle.

On fait une théorie classique du premier gradient pour les contraintes [3]. Pour l'adhérence on peut parler d'une théorie du zéro gradient. Une théorie du premier gradient ferait dépendre $\hat{\mathcal{T}}_i$ du gradient de γ. C'est une théorie plus fine qui suppose une influence mutuelle entre les points de la surface de contact. Nous ne l'abordons pas ici.

4. EFFORTS EXTERIEURS

Ils sont définis par leur puissance virtuelle $\hat{\mathcal{T}}_e$, forme linéaire sur $\mathcal{B}\times\mathcal{V}\times\mathcal{V}_s$:

$$\forall\mathcal{D}\subset\Omega,\quad \hat{\mathcal{T}}_e = \int_{\mathcal{D}} \vec{f}\cdot\vec{v}d\Omega + \int_{\partial\mathcal{D}} \vec{T}\cdot\vec{v}d\Gamma + \int_{\partial\mathcal{D}\cap\Gamma_1} \{A\gamma + \vec{g}_s\vec{v}_s\}d\Gamma$$

Sur Γ_1 les efforts extérieurs A, $\vec{T}$ sont imposés à la structure, l'effort $\vec{g}_s$ est lui imposé au support.

5. PRINCIPE DES PUISSANCES VIRTUELLES. EQUATIONS D'EQUILIBRE

Pour simplifier nous nous limitons au cas quasi-statique. Le principe des puissances virtuelles est

$$\forall\mathcal{D}\subset\Omega,\quad \forall(\gamma,\vec{v},\vec{v}_s)\in\mathcal{B}\times\mathcal{V}\times\mathcal{V}_s,\ \hat{\mathcal{T}}_e + \hat{\mathcal{T}}_i = 0.$$

On obtient d'abord les propriétés classiques [3],

$$\sigma_{ij,j} + f_i = 0 \text{ dans } \mathcal{D} \text{ et } \sigma_{ij}n_j = T_i \text{ sur } \partial\mathcal{D}-\Gamma_1,$$

où les n_i sont les cosinus directeurs de la normale extérieure à $\mathcal{D}$, et

$$\forall(\gamma,\vec{v},\vec{v}_s),\ \int_{\partial\mathcal{D}\cap\Gamma_1}\{-\sigma_{ij}n_jv_i + \vec{T}\cdot\vec{v} + \vec{g}_s\vec{v}_s - \vec{Q}(\vec{v}-\vec{v}_s) + \gamma(A-F)\}d\Gamma = 0,$$

qui donne les équations nouvelles sur $\partial\mathcal{D}\cap\Gamma_1$,

$$F = A,\ \sigma_{ij}n_j = T_i - Q_i,\ \vec{g}_s + \vec{Q} = 0.$$

Dans la théorie que nous exposons, les efforts intérieurs F et $\vec{Q}$ sont déterminés dès que l'on connaît les efforts extérieurs. Les égalités précédentes donnent aussi la signification physique des efforts F et $\vec{Q}$: F est l'effort extérieur d'adhésion et $-\vec{Q}$ l'effort extérieur appliqué au support.

Si l'on suppose $\vec{T} = 0$ sur Γ_1 (il n'y a pas d'action extérieure sur la structure sur Γ_1), on a $\sigma\cdot\vec{n} = -\vec{Q} = \vec{g}_s$ qui exprime que $\vec{g}_s$ est l'action du support sur la structure, c'est-à-dire la réaction du support. Pour la suite on suppose que $\vec{T} = 0$ et $A = 0$ sur Γ_1.

6. CONSERVATION DE L'ENERGIE

On introduit en plus de l'énergie spécifique volumique classique e, une énergie surfacique E. L'énergie interne est alors

$$\mathcal{E}(\mathcal{D}) = \int_{\mathcal{D}} \rho\, e\, d\Omega \quad \int_{\partial\mathcal{D}\cap\Gamma_1} E\, d\Gamma,$$

où ρ est la masse volumique. L'équation de la conservation de l'énergie est

$$\forall \mathcal{D} \subset \Omega, \quad \frac{d\mathcal{E}}{dt}(\mathcal{D}) = \mathcal{T}_e + \dot{Q} = \dot{Q} - \mathcal{T}_i,$$

où $\dot{Q}$ est le taux de la chaleur reçue,

$$Q(\mathcal{D}) = \int_{\mathcal{D}} r\, d\Omega - \int_{\partial\mathcal{D}} \vec{q}\cdot\vec{n}\, d\Gamma + \int_{\partial\mathcal{D}\cap\Gamma_1} R\, d\Gamma,$$

où r est le taux de production de chaleur volumique, R le taux de production surfacique de chaleur sur Γ_1 et $\vec{q}$ le vecteur courant de chaleur. Elle donne l'équation classique dans Ω [3],

$$\rho \frac{de}{dt} + q_{i,i} = \sigma_{ij} D_{ij} + r,$$

et sur Γ_1

$$\frac{dE}{dt} = F\dot{\beta} + \vec{Q}(\vec{u} - \vec{u}_s) + R.$$

7. SECOND PRINCIPE DE LA THERMODYNAMIQUE

7.1. Rappel. Nous allons utiliser les résultats classiques de thermodynamique [3]. L'inégalité fondamentale exprime qu'à tout instant t et pour tout système $\mathcal{D}$

$$\forall \mathcal{D} \subset \Omega, \quad \frac{d\mathcal{S}}{dt} \geq \int_{\mathcal{D}} \frac{r\, d\Omega}{T} + \int_{\partial\mathcal{D}\cap\Gamma_1} \frac{R\, d\Gamma}{T} - \int_{\partial\mathcal{D}} \frac{\vec{q}\cdot\vec{n}}{T}\, d\Gamma,$$

où T est la température absolue.

On suppose ici que $\mathcal{S} = \int_{\mathcal{D}} \rho\, s\, d\Omega + \int_{\partial\mathcal{D}\cap\Gamma_1} S\, d\Gamma$ où s est l'entropie surfacique. Par un calcul classique utilisant les équations de conservation de l'énergie, on obtient l'inégalité de Clausius-Duhem

$$(3) \qquad \forall \mathcal{D} \subset \Omega, \quad \int_{\mathcal{D}} \rho \frac{d\psi}{dt} d\Omega + \int_{\partial\mathcal{D}\cap\Gamma_1} \frac{d\Psi}{dt} d\Gamma \leq \int_{\mathcal{D}} \{\sigma_{ij} D_{ij} - \rho s \frac{dT}{dt} - \frac{\vec{q}\cdot \text{grad}\, T}{T}\} d\Omega$$
$$+ \int_{\partial\mathcal{D}\cap\Gamma_1} \{F\dot{\beta} + \vec{Q}(\dot{\vec{u}} - \dot{\vec{u}}_s) - S \frac{dT}{dt}\} d\Gamma,$$

où l'on a défini les énergies libres ψ et Ψ par $\psi = e - sT$, $\Psi = E - ST$.

Les *inégalités de Clausius-Duhem* sont vérifiées par toute *évolution réelle* du système $\beta, \vec{u}, t, \dots$. On retient l'axiome de l'état local [3] : à tout instant t, l'énergie libre, l'entropie sont fonctions de n paramètres, les variables normales χ_i et ont les mêmes expressions qu'en thermostatique.

7.2. <u>Application à la théorie de l'adhérence</u>. Pour la suite de l'exposé, nous faisons l'hypothèse des petites perturbations [3]. Nous supposons de plus que la température T, les petites déformations $\varepsilon_{ij}(\vec{u}) = \frac{1}{2}(u_{i,j} + u_{j,i})$, l'écart $\bar{u} = \vec{u} - \vec{u}_s$ entre les déplacements de la structure et du socle, l'intensité d'adhérence β, forment un système de variables normales. Les énergies libres sont fonction de ces fonctions. Ce sont des applications qui aux fonctions $\varepsilon, \bar{u}, \beta$ et T font correspondre les fonctions ψ et Ψ. On dit que ce sont des énergies libres globales. Leur valeur en un point x dépend donc des fonctions $\varepsilon, \bar{u}, \beta$ et de T(x) d'après la définition des énergies libres. Si l'on veut que ces expressions soient locales (hypothèses que l'on retient pour la suite), on suppose que

$$\forall x \in \Omega, \quad \psi\big(\bar{u}, \beta, \varepsilon, T(x)\big)(x) = \bar{\psi}\big(\varepsilon(x), T(x)\big)$$

$$\forall x \in \Gamma_1, \quad \Psi\big(\bar{u}, \beta, \varepsilon, T(x)\big)(x) = \bar{\Psi}\big(\bar{u}(x), \beta(x), T(x)\big)$$

où $\bar{\psi}$ et $\bar{\Psi}$ sont des fonctions réelles de variables réelles.

L'inégalité de Clausius-Duhem est vérifiée par les évolutions réelles, donc par des β et $\bar{u}$ qui vérifient les liaisons internes (1) et (2) :

(1) $0 \leq \beta \leq 1$, (2) $\beta \bar{u} = 0$.

On peut donc définir à loisir les énergies libres si les liaisons internes ne sont pas vérifiées sans affecter l'inégalité de Clausius Duhem. Nous profitons de cette liberté pour inclure les liaisons internes dans le potentiel $\bar{\psi}$ en lui donnant la valeur $+\infty$ si (1) ou (2) n'est pas vérifiée. La dérivée $\frac{d\bar{\psi}}{dt}$ doit toujours s'entendre comme dérivée particulaire, c'est-à-dire en suivant les points matériels qui vérifient à chaque instant les liaisons internes. Sa valeur n'est donc pas affectée par la modification apportée au potentiel $\bar{\psi}$.

La fonction $\bar{\psi}$ garde là où les liaisons internes sont vérifiées toutes les propriétés des potentiels thermodynamiques : $\bar{\psi}$ est convexe sur toute partie convexe où elle est finie ; elle est dérivable par rapport à T avec $\frac{\partial\bar{\psi}}{\partial T} = -S$ en tout point où elle est finie.

Plus précisément, les fonctions de $\bar{u}, T : (\bar{u}, T) \rightarrow \bar{\psi}(\bar{u}, \beta, T)$ et de $\beta, T : (\beta, T) \rightarrow \bar{\psi}(\bar{u}, \beta, T)$ sont convexes et là où elles sont finies $\bar{\psi}$ est dérivable par rapport à T. La fonction $(\bar{u}, \beta, T) \rightarrow \psi(\bar{u}, \beta, T)$ n'est pas par contre une fonction convexe de $\bar{u}$, β et T.

Avant de préciser les potentiels $\bar{\psi}$ et $\bar{\psi}$ introduisons une notion mathématique qui nous paraît utile, généralisant la notion de sous-différentiel [4] aux fonctions $\bar{\psi}$ que nous aurons à envisager.

7.3. <u>Sous-différentiabilité locale</u>. Soit Φ une application d'un espace vectoriel topologique X en dualité avec $X^{\star}$ dans $\mathbb{R} \cup \{+\infty\}$. On dit que Φ est localement sous-différentiable au point $x \in X$ s'il existe un voisinage $\mathcal{W}(x)$ et $x^{\star} \in X^{\star}$ tels que

(4) $\forall z \in \mathcal{W}(x), \quad \Phi(z) \geq \Phi(x) + \langle x^{\star}, z-x \rangle$.

Remarquons que cette notion n'impose pas que Φ soit convexe. Donnons un exemple (figure 2)

$$X = X^{\star} = \mathbb{R}^2,$$

$$\Phi(0,u) = \frac{1}{2} k u^2,$$

$$\Phi(\beta,0) = -w\beta \quad \text{si} \quad 0 \leqslant \beta \leqslant 1,$$

$$\Phi(\beta,u) = +\infty \quad \text{si} \quad \beta u \neq 0 \quad \text{et} \quad \beta \notin [0,1] \; .$$

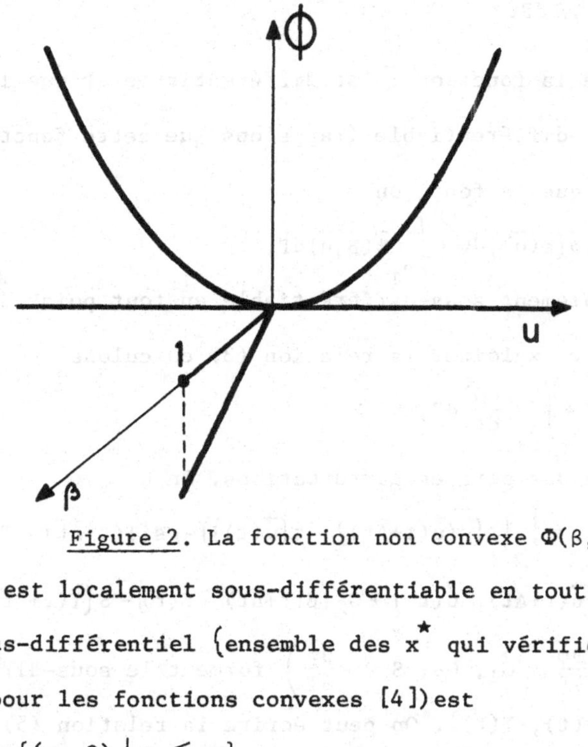

Figure 2. La fonction non convexe $\Phi(\beta,u)$.

La fonction Φ est localement sous-différentiable en tout point où elle est finie. Son sous-différentiel {ensemble des $x^{\star}$ qui vérifient (4)} noté $\partial\Phi(x)$ (comme pour les fonctions convexes [4]) est

$$\partial\Phi(0,0) = \{(\alpha_1,0) \mid \alpha_1 \leqslant -w\},$$

$$\partial\Phi(\beta,0) = \{(-w,\alpha_2) \mid \alpha_2 \in \mathbb{R}\}, \text{ si } 0 < \beta < 1,$$

$$\partial\Phi(1,0) = \{(\alpha_1,\alpha_2) \mid \alpha_1 \geqslant -w, \ \alpha_2 \in \mathbb{R}\},$$

$$\partial\Phi(0,u) = \{(\alpha_1,ku) \mid \alpha_2 \in \mathbb{R}\}, \text{ si } u \neq 0.$$

A partir de maintenant on suppose que le socle est rigide et fixe $\vec{u}_s = 0$ et que la structure est fixée sur la partie Γ_0 ($\vec{u}|\Gamma_0 = 0$). On pose

$$\mathcal{V}_0 = \{\vec{v} \mid \vec{v} \in \mathcal{V} ; \ \vec{v}|\Gamma_0 = 0\}$$

qui est en dualité avec l'espace formé des éléments $(\sigma,\vec{Q})$ par la forme

$$\int_\Omega \sigma_{ij}\varepsilon_{ij}(\vec{v})d\Omega + \int_{\Gamma_1} \vec{Q}\vec{v}\,d\Gamma.$$

De même $\mathcal{B}$ est en dualité avec son dual par la forme

$$\int_{\Gamma_1} F\gamma\, d\Gamma,$$

et l'espace des températures $\mathcal{T}$ en dualité avec son dual par la forme

$$\int_\Omega \rho\, Ts\, d\Gamma + \int_{\Gamma_1} TS\, d\Gamma.$$

On suppose que la fonction $\overline{\psi}$ est différentiable et que la fonction $\overline{\psi}$ est localement sous-différentiable (rappelons que cette fonction n'est pas convexe). On admet que la fonction

$$\mathcal{A}(\vec{u},\beta,T) = \int_\Omega \rho\overline{\psi}\big(\varepsilon(\vec{u})\big)d\Omega + \int_{\Gamma_1} \overline{\psi}(\beta,\vec{u})d\Gamma,$$

est-elle aussi localement sous-différentiable en tout point de $\mathcal{V}\times\mathcal{B}\times\mathcal{T}$ où elle est finie. Pour exploiter la relation (3) calculons

$$\frac{\partial\mathcal{A}}{\partial t} = \int_\Omega \rho\,\frac{d\overline{\psi}}{dt}\,d\Omega + \int_{\Gamma_1} \frac{d\overline{\psi}}{dt}\,d\Gamma,$$

d'après l'hypothèse des petites perturbations. On a

$$(5)\quad \mathcal{A}(t+\Delta t) - \mathcal{A}(t) = \int_\Omega \{\tau\big(\varepsilon(\vec{u}(t+\Delta t)) - \varepsilon(\vec{u}(t))\big) - \rho s\big(T(t+\Delta t) - T(t)\big)\}d\Omega$$
$$+ \int_{\Gamma_1} \{\vec{G}_1\cdot\big(\vec{u}(t+\Delta t) - \vec{u}(t)\big) + G_2\big(\beta(t+\Delta t) - \beta(t)\big) - S\big(T(t+\Delta t) - T(t)\big)\}d\Gamma$$

où $\left(\tau = \frac{1}{\rho}\frac{\partial\overline{\psi}}{\partial\varepsilon},\ s = -\frac{\partial\overline{\psi}}{\partial T},\ \vec{G}_1,\ G_2,\ S = -\frac{\partial\overline{\psi}}{\partial T}\right)$ forment le sous-différentiel de $\mathcal{A}$ au point $\big(\vec{u}(t),\ \beta(t),\ T(t)\big)$. On peut écrire la relation (5) pour Δt assez petit car les fonctions $\vec{u}(t)$, $\beta(t)$, $T(t)$ sont supposées continues en t. En divisant par Δt positif et négatif et en passant à la limite on trouve

$$\frac{d\mathcal{A}}{dt} = \int_\Omega \{\tau\varepsilon(\dot{\vec{u}}) - \rho s\,\dot{T}\}d\Omega + \int_{\Gamma_1} \{\vec{G}_1\cdot\dot{\vec{u}} + G_2\dot{\beta} - S\dot{T}\}d\Gamma.$$

La relation (3) donne

$$\int_\Omega \{(\sigma_{ij}-\tau_{ij})\varepsilon_{ij}(\dot{\vec{u}}) - \frac{\vec{q}\cdot\text{grad}T}{T} - \rho s\dot{T}\}d\Omega + \int_{\Gamma_1} \{(\vec{Q}-\vec{G}_1)\dot{\vec{u}} + (F-G_2)\dot{\beta} - S\dot{T}\}d\Gamma \geq 0,$$

qui doit être vérifiée pour toute évolution réelle du système. La définition du sous-différentiel de $\mathcal{A}$ forme les lois d'état.

Pour simplifier on suppose les problèmes thermique et mécanique découplés (évolution isotherme par exemple) et l'on omet la température dans ce qui suit. La relation de Clausius-Duhem (3) exprime alors que la dissipation $\widetilde{D}$ est positive.

$$\widetilde{D} = \int_\Omega (\sigma_{ij} - \tau_{ij})\varepsilon_{ij}(\dot{\vec{u}})d\Omega + \int_{\Gamma_1} \{(\vec{Q}-\vec{G}_1)\dot{\vec{u}} + (F-G_2)\dot{\beta}\}d\Gamma \geqslant 0.$$

Les lois d'état s'écrivent alors

$$(\tau, \vec{G}_1, G_2) \in \partial\mathcal{A}(\vec{u}, \beta).$$

Cette relation exprime aussi que $\mathcal{A}(\vec{u},\beta) < +\infty$ puisque $\mathcal{A}$ est dérivable par rapport à $\varepsilon(\vec{u})$. Les liaisons internes (1) et (2) étant vérifiées si et seulement si $\mathcal{A}$ est finie, on remarque encore ici qu'elles font partie des lois de comportement.

On doit adjoindre aux lois d'état les lois complémentaires pour avoir toutes les lois de comportement.

8. LOIS DE COMPORTEMENT : LOIS COMPLEMENTAIRES

Nous donnons deux exemples, l'un le comportement élastique a une dissipation nulle $\widetilde{D} = 0$, l'autre le comportement standard généralisé [7], [8] a une dissipation non nulle. Il recouvre les comportements classiques comme la plasticité, la viscosité, ...

8.1. Comportement élastique (dissipation nulle). On dit que le comportement est élastique si

$$\sigma = \tau, \quad \vec{G}_1 = \vec{Q}, \quad G_2 = F.$$

Pour les contraintes, on retrouve la loi de comportement élastique classique.

8.2. Comportement standard généralisé (dissipation non nulle). Pour définir le comportement standard généralisé [7], [8] on introduit un pseudopotentiel de dissipation défini par des densités volumique $\varphi(\dot{\varepsilon})$ et surfacique $\Phi(\dot{\bar{u}},\dot{\beta})$.Les fonctions φ et Φ sont convexes, positives et nulles à l'origine. Le comportement est standard généralisé si

$$\sigma - \tau \in \partial\varphi\left(\varepsilon(\dot{\vec{u}})\right),\ (\vec{Q}-\vec{G}_1, F-G_2) \in \partial\Phi(\dot{\vec{u}},\dot{\beta}).$$

Les études expérimentales actuelles ne semblent pas s'opposer à une loi de comportement standard généralisée pour divers matériaux (par exemple, contact verre polyuréthane [5]).

Dans la suite nous nous consacrons à l'étude de la loi de comportement élastique.

9. EXEMPLE DE COMPORTEMENT ELASTIQUE

On choisit

$$\rho\bar{\psi}(\varepsilon) = \frac{1}{2}(\lambda\varepsilon_{kk}^2 + 2\mu\varepsilon_{ij}\varepsilon_{ij}) = \frac{1}{2}a(\vec{v},\vec{v})$$

avec $a(\vec{v},\vec{w}) = \lambda\varepsilon_{kk}(\vec{v})\varepsilon_{\ell\ell}(\vec{w}) + 2\mu\varepsilon_{ij}(\vec{v})\varepsilon_{ij}(\vec{w})$, où λ et μ sont les paramètres de Lamé de la théorie de l'élasticité classique. On définit l'ensemble des variables cinématiques admissibles

$$K = \{(\gamma,\vec{v}) \mid \gamma \in \mathcal{B}\ ;\ \vec{v} \in \mathcal{V}_0\ ;\ 0 \leq \gamma \leq 1\ ;\ \gamma\vec{v} = 0\ ;\ \vec{v}\cdot\vec{N} = v_N \leq 0 \text{ sur } \Gamma_1\}.$$

La dernière condition $\vec{v}\cdot\vec{N} \leq 0$ exprime qu'il n'y a pas interpénétration de la structure et du socle ($\vec{N}$ est la normale extérieure à Ω sur Γ_1).

On choisit encore

$$\bar{\psi}(\vec{v},\gamma) = -w\gamma + I_K(\vec{v},\gamma)$$

où I_K est l'indicatrice de l'ensemble K (l'indicatrice d'un ensemble [4] K est définie par $I_K(x) = 0$ si $x \in K$ et $I_K(x) = +\infty$ si $x \notin K$). La constante positive w est l'énergie d'adhésion de Dupré [9]. On note encore que les liaisons internes et les conditions aux limites cinématiques sont incluses dans $\bar{\psi}$.

Remarque. On peut imaginer d'autres expressions pour $\bar{\psi}$, par exemple

$$\overline{\psi}(\vec{v},\gamma) = - w\gamma + \frac{1}{2}k\vec{v}^2 + I_K(\vec{v},\gamma),$$

qui modélise une interaction élastique entre le socle et la structure.

10. INTERPRETATION DE LA LOI DE COMPORTEMENT

Examinons les implications du choix de $\overline{\psi}$ et $\overline{\psi}$ Pour simplifier nous nous plaçons dans la situation de l'équilibre et nous nous limitons au comportement sur la partie Γ_1. On a

$$\forall(\gamma,\vec{v}) \in \mathcal{W} \cap K,$$

$$\int_\Omega \{\frac{1}{2}a(\vec{v},\vec{v}) - \frac{1}{2}a(\vec{u},\vec{u})\}d\Omega - \int_{\Gamma_1} w(\gamma-\beta)d\Gamma \geqslant \int_\Omega \sigma_{ij}\varepsilon_{ij}(\vec{v}-\vec{u})d\Omega + \int_{\Gamma_1}\{\vec{Q}(\vec{v}-\vec{u}) + F(\gamma-\beta)\}d\Gamma.$$

Comme $\sigma_{ij} = \lambda\varepsilon_{kk}(\vec{u})\delta_{ij} + 2\mu\varepsilon_{ij}(\vec{u})$, on a

$$\sigma_{ij}\varepsilon_{ij}(\vec{v}-\vec{u}) = a(\vec{u},\vec{v}-\vec{u}),$$

et

$$\int_\Omega \frac{1}{2}a(\vec{v}-\vec{u},\vec{v}-\vec{u})d\Omega - w\int_{\Gamma_1}(\gamma-\beta)d\Gamma \geqslant \int_{\Gamma_1}\vec{Q}(\vec{v}-\vec{u})d\Gamma, \tag{6}$$

car $F = A = 0$.

10.1. Interprétation de la loi de comportement. Cas du contact ($\vec{u} = 0$.

Comme $F = 0$ on a $\beta = 1$ (voir l'exemple du paragraphe 7.3). Soit $\mathcal{O}$ un voisinage de $x \in \Gamma_1$ où $\vec{u} = 0$. Prenons $\gamma = \beta$ et $\vec{v} = \vec{u}$ en dehors de $\mathcal{O}$ et $\gamma = 0$ sur $\mathcal{O}$. On a alors

$$\int_\Omega a(\vec{u},\vec{v}-\vec{u})d\Omega = -\int_\Omega \sigma_{ij,j}(v_i-u_i)d\Omega + \int_{\Gamma_1}\sigma_{ij}n_j(v_i-u_i)d\Gamma = \int_\Omega \vec{f}(\vec{v}-\vec{u})d\Omega - \int_{\Gamma_1}\vec{Q}(\vec{v}-\vec{u})d\Gamma.$$

Posons

$$\mathcal{F}[\vec{v}] = \frac{1}{2}\int_\Omega \frac{1}{2}a(\vec{v},\vec{v})d\Omega - \int_\Omega \vec{f}\,\vec{v}\,d\Omega - \int_{\Gamma_2}\vec{g}\cdot\vec{v}\,d\Gamma.$$

On a alors d'après (6)

$$\mathcal{F}[\vec{v}] - \mathcal{F}[\vec{u}] + \int_{\mathcal{O}} w\,d\Gamma \geqslant 0.$$

En divisant par mes$(\mathcal{O})$, on obtient

$$w \geqslant \limsup_{\text{mes}(\mathcal{O})\to 0} \frac{\mathcal{F}[\vec{u}] - \mathcal{F}[\vec{v}]}{\text{mes}(\mathcal{O})} = G(x). \tag{7}$$

La quantité $G(x)$ est le taux de restitution de l'énergie. La relation précédente est souvent prise comme base de la théorie de la rupture [10]. Ici c'est une conséquence des lois de comportement.

La relation (6) peut encore être utilisée pour montrer que $-Q_N$ réaction normale n'est pas trop grande si c'est une traction. De même la réaction tangentielle n'est pas trop grande [11].

10.2. Interprétation de la loi de comportement. Cas du décollement $\vec{u} \neq 0$. En prenant $\gamma = \beta = 0$ dans (6), on montre facilement que $-\vec{Q}_{\vec{T}}$ (réaction tangentielle est nulle) et que $-Q_N$ réaction normale est négative avec $Q_N u_N = 0$. On montre encore que

$$G(x) \geqslant w,$$

autre relation classique de la théorie de la rupture [10]. Cette relation exprime encore que là où il y a décollement la déformation est assez grande.

11. EQUILIBRE D'UNE STRUCTURE ELASTIQUE. EQUATIONS DU PROBLEME

Nous avons déjà considéré la structure chargée par les forces de volume $\vec{f}$, de surface $\vec{g}$ sur Γ_2, fixée sur Γ_0 et en contact unilatéral avec adhérence sur Γ_1 (figure 1).

Les équations sont celles que nous avons écrites progressivement dans les paragraphes précédents. Les inconnues sont les fonctions $\vec{u}, \beta, \sigma, \vec{Q}, F, \vec{g}_s$. Les données sont $\Omega, \Gamma_0, \Gamma_1, \Gamma_2, \vec{g}, \vec{f}$ et les fonctions $\rho\overline{\psi}$ et $\overline{\psi}$. Les équations sont

$$\sigma_{ij,j} + f_i = 0 \text{ dans } \Omega, \quad \sigma_{ij} n_j = g_i \text{ sur } \Gamma_2,$$

$$F = 0, \quad \sigma_{ij} n_j = -Q_i, \quad \vec{g}_s + \vec{Q} = 0 \text{ sur } \Gamma_1,$$

$$(\sigma, \vec{Q}, F) \in \partial\mathcal{A}(\vec{u}, \beta). \tag{8}$$

12. EQUILIBRE D'UNE STRUCTURE ELASTIQUE. FORMULATION VARIATIONNELLE.

On définit l'énergie potentielle

$$\mathcal{F}[\gamma,\vec{v}] = \frac{1}{2}\int_\Omega a(\vec{v},\vec{v})\,d\Omega - \int_{\Gamma_1} w\,\gamma\,d\Gamma - \int_\Omega \vec{f}\cdot\vec{v}\,d\Omega - \int_{\Gamma_2} \vec{g}\cdot\vec{v}\,d\Gamma.$$

On montre

<u>THEOREME 1</u>. Si les fonctions $\vec{u},\beta,\sigma,\vec{Q}$ sont régulières, les équations (8) sont équivalentes à trouver $(\beta,\vec{u})$ qui vérifient

$$(\beta,\vec{u}) \in K, \quad \exists \mathcal{W}(\vec{\beta},\vec{u}) \text{ voisinage de } (\beta,\vec{u}) \text{ dans } \mathcal{B}\times\mathcal{V} \text{ tel que}$$

(9) $\quad \forall(\gamma,\vec{v}) \in K \cap \mathcal{W}, \quad \mathcal{F}[\beta,\vec{u}] \leqslant \mathcal{F}[\gamma,\vec{v}]$

<u>Remarque</u>. On dit que $(\beta,\vec{u})$ est une position d'équilibre relatif ou un minimum relatif. L'équilibre est absolu si $\mathcal{W} = \mathcal{B}\times\mathcal{V}$.

<u>Démonstration</u>. Supposons les équations (8) vérifiées. Soit $(\gamma,\vec{v}) \in K$ on a d'après la définition du sous-différentiel

$$\forall(\gamma,\vec{v}) \in \mathcal{W}(\beta,\vec{u}), \quad \int_\Omega \frac{1}{2}a(\vec{v},\vec{v})\,d\Omega - w\int_{\Gamma_1}\gamma\,d\Gamma - \left\{\int_\Omega \frac{1}{2}a(\vec{u},\vec{u})\,d\Omega - w\int_{\Gamma_1}\beta\,d\Gamma\right\}$$

$$\geqslant \int_\Omega \sigma_{ij}\varepsilon_{ij}(\vec{v}-\vec{u})\,d\Omega + \int_{\Gamma_1} \vec{Q}(\vec{v}-\vec{u})\,d\Omega \quad \int_{\Gamma_1} F(\gamma-\beta)\,d\Gamma$$

qui donne la formule (9) en utilisant les équations d'équilibre.

Réciproquement supposons que (9) soit vérifiée. On définit σ par $\sigma_{ij} = \lambda\varepsilon_{kk}(\vec{u})\delta_{ij} + 2\mu\varepsilon_{ij}(\vec{u})$ et l'on vérifie classiquement les équations d'équilibre dans Ω puis l'on définit $\vec{Q}$ par $Q_i = -\sigma_{ij}n_j$. On a alors

$$\int_\Omega \frac{1}{2}a(\vec{v},\vec{v})\,d\Omega - w\int_{\Gamma_1}\gamma\,d\Gamma - \left\{\int_\Omega \frac{1}{2}a(\vec{u},\vec{u}) - w\int_{\Gamma_1}\beta\,d\Gamma\right\} \geqslant \int_\Omega \sigma_{ij}\varepsilon_{ij}(\vec{v}-\vec{u})\,d\Omega$$

$$+ \int_{\Gamma_1} \{0\times(\gamma-\beta) + \vec{Q}(\vec{v}-\vec{u})\}\,d\Gamma, \text{ pour } (\gamma,\vec{v}) \in \mathcal{W}(\beta,\vec{u}) \cap K,$$

qui est la relation $(\sigma,\vec{Q},0) \in \partial\mathcal{A}(\vec{u},\beta)$. On obtient ensuite facilement les autres relations de (8).

En précisant les espaces on peut démontrer le théorème d'existence suivant :

<u>THEOREME 2</u>. Si l'ouvert Ω est régulier, $\mathrm{mes}(\Gamma_0) > 0$, $\mathrm{mes}(\Gamma_1) > 0$, $\mathcal{V} = \{\vec{v} \mid v_i \in H^1(\Omega)\}$, $\mathcal{B} = L^2(\Gamma_1)$, $f_i \in L^2(\Omega)$, $g_i \in L^2(\Gamma_2)$, le problème (9) possède

au moins une solution qui vérifie $\beta = 1$ là où $\vec{u} = 0$ sur Γ_1.

Remarque. Cette solution peut ne pas être unique puisque K n'est pas convexe.

Démonstration. Elle consiste à chercher un minimum absolu de $\mathcal{F}(\mathcal{W} = \mathcal{B}\times\mathcal{V})$. Montrons que K est fermé pour $\mathcal{B}\times\mathcal{V}$ muni de la topologie faible de $L^2(\Gamma_1) \times H^1(\Omega)$. Soit donc $(\gamma_n, \vec{v}_n)$ qui tend faiblement vers $(\gamma, \vec{v})$. On sait que $\vec{v}_n$ tend fortement dans $L^2(\Gamma_1)$ vers $\vec{v}$ donc que $\gamma_n \vec{v}_n$ tend faiblement vers $\gamma\vec{v}$ qui est donc nul. On a donc $(\gamma, \vec{v}) \in K$.

La fonction $\mathcal{F}[\gamma, \vec{v}]$ est faiblement semi-continue inférieurement et est coercive ($\lim \mathcal{F}[\gamma, \vec{v}] = +\infty$ si $\|(\gamma, \vec{v})\| \to \infty$) puisque $\operatorname{mes}(\Gamma_0) > 0$. On peut donc se contenter de chercher la borne inférieure de $\mathcal{F}$ sur un ensemble borné donc faiblement compact. Toute fonction semi-continue inférieurement sur un ensemble compact atteignant sa borne inférieure, le problème a au moins une solution.

Supposons que $\vec{u} = 0$ presque partout sur l'ensemble Z de mesure non nulle de Γ_1, prenons $\vec{v} = \vec{u}$, $\gamma = \beta$ sur $\complement Z$, la relation (9) donne alors

$$- w \int_Z (\gamma - \beta)\, d\Gamma \geq 0$$

ce qui donne $\beta = 1$ presque partout sur Z.

13. PROBLEME D'ADHERENCE CLASSIQUE

La dernière propriété de la solution $(\beta, \vec{u})$ montre que l'adhérence est soit nulle, soit totale. On a d'ailleurs déjà noté cette propriété lors de l'interprétation des lois de comportement et d'équilibre (paragraphe 10).

Posons

$$K_1 = \{\vec{v} \mid \vec{v} \in \mathcal{V}\,;\ \vec{v} = 0 \text{ sur } \Gamma_0\ ;\ v_N \leq 0 \text{ sur } \Gamma_1\}$$

$$\mathcal{F}_1[\vec{v}] = \int_\Omega \rho\bar{\psi}(\varepsilon(\vec{v}))\, d\Omega - w\mathcal{H}(\vec{v}) - \int_\Omega \vec{f}\cdot\vec{v}\, d\Omega - \int_{\Gamma_2} \vec{g}\cdot\vec{v}\, d\Gamma$$

où $\mathcal{H}(\vec{v})$ est l'aire de la surface de contact :

$$\mathcal{H}(\vec{v}) = \text{mes}\{x \mid x \in \Gamma_1 \ ; \ \vec{v}(x) = 0\}.$$

Le problème (9) est alors équivalent à trouver $\vec{u}$ tel que

(10) $\vec{u} \in K_1, \quad \forall \vec{v} \in K_1, \ \mathcal{F}_1[\vec{v}] \geq \mathcal{F}_1[\vec{u}].$

On retrouve ici la formulation classique des problèmes de contact avec adhérence [5], [9] qui est présentée à partir de l'énergie potentielle $\mathcal{F}_1$. On peut d'ailleurs étudier directement ce problème en montrant que l'application $\vec{v} \longrightarrow \mathcal{H}(\vec{v})$ est faiblement semi-continue supérieurement.

14. UN EXEMPLE. RESSORT ADHERANT A SON SUPPORT

On peut faire la même théorie pour les milieux curvilignes. Certains paradoxes inattendus apparaissent. Nous les mettons en évidence sur un exemple : dans certaines circonstances la théorie de l'adhérence classique montre qu'il n'est pas possible de décoller deux solides qui adhèrent ! Les possibilités offertes par la théorie présentée ci-dessus permettent de surmonter cette difficulté et de se retrouver en accord avec l'expérience.

Considérons un ressort collé sur un socle rigide (figure 3) et soumis à l'action de la force d'intensité f. On suppose que l'effort extérieur A est nul. Le ressort est considéré comme une structure occupant un point de l'espace. Son déplacement comme toutes les quantités qui interviennent dans la théorie sont des scalaires.

Nous envisageons de chercher les diverses positions d'équilibre du ressort sous l'action de la force f. Les inconnues du problème sont les scalaires : u, déplacement du ressort ; σ, effort intérieur dans le ressort (tension) ; β, intensité d'adhésion ; F, énergie d'adhésion, -Q , effort de réaction du contact unilatéral. On suppose le ressort élastique

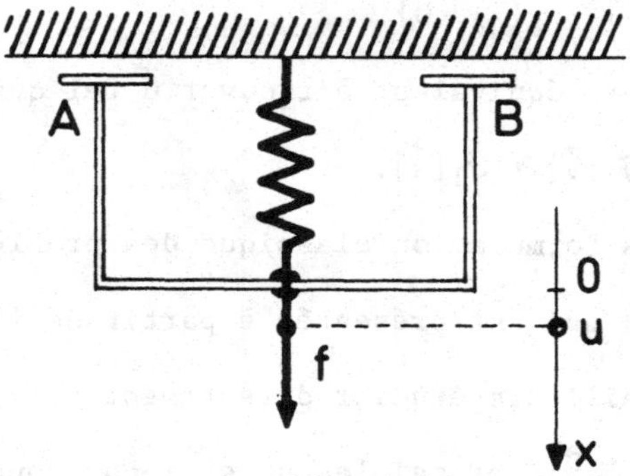

Figure 3. Le ressort adhère au support par l'intermédiaire de deux plaques A et B.

de rigidité k. L'énergie d'adhésion de Dupré est w.

Les équations du problème sont

$$\sigma = f, \quad F = 0, \quad \sigma = -Q,$$

$$(\sigma, Q, F) \in \partial \mathcal{A}(\beta, u),$$

avec

$$\mathcal{A}(\gamma, v) = \frac{1}{2} k v^2 - w\gamma + I_K(\gamma, v),$$

et

$$K = \{(\gamma, v) \mid \gamma \in \mathbb{R} \ ; \ v \in \mathbb{R} \ ; \ 0 \leq \gamma \leq 1 \ ; \ \gamma v = 0 \ ; \ v \geq 0\}.$$

On se donne f et l'on cherche u, β, F, σ, Q. Deux minima absolus sont possibles pour

$$\mathcal{F}[\gamma, v] = \frac{1}{2} k v^2 - f v - w\gamma,$$

sur K (figure 4).

Si $u = 0$ est minimum absolu, on a $f \leq \sqrt{2kw}$ et

$$0 = G(x) \leq w.$$

Si $u = \frac{f}{k}$ est minimum absolu on a, $f \geq \sqrt{2kw}$ et

$$G(x) = \frac{1}{2} k u^2 = \frac{1}{2} \frac{f^2}{k} \geq w.$$

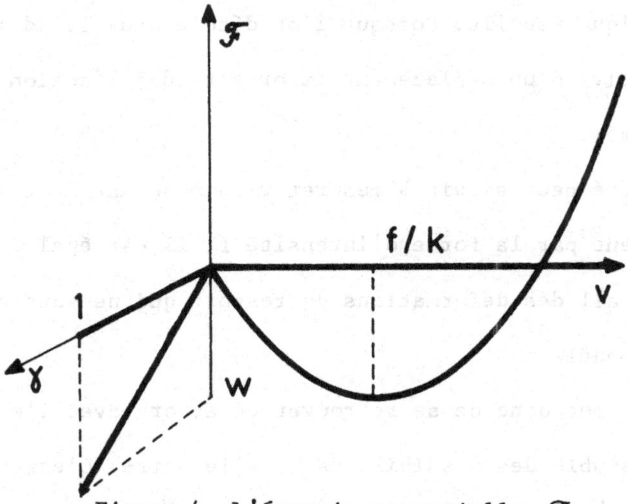

Figure 4. L'énergie potentielle $\mathcal{F}$.

La position u = 0 est position d'équilibre absolu ($\mathcal{W}=\mathcal{B}\times\mathcal{V}$) tant que $Q \leqslant \sqrt{2kw}$, valeur pour laquelle $u = \frac{Q}{k}$ devient position d'équilibre absolu et le reste pour $Q \geqslant \sqrt{2kw}$. Il est évident que $(u = 0, \beta = 1)$ reste position d'équilibre relatif $\mathcal{W} \neq \mathcal{B}\times\mathcal{V}$ pour toute valeur de f. Il semble donc que l'on ne puisse décoller le ressort de son support !

On peut surmonter cette difficulté en élargissant la classe des efforts extérieurs appliqués au ressort. Supposons qu'en plus de la force d'intensité f, on lui impose un travail d'adhésion négatif d'intensité fA (A<0) représentant le travail de déformation du solide tridimensionnel constituant le ressort. Ce travail n'est pas pris en compte par la modélisation usuelle des ressorts. L'équation d'équilibre F = 0 devient F = fA et l'on obtient facilement que u = 0 est position d'équilibre relatif tant que

$$- fv + \frac{1}{2} kv^2 - (w+fA)(\gamma-1) \geqslant 0,$$

pour $(\gamma, v) \in K$ et voisin de (1,0). Ceci est vérifié tant que

$$f \leqslant \frac{w}{-A} .$$

La position d'équilibre est ensuite $u = \frac{f}{k}$. Ceci correspond parfaitement à l'expérience physique usuelle. Lorsque l'on décole deux solides, le décollement est suivi d'un déplacement important dû à l'action des efforts extérieurs appliqués.

Cette propriété peut servir à mesurer w en mesurant le travail fourni avant le décollement par la force d'intensité f, il est égal à $-fA$ et représente le travail des déformations du ressort qui ne sont pas prises en compte par le modèle.

La théorie permet donc de se retrouver en accord avec l'expérience en utilisant l'ensemble des possibilités qu'elle offre. L'effort extérieur A est raisonnable car l'on sait bien qu'il faut "forcer" pour décoller deux pièces collées. Le travail de cet effort est $-fA$. Il est fourni sans déplacement notable des deux pièces. Il correspond a un travail produit sans que les déplacements cinématiques choisis (ici le déplacement du ressort) varient.

15. EXEMPLE DE COMPORTEMENT DISSIPATIF

Considérons le même ressort qu'au paragraphe précédent et supposons qu'il y a une dissipation représentant la dissipation dans la colle liant le ressort à son support. On prend

$$\varphi(\dot{\beta},\dot{\vec{u}}) = \frac{\bar{k}}{2}\dot{\beta}^2 + I_{K_2}(\dot{\beta})$$

avec

$$K_2 = \{\dot{\beta} \mid \dot{\beta} \leqslant 0\}.$$

On représente ici une colle qui rompt et n'adhère plus après rupture puisque β ne peut jamais augmenter. C'est le cas pour un papier à tapisser usagé que l'on décolle d'un mur. L'effort intérieur F est la somme d'un effort réversible ou élastique F_e et d'un effort irréversible F_i.

Les équations sont

$$\sigma = f, \quad f = -Q, \quad F = A,$$

$$(\sigma, Q, F_e) \in \partial\mathcal{A}(\beta, \vec{u}), \quad F_i \in \partial\varphi(\dot{\beta}), \quad F = F_i + F_e,$$

avec la condition initiale $\beta(0) = 1$ (ressort collé).

Les données du problème sont f et A et l'on cherche les fonctions du temps $\beta(t)$ et $u(t)$. Supposons qu'à un instant t l'on connaisse l'effort irréversible $F_i = A - F_e$, $\beta(t)$ et $u(t)$ sont alors tels qu'à cet instant

$$(\beta, \vec{u}) \in K, \quad \forall (\gamma, v) \in K \cap \mathcal{W}(\beta, \vec{u}), \quad \mathcal{F}[\beta, \vec{u}] \leq \mathcal{F}[\gamma, \vec{v}],$$

avec

$$\mathcal{F}[\gamma, v] = \frac{1}{2} k v^2 - f v - (w + A - F_i)\gamma .$$

On voit alors facilement que

1) si $w + A \geq 0$, $\beta(t) = 0$, $u(t) = 0$ pour tout t, il n'y a pas décollement car on ne tire pas assez fort,

2) si $w + A < 0$, il y a décollement au bout du temps $t_0 = \frac{-\bar{k}}{w+A}$. Entre les instants 0 et t_0, l'intensité d'adhésion décroît linéairement de 1 à 0 ($\bar{k}\dot{\beta} = w + A$). Les liaisons se rompent progressivement sans qu'aucun mouvement de la structure ne se produise. Cette dissipation rend compte d'un fait expérimental : pour séparer deux pièces collées il faut tirer avec force pendant un certain temps, la rupture n'étant pas immédiate. La figure 5 représente l'évolution de $\beta(t)$ et $u(t)$ lorsque $w + A < 0$.

16. CONCLUSION

Les méthodes de la mécanique des milieux continus classique permettent de rendre compte de l'adhérence en introduisant une nouvelle variable cinématique : l'intensité d'adhésion β. Elles permettent d'expliquer un paradoxe de la théorie classique de l'adhésion appliquée aux milieux curvilignes en introduisant un effort extérieur nouveau représentant le tra-

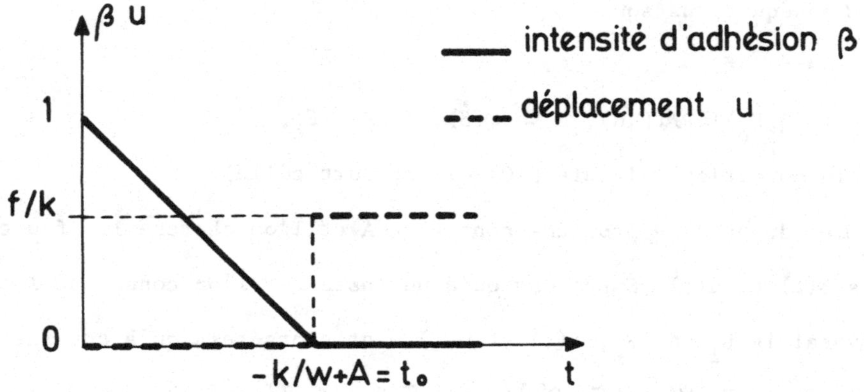

Figure 5. L'intensité d'adhésion β et le déplacement du ressort fonction du temps. Lorsque l'effort A est assez grand (A+w<0). Il y a décollement au temps t_0.

vail de déformations ignorées par la schématisation cinématique.

D'autres points peuvent être abordés : estimation de la force nécessaire à la séparation de deux solides ($\Gamma_0 = \phi$) [1], [2], [11], approximations numériques [12]. On peut à ce propos noter que la méthode présentée dans [13] est une approximation du problème (10). On peut aussi bâtir une théorie du premier gradient pour l'intensité d'adhésion β [11].

REFERENCES

[1] FREMOND, M., Adhérence des Solides. *Comptes Rendus de l'Académie des Sciences*, Série II, 295, Paris, 1982, p. 769-772.

[2] FREMOND, M., Equilibre de structures qui adhèrent à leur support. *Comptes Rendus de l'Académie des Sciences*, Série II, 295, Paris, 1982, p. 913-915.

[3] GERMAIN, P., *Mécanique des Milieux Continus*, MASSON, Paris, 1973.

[4] MOREAU, J.J., Fonctionnelles convexes. *Séminaire sur les équations aux dérivées partielles. Collège de France*, Paris, 1966.

[5] MAUGIS, D., M. BARQUINS, Fracture Mechanics and Adherence of Viscoelastic Solids dans *Adhésion and Adsorption of Polymers*, Plenum Publ. Corp. New York, 1980, part A, p. 203.

[6] GERMAIN, P., Sur l'application de la méthode des puissances virtuelles en mécanique des milieux continus. *Comptes Rendus de l'Académie des Sciences*, Série A, 274, Paris, 1972, p. 1051-1055.

[7] MOREAU J.J., Fonctions de résistance et fonctions de dissipation *(Séminaire d'Analyse Convexe,* Montpellier, 1971).

[8] HALPHEN, B., Q.S. NGUYEN, Sur les lois de comportement élastovisco-plastique à potentiel généralisé. *Comptes Rendus de l'Académie des Sciences,* Série A, 277, Paris, 1973, p. 319.

[9] MAUGIS, D., Adherence of solids dans *Microscopic Aspects of Adhesion and Lubrification,* J.M. Georges éd. ELSEVIER, 1982, p. 221.

[10] BUI, H.D., *Mécanique de la rupture fragile,* MASSON, Paris, 1978.

[11] FREMOND, à paraître.

[12] FREMOND, M., Adhésion et contact unilatéral. *International Symposium on Contact Mechanics and Wear of Rail/Wheel Systems,* VANCOUVER, (Canada), 6-9 July 1982.

[13] ASCIONE, L., D. BRUNO and A. GRIMALDI, *Plate delamination : an unilateral approach.* (dans ce volume).

ON THE CONTACT PROBLEM IN ELASTOSTATICS*

J.J. Kalker
Department of Mathematics and Informatics
Delft University of Technology

ABSTRACT

Some aspects of the contact between elastic bodies are presented. Starting with the formulation of the contact problem we proceed to give algorithms for contact formation and frictional contact, with emphasis on rolling, which is the most general form of contact with friction. Numerical results are shown, and a comparison with experiments is made.

*) This article is based on Ref. [10], with the kind permission of the International Center for Transportation Studies, Rome, Italy.

FORMULATION OF THE CONTACT PROBLEM

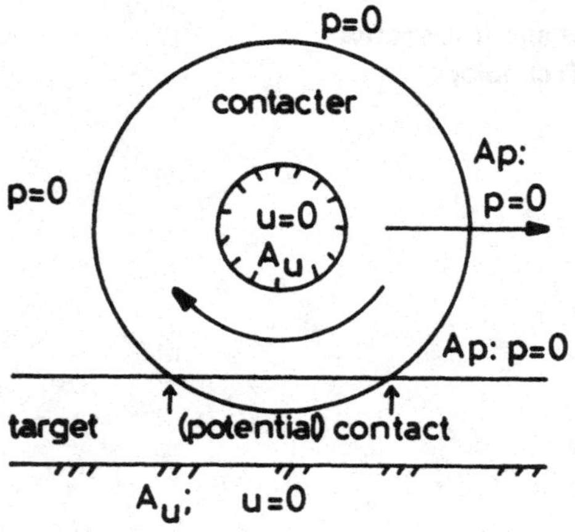

Fig. 1 - contacter and target in contact.

Consider two bodies in contact, called the contacter and the target, see Fig. 1. On part of the boundary, denoted by A_u, contacter and target are clamped, i.e. the surface displacement $\underline{u}$ is prescribed as $\underline{0}$. In another part of the boundary, denoted by A_p, the surface traction vanishes. In the remainder of the surface contacter and target are, or can be in contact: the potential contact area A_c. In the potential contact the laws of contact mechanics prevail.

In Fig. 1 contacter and target are shown in the unstressed state corresponding to $\underline{u} = \underline{0}$ in A_u. This state we call the reference state. Note that the reference states of contacter and target may move with respect to each other in the course of time. As a consequence of the fact that $\underline{u} = \underline{0}$ in A_u, contacter and target, which are unstressed, penetrate. This penetration is counteracted by the elastic deformation, which causes an elastic field to appear in contacter and target, so that a surface load comes into being where contacter and target touch. Contacter and target are now said to be in the deformed state.

We distinguish two aspects of the problem:

1. Contact formation.

2. Friction.

Contact formation

In the reference state contacter and target intersect, see Fig. 2.

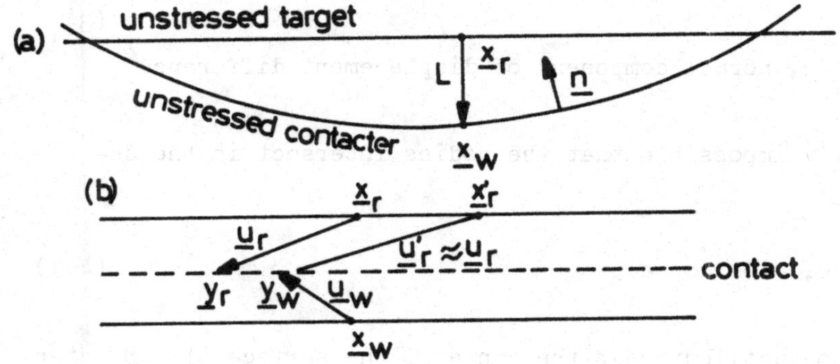

Fig. 2. - Contact formation a: unstressed state, b: deformation.

Two surface points $\underline{x}_w$ and $\underline{x}_r$ of contacter and target are called <u>opposing</u> when they lie on the same inner normal $\underline{n}$ to the contacter. The distance $L(\underline{x}_r)$ in the reference state is measured along this normal. It is taken negative where contacter and target intersect, and positive or zero otherwise. We have that:

$$L = (\underline{x}_w - \underline{x}_r)\underline{n} \quad \text{(inner product is meant)} \tag{1}$$

Now the bodies are deformed, so that a displacement is added to $\underline{x}_w$ and $\underline{x}_r$, and they come to lie in

$$\underline{y}_w = \underline{x}_w + \underline{u}_w \ , \ \underline{y}_r = \underline{x}_r + \underline{u}_r \qquad (2)$$

In general, $\underline{y}_r$ will not coincide with $\underline{y}_w$, see Fig. 2b.
Owing to the assumed smallness of the displacements and their gradients, the distance between the bodies in the deformed state at $\underline{x}_r$ may be approximated by:

$$d = (\underline{y}_w - \underline{y}_r)\underline{n} = (\underline{x}_w - \underline{x}_r)\underline{n} + (\underline{u}_w - \underline{u}_r)\underline{n} = L + u_n$$

$$u_n \overset{\text{def}}{=} (\underline{u}_w - \underline{u}_r)\underline{n}, \text{ normal component of displacement difference} \qquad (3)$$

Now it is physically impossible that the bodies intersect in the deformed state, hence

$$d = L + u_n \geq 0, \ = 0\text{: contact} \qquad (4.1)$$

Further, outside contact, but near the contact, the surface of contacter and target is free of traction $\underline{p}$(= load per unit area exerted on the surface of the bodies):

$$\underline{p} = 0 \quad \text{outside but near contact} \qquad (4.2)$$

It is experimentally verified that ordinarily the normal contact force is compressive:

$$\underline{p}_w \cdot \underline{n} \geq 0 \quad \text{inside contact} \qquad (4.3)$$

Finally it follows from Newton's Third Law that

$$\underline{p}_r = \underline{p}_w \quad \text{in contact} \qquad (4.4)$$

(4.1,2,3,4,) constitute the conditions of contact formation.
Note that (4.1) and (4.3) are <u>inequalities</u>.

Friction

We assume Coulomb friction. Let $\underline{p}$ be the load per unit area exerted on the contacter at the contact, and $\underline{w}$ is the slip, defined as the local velocity of sliding of the contacter with respect to the target. p_n is the normal component of $\underline{p}$, where $\underline{n}$ is the inner normal on the contacter:

$$p_n \overset{\text{def}}{=} \underline{p} \cdot \underline{n} \geq 0, \quad \text{see (4.3)} \quad \text{normal pressure} \tag{5.1}$$

and $\underline{p}_t$ is the tangential component of $\underline{p}$, defined by

$$\underline{p}_t = \underline{p} - (\underline{p} \cdot \underline{n})\underline{n} = \underline{p} - p_n\underline{n} \quad \text{tangential traction} \tag{5.2}$$

Then we have that

$$\begin{aligned} &|\underline{p}_t| \geq fp_n \qquad f : \text{coefficient of friction;} \\ &\text{if } \underline{\dot{w}} \neq \underline{0} : \underline{p}_t = -(fp_n)\underline{\dot{w}}/|\underline{\dot{w}}| \qquad \underline{\dot{w}} : \text{slip} \end{aligned} \tag{6}$$

This is Coulomb's law in local form. When the slip $\underline{\dot{w}}$ vanishes (this happens in the region of the contact area called area of adhesion) the local tangential traction falls normally below fp_n, while in the remainder of the contact area $\underline{\dot{w}} \neq \underline{0}$ (the area of slip) the local traction equals (fp_n) in magnitude, and is opposedly directed to the slip.

Synthesis

We get the following formulation of the contact problem:

$$\begin{aligned} &\underline{u} = \underline{0} \text{ in } A_u && \text{(See Fig. 1)} \\ &\underline{p} = \underline{0} \text{ in } A_p && \text{(See Fig. 1)} \\ &p = \underline{0} \text{ outside but near the contact} && \end{aligned} \tag{7}$$

$$d = L + u_n \geq 0 \text{ in and near contact } , = 0 \text{ in contact} \quad (7)$$

$$p_n \overset{\text{def}}{=} \underline{p} \cdot \underline{n} \geq 0$$

$$|p_t| \leq fp_n \qquad \text{inside contact}$$

$$\text{if } \underline{\dot{w}} \neq \underline{0}:\ p_t = -(fp_n)\underline{\dot{w}} / |\underline{\dot{w}}|$$

This is a complicated boundary value problem. Indeed, up to now a general proof of the existence and uniqueness of the elastic field of this problem has not been given. This is an indication that there are difficulties in the formulation; indeed, in the numerical evaluation such difficulties have been encountered by myself and by Kikuchi [22]. We return to this at the end of this section. The problem may be split into two partial problems in which such difficulties do not occur.
They are:

N $\underline{u} = \underline{0}$ in A_u

$\underline{p} = \underline{0}$ in A_p

$\underline{p}_t$ is given in contact — normal

$d = L + u_n = 0$ in contact, ≥ 0 outside — contact (8)

$p_n \geq 0$ in contact — problem

F $\underline{u} = 0$ in A_u

$\underline{p} = 0$ in A_p — frictional

p_n is given in contact — contact (9)

$|\underline{p}_t| \leq fp_n$ — problem

if $\underline{\dot{w}} \neq 0$: $\underline{p}_t = -(fp_n)\underline{\dot{w}} / |\underline{\dot{w}}|$

The problem N - (8) possesses a unique solution [2], in problem (9) existence and uniqueness are guaranteed if the problem is a so-called shift [3]: that is, two bodies are pressed together, and they are displaced and rotated with respect to each other over a distance comparable to the displacement, while the normal pressure at each point of the contact is kept constant, see Fig. 3. The local shift of the contacter with respect to the target denoted by $\underline{w}$, is given by

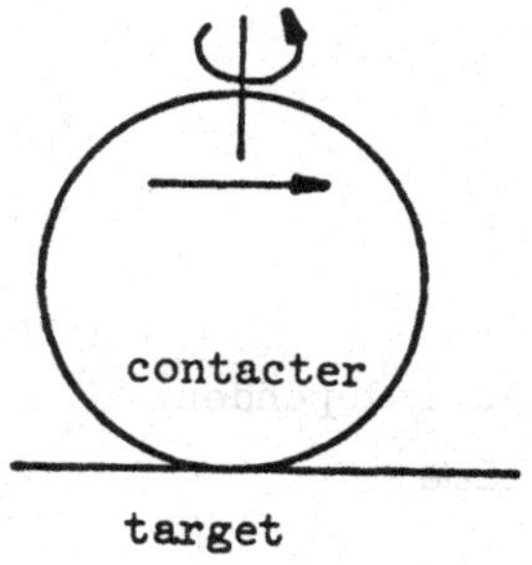

Fig. 3 - The shift.

$$\underline{w}(\underline{x}) = \{\underline{s}+\underline{u}_{t,w}-\underline{u}_{t,r}\}|_{\underline{x}} \tag{10}$$

where $\underline{s}$ is the displacement at $\underline{x}$ of the reference state of the contacter with respect to the reference state of the target, while $\underline{u}_{t,r}$ and $\underline{u}_{t,w}$ are the tangential components of the surface displacement of target and contacter, all taken at the position $\underline{x}$.

Generalisation of the shift is the evolution, which is defined as a time sequence of shifts. As far as the author is aware, no existence-uniqueness proof has been given for evolutions which are a continuous time sequence of shifts.

Rolling is an example of such a continuous time sequence. In discretised form, by which we mean that we approximate the evolution by a finite number of shifts, we have for the local displacement of the contacter with respect to the target at the time t, when the previous shift took place at the time $(t-\tau)$:

$$\dot{\underline{w}}(\underline{x})\tau = \dot{\underline{s}}\tau+\{\underline{u}_{t,w}(\underline{x},t)-\underline{u}_{t,r}(\underline{x},t)\}-\{\underline{u}_{t,w}(\underline{x},t-\tau)-\underline{u}_{t,r}(\underline{x},t-\tau)\} \tag{11}$$

Here, $\{\underline{u}_{t,w}(\underline{x},t)-\underline{u}_{t,r}(\underline{x},t)\}$ is the displacement of the particle $\underline{x}$ of the contacter at the time t with respect to the position of the opposing particle $\underline{x}$ of the target at the time t. $\{\underline{u}_{t,w}(\underline{x},t-\tau)-\underline{u}_{t,r}(\underline{x},t-\tau)\}$ is similarly defined for the instant $(t-\tau)$. Note that we speak of particles at the opposing reference positions $\underline{x}$, as we may, since displacements and displacement gradients are small, see the discussion in sec. "Contact

Formation". $\underline{\dot{s}}\tau$ is the change of relative position of the reference states of contacter and target during the time interval $(t-\tau,t)$. $\underline{\dot{s}}$ has the dimension of a velocity; it is a rigid translation and a rigid rotation, and has the form, if (x,y) are the Cartesian coordinates in the plane of contact,

$$\underline{\dot{s}} = V(\upsilon_x - \phi y,\ \upsilon_y + \phi x),$$

with V: rolling velocity, (12)

$\upsilon_x, \upsilon_y, \phi$: longitudinal, lateral, spin creepages independent of position, possibly dependent on time

In order to simulate continuous time sequences of shifts (continuous evolutions), one should take τ smaller and smaller and apply a limiting process. Here arises a difficulty, which we know from the numerical integration of ordinary differential equations, namely that a smaller time step τ yield a higher accuracy, but needs more calculating time. A larger time step has the drawback that the numerical process may converge to a very inaccurate solution. In our case we solve the difficulty by choosing τ, calculating for τ and $\frac{1}{2}\tau$, and extrapolating to $\tau = 0$. Especially in the cases that spin creepage is dominant, this has a beneficial effect.

In the program Duvorol (1978) we find the solution of steady state rolling as the steady-state limit of an evolution in which the creepages and the normal pressure and contact area are kept constant. Existence of such a limiting state has not been proved, but work with the program Duvorol indicates that the limit does indeed exist.

In the program Contact (1982) the steady state is calculated directly. Here also we worked with a discretised evolution, as follows. We observe that $\underline{u}_t(\underline{x},t)$ and $\underline{u}_t(x,t-\tau)$ are caused by the same pressure distribution, see Fig. 4.:

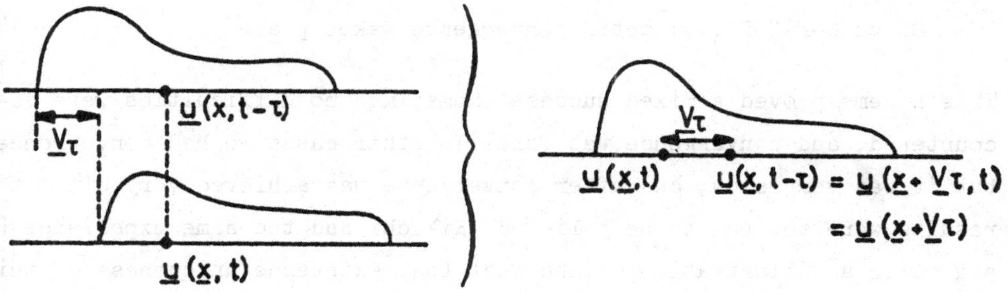

Fig. 4 - $\underline{u}(\underline{x},t)$ and $\underline{u}(\underline{x},t-\tau) = \underline{u}(\underline{x} + \underline{V}\tau,t)$.

The property that $\underline{u}(\underline{x},t-\tau) = \underline{u}(\underline{x}+\underline{V}\tau,t) = \underline{u}(\underline{x}+\underline{V}\tau)$ (t is immaterial) is exploited:

$$\underline{\dot{w}}(\underline{x})\tau = \underline{\dot{s}}\tau+\{\underline{u}_{t,w}(\underline{x})-\underline{u}_{t,r}(\underline{x})\}-\{\underline{u}_{t,w}(\underline{x}+\underline{V}\tau)-\underline{u}_{t,r}(\underline{x}+\underline{V}\tau)\} \tag{13}$$

from which explicit time t has vanished. Here also an extrapolation over τ is performed. Again no existence/uniqueness proof has been given when (13) is inserted into (9). The program Contact has been built on it, and whereas in the great majority of cases no difficulties are encountered, there are situations in which the process fails and Contact has to shift back to a Duvorol-type solution. When the method "(13) into (9)" works, it is generally much faster than Duvorol.

In the problem N and F, see (8) and (9), the tangential traction $\underline{p}_t$ must be given (N) or the normal pressure is supposedly known (F). In general, the tangential traction affects the deformed distance d. As d = 0 in the contact area, the normal pressure is affected, and hence, at any rate through the friction mechanism, the tangential traction changes. Hence the normal pressure changes, etc. In order to calculate the general contact problem (7) the program Contact follows a proposal by Panagiotopoulos [4]:

Solve N-F-N-F .... until convergence takes place (14)

This scheme proved a mixed success. Sometimes no difficulties were encountered, and convergence was fast. In other cases we had convergence, but it was very slow, and after convergence was achieved very often the results were too bad to be believed. Kikuchi had the same experience [22], and it is an illustration of the fact that existence/uniqueness of this problem can be proved only conditionally. An example of this will be treated later on in this paper.

In some technologically very important cases, the normal problem is independent of the tangential traction. Then the process (14) stops after a single cycle N.F.

This is called the Johnson process (5); Johnson even proposed to use it in the general, "Panagiotopoulos", case as an approximation.

THE HALF-SPACE APPROXIMATION

The contact between contacter and target is often small with respect to the dimension of those bodies. Then, a good approximation is possible, as follows.

In elasticity theory a very convenient abstraction is the point load. This is the limit of loads of fixed resulting magnitude and direction, whose area of application dwindles to a point. Another very convenient abstraction is the elastic half-space which is an elastic body that in a Cartesian coordinate system (x_1,x_2,x_3), occupies, say, the region $\{(x_1,x_2,x_3)|x_3 \geq 0\}$. The stresses due to a point load acting on the boundary of a half-space fixed at infinity die out, away from the point of application, $O(1/r^2)$, where r is the distance of the point of observation to the point of application of the load.

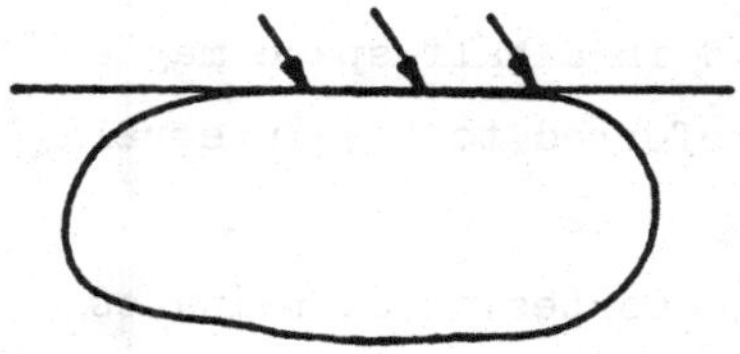

Fig. 5 - An elastic body embedded in a half-space.

Now consider an elastic body embedded in a half-space, see Fig. 5. We simulate the contact by a number of point loads lying close together in a small area of application. We consider the elastic field in the half-space, and in particular the stress on the surface of the rail head. On the upper part of the elastic body the surface stress of elastic body and half-space coincide. The remainder of the elastic body is a few contact diameters away from the contact area, hence by the fast decay of the stress in the half-space the stress on it is small. We will

neglect that stress, which means that as far as the elastic field is concerned we approximate the elastic body by a half-space.
It should be stressed that the boundary conditions, involving among others the undeformed distance L, see eq.(15), are calculated for the real bodies, and that only the elastic field is calculated as if contacter and target were half-spaces.
The reason for the half-space approximation is simplification:

1. The elastic field in all forms and types of contacter and target, whatever their relative position, is calculated for one and the same geometry.
2. The elastic field in an elastic half-space may be explicity expressed in the surface loads, see sec. "The integral interpretation of Boussinesq-Cerruti".
   Knowing that for many purposes, such as vehicle simulation, half-space contact programs are already too slow, it will be clear that in this field half-space calculations are the most accurate that can be achieved.

The integral representation of Boussinesq-Cerruti

In 2 point 2 we mentioned that the elastic field in a half-space may be explicitly expressed in the surface loads. We referred to the integral representation of Boussinesq and Cerruti.
In the upper half-space (contacter) we introduce Cartesian coordinates (x_1,x_2,x_3), where the axis 1 and 2 lie in the surface of the half-space, and the axis 3 points normally inwards. Then it was shown by Boussinesq and Cerruti (see A.E.H. Love [1], pg.192, 243) that

$$\underline{u}(\underline{x}) = \frac{1+\tilde{\nu}}{\pi\tilde{E}} \iint_C H(\underline{x}-\underline{x}')\,p(\underline{x}')\,dx_1'dx_2' \qquad \tilde{\nu}: \text{Poisson's ratio (contacter)}$$

$$\tilde{E}: \text{Young's modulus (contacter)}$$

C: contact area; $|\underline{y}| = \sqrt{y_1^2 + y_2^2 + y_3^2}$;

with: $\underline{u} = (u_1,u_2,u_3)^T$, surface displacement
$\underline{p} = (p_1,p_2,p_3)^T$, surface load (/unit area)
$\underline{x} = (x_1,x_2,0)^T$, observation point
$\underline{x}' = (x_1',x_2',0)^T$, application point (15)

$$\text{and}\quad H(\underline{y}) = (H_{ij}) = \begin{bmatrix} \frac{1-\tilde{\nu}}{|\underline{y}|} + \frac{\tilde{\nu}y_1^2}{|\underline{y}|^3}, & \frac{\tilde{\nu}y_1y_2}{|\underline{y}|^3}, & -\frac{(1-2\tilde{\nu})y_1}{|\underline{y}|^2} \\ \frac{\tilde{\nu}y_1y_2}{|\underline{y}|^3}, & \frac{1-\tilde{\nu}}{|\underline{y}|} + \frac{\tilde{\nu}y_2^2}{|\underline{y}|^3}, & -\frac{(1-2\tilde{\nu})y_2}{|\underline{y}|^2} \\ \frac{(1-2\tilde{\nu})y_1}{|\underline{y}|^2}, & \frac{(1-2\tilde{\nu})y_2}{|\underline{y}|^2}, & \frac{1-\tilde{\nu}}{|\underline{y}|} \end{bmatrix}$$

The load on the lower half-space (target) is the reaction to the load on the contacter, see (16). We denote quantities of the target by an overbar. We have

$$\bar{\underline{p}}(\underline{x}) = -\,\underline{p}(\underline{x}) \tag{16}$$

The displacement in the lower half-space is as above, but with components in the $(\bar{1},\bar{2},\bar{3})$ coordinate system, of which the $\bar{1},\bar{2}$ axes coincide with the 1,2 axes, but the $\bar{3}$-axis points normally inward into the lower half-space, hence

$$\bar{3}\text{ - component of a vector} = -(3\text{-component of the same vector}) \tag{17}$$

We find the analogue of (15) for the lower half-space by replacing all vectors by their overbar version, and also the components should be given an overbar. Finally, Poisson's ratio and Young's modulus should be adapted: $\bar\nu$, $\bar E$.

Now, in contact formation, we need the normal displacement difference $u_3-u_3 = (u_3+\bar u_{\bar 3})$; for the slip, we are interested in the tangential components of the displacement difference $u_i-\bar u_i$; $i = 1,2$:

$$\begin{aligned} &u_1-\bar u_1 = u_1-\bar u_{\bar 1};\ u_2-\bar u_2 = u_2-\bar u_{\bar 2};\ u_3-\bar u_3 = u_3+\bar u_{\bar 3} \\ &\bar p_{\bar 1} = -\,p_1,\ \bar p_{\bar 2} = -\,p_2;\ \bar p_{\bar 3} = p_3 \end{aligned} \tag{18}$$

That is,

$$\begin{aligned} u_3-\bar u_3 = u_3+\bar u_{\bar 3} = \frac{1}{\pi}\iint_C &\left\{\frac{1}{|\underline{x}-\underline{x}'|}\left(\frac{1-\tilde\nu^2}{\tilde E}\,p_3 + \frac{1-\bar\nu^2}{\bar E}\,\bar p_{\bar 3}\right.\right. + \\ &+ \frac{x_1-x_1'}{|\underline{x}-\underline{x}'|^2}\left(\frac{(1+\tilde\nu)(1-2\tilde\nu)}{\tilde E}\,p_1 + \frac{(1+\bar\nu)(1-2\bar\nu)}{\bar E}\,\bar p_{\bar 1}\right. + \\ &\left.+ \frac{x_2-x_2'}{|\underline{x}-\underline{x}'|^2}\left(\frac{(1+\tilde\nu)(1-2\tilde\nu)}{\tilde E}\,p_2 + \frac{(1+\bar\nu)(1-2\bar\nu)}{\bar E}\,\bar p_{\bar 2}\right)\right\}dx'_1 dx'_2 \end{aligned} \tag{a}$$

$$\begin{aligned} = \frac{1}{\pi}\iint_c &\left\{\left(\frac{1-\tilde\nu^2}{\tilde E} + \frac{1-\bar\nu^2}{\bar E}\right)\frac{p_3}{|\underline{x}-\underline{x}'|} + \right. \\ &\left.+ \left(\frac{(1+\tilde\nu)(1-2\tilde\nu)}{\tilde E} - \frac{(1+\bar\nu)(1-2\bar\nu)}{\bar E}\right)\frac{(x_1-x_1')p_1+(x_2-x'_2)p_2}{|\underline{x}-\underline{x}'|^2}\right\}dx_1\ dx_2 \end{aligned}$$

$$\begin{aligned} u_1-\bar u_1 = u_1-\bar u_{\bar 1} = \frac{1}{\pi}\iint_c &\left\{\left(\frac{1-\tilde\nu^2}{\tilde E} + \frac{1-\bar\nu^2}{\bar E}\right)\frac{p_1}{|\underline{x}-\underline{x}'|} + \right. \\ &+ \left(\frac{(1+\tilde\nu)\tilde\nu}{\tilde E} + \frac{(1+\bar\nu)\bar\nu}{\bar E}\right)\left(\frac{p_1(x_1-x_1')^2}{|\underline{x}-\underline{x}'|^3} + \frac{p_2(x_1-x_1')(x_2-x_2')}{|\underline{x}-\underline{x}'|^3}\right) + \\ &\left.- \left(\frac{(1+\tilde\nu)(1-2\tilde\nu)}{\tilde E} - \frac{(1+\bar\nu)(1-2\bar\nu)}{\bar E}\right)\frac{p_3(x_1-x_1')}{|\underline{x}-\underline{x}'|^2}\right\}dx_1'\ dx_2' \end{aligned} \tag{b}$$

$u_2 - \bar{u}_2$: interchange 1 and 2 in (b). (c)

Symbolically, the displacement difference $(\underline{u}-\underline{\bar{u}})$ is given by

$$\underline{u}(\underline{x})-\underline{\bar{u}}(\underline{x}) = \frac{1}{\pi G} \iint_C K(\underline{x}-\underline{x}')\underline{p}(\underline{x}')dx_1', \; dx_2' \; ,$$

$$\text{with } K(\underline{y}) = \begin{bmatrix} \frac{1-\nu}{|\underline{y}|} + \frac{\nu y_1^2}{|\underline{y}|^3} & , \frac{\nu y_1 y_2}{|\underline{y}|^3} & , \frac{K y_1}{|\underline{y}|^3} \\ \frac{\nu y_1 y_2}{|\underline{y}|^3} & , \frac{1-\nu}{|\underline{y}|} + \frac{\nu y_2^2}{|\underline{y}|^3} & , -\frac{K y_2}{|\underline{y}|^2} \\ \frac{K y_1}{|\underline{y}|^2} & , \frac{K y_2}{|\underline{y}|^2} & , \frac{1-\nu}{|\underline{y}|} \end{bmatrix} \tag{20}$$

where

$$\frac{1}{G} \overset{\text{def}}{=} \frac{1+\tilde{\nu}}{\tilde{E}} + \frac{1+\bar{\nu}}{\bar{E}}; \text{ G: combined modulus of rigidity}$$

$$\frac{\nu}{G} \overset{\text{def}}{=} \frac{(1+\tilde{\nu})\tilde{\nu}}{\tilde{E}} + \frac{(1+\bar{\nu})\bar{\nu}}{\bar{E}}; \; \nu\text{: combined Poisson's ratio}$$

$$\frac{K}{G} \overset{\text{def}}{=} \frac{(1+\tilde{\nu})(1-2\tilde{\nu})}{\tilde{E}} - \frac{(1+\bar{\nu})(1-2\bar{\nu})}{\bar{E}} \; ; \text{ K: difference parameter;}$$

when $\bar{E}=\tilde{E}$, $\bar{\nu}=\tilde{\nu}$ as in the steel-on-steel contact of a railway wheel on a rail, then $G=\frac{1}{2}\tilde{E}/(1+\tilde{\nu})$, modulus of rigidity of steel$= 0.82\text{x}10^{11}\text{N/m}^2$. $\nu=\tilde{\nu}=\bar{\nu}= 0.28$, $K=0$.

Discretisation

In our numerical algorithms we discretise the contact area and its surroundings by a set of rectangular meshes, see Fig. 6. We consider the centers of the meshes as their representative points. We take the pressure constant in each rectangular mesh. We can calculate exactly the displacement difference in an arbitrary surface point of the half-spaces, due to the pressure distribution over one mesh. In order to see that, we observe that we need the following six integrals [6]:

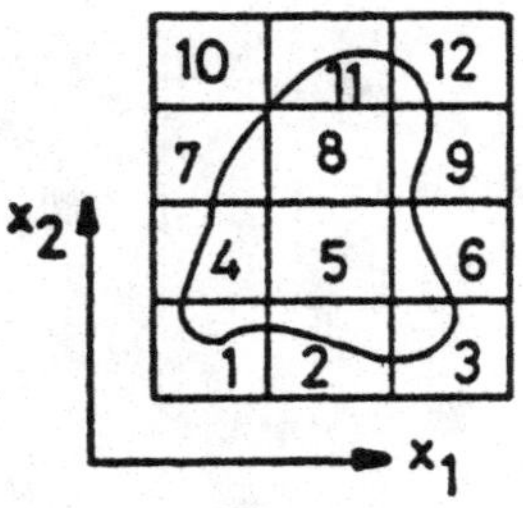

Fig. 6 - Discretisation of the contact area.

$$I_1 = \iint_M \frac{dx_1' \, dx_2'}{|\underline{x}-\underline{x}'|}$$

$$I_2 = \iint_M \frac{(x_1-x_1')^2}{|\underline{x}-\underline{x}'|^3} dx_1' \, dx_2'$$

$$I_3 = \iint_M \frac{(x_1-x_1')(x_2-x_2')}{|\underline{x}-\underline{x}'|^3} dx_1' \, dx_2'$$

$$I_4 = \iint_M \frac{(x_2-x_2')^2}{|\underline{x}-\underline{x}'|^3} dx_1' \, dx_2' \tag{4}$$

$$I_5 = \iint_M \frac{x_1-x_1'}{|\underline{x}-\underline{x}'|^2} dx_1' \, dx_2'$$

$$I_6 = \iint_M \frac{x_2-x_2'}{|\underline{x}-\underline{x}'|^2} dx_1' \, dx_2' \qquad I_j = I_j(\underline{x},\underline{x}_p,\Delta\underline{x}_p), \; j = 1,\dots,6$$

M is the rectangle with vertices $(x_{1p} \pm \Delta x_1/2,\; x_{2p} \pm \Delta x_2/2)$,

Δx_1, Δx_2: length of the sides of the discretisation rectangles. These sides are parallel to the x_1, x_2 axes.

Using the following abbreviated notations

$$[\dots.]_i = [\dots.]_{x_i-x_{ip}-\Delta x_i/2}^{x_i-x_{ip}+\Delta x_i/2} \tag{22}$$

we have that

$$\iint_M \frac{\partial^2 f}{\partial x_1' \partial x_2'} dx_1' \, dx_2' = [[f(x_1',x_2')]_1]_2$$

and it may be verified that, with $sh^{-1} z \overset{def}{=} \ln(z + \sqrt{z^2+1})$,

$$\frac{\partial^2}{\partial x_1' \partial x_2'} \{x_2' \, sh^{-1}(x_1'/|x_2'|) + x_1' \, sh^{-1}(x_2'/|x_1'|)\} = (x_1'^2 + x_2'^2)^{-1/2}$$

$$\frac{\partial^2}{\partial x_1' \partial x_2'} \{x_2' \, sh^{-1}(x_1'/|x_2'|)\} = x_1'^2 (x_1'^2 + x_2'^2)^{-3/2}$$

$$\frac{\partial^2}{\partial x_1' \partial x_2'} \{-(x_1'^2 + x_2'^2)^{1/2}\} = x_1' x_2' (x_1'^2 + x_2'^2)^{-3/2} \qquad (23)$$

$$\frac{\partial^2}{\partial x_1' \partial x_2'} \{1/2 \, x_2' \ln(x_1'^2 + x_2'^2) + x_1' \tan^{-1}(x_2'/x_1')\} = x_1'(x_1'^2 + x_2'^2)^{-1}$$

By means of (22) and (23) we can readily calculate the integrals (21). From the above integrals we can form the discretisations suitable for the normal and the frictional problem.

The discretisation for the normal contact problem

We number the elements of the rectangular net consecutively from 1 to N, see Fig. 6. The centers of the elements are denoted by $\underline{x}_I$. Then, by (20), the definition (21) of the I_j and the evaluation (23), the normal displacement difference $(u_3 - u_3)$ at $\underline{x}$ due to a load/unit area which is constant over the meshes of the net, with components (p_{Ii}), i = 1,2,3 is given by

$$(u_3-\bar{u}_3)(\underline{x}) = \frac{1}{\pi G}\sum_{I=1}^{N}\{K\, p_{I1} I_5(\underline{x},\underline{x}_I,\Delta\underline{x}_I)+Kp_{I2}I_6(\underline{x}.\underline{x}_I,\Delta\underline{x}_I)$$
$$+ (1-\nu)p_{I3}I_1(\underline{x},\underline{x}_I,\Delta\underline{x}_I)\} \tag{24}$$

Note that p_{I1}, p_{I2} must be prescribed, unless K = 0, as is the case when the elastic constants are equal, which happens, for instance, in wheel-rail contact. It is seen from (24) that the tangential traction then does not influence the deformed distance d, see (3). The discretised $(u_3-\bar{u}_3)$ are best defined in the centers of the elements.
We call $(u_3-\bar{u}_3)(\underline{x}_J) \overset{def}{=} (u_{J3}-\bar{u}_{J3})$, and we have

$$(u_{J3}-\bar{u}_{J3}) = \frac{1}{\pi G}\sum_{I=1}^{N}\{Kp_{I1}I_5(\underline{x}_J,\underline{x}_I,\Delta\underline{x}_I) + Kp_{I2}I_6(\underline{x}_J,\underline{x}_I.\Delta\underline{x}_I) +$$
$$+ (1-\nu)p_{I3}I_1(\underline{x}_J,\underline{x}_I,\Delta\underline{x}_I)\} \tag{25}$$

$$u_{J3} = u_3(\underline{x}_J),\ \bar{u}_{J3} = \bar{u}_3(\underline{x}_J)\ ;\ I,J = 1, \ldots, N.$$

The discretisation for the shift

We start with (10),

$$\underline{w}(\underline{x}) = (\underline{s}+\underline{u}_{t,w}-\underline{u}_{t,r})|_{\underline{x}} \quad \underline{w}\text{: local shift}$$

$$\underline{u}_t = \underline{u}_{t,w}\text{: tangential component of the displacement of the contacter} \tag{10}$$

$$\bar{\underline{u}}_t = \underline{u}_{t,r}\text{: tangential component of the displacement of the target.}$$

The local shift due to a load/unit area (p_{I1},p_{I2},p_{I3}), I = 1, ..., N acting on the contacter which is constant over each discretisation mesh, is

$$\begin{bmatrix} w_1(\underline{x}) \\ w_2(\underline{x}) \end{bmatrix} = \begin{bmatrix} s_1(\underline{x}) \\ s_2(\underline{x}) \end{bmatrix} +$$

$$+ \frac{1}{\pi G} \sum_{I=1}^{N} \begin{bmatrix} (1-\nu)I_1+\nu I_2, \nu I_3 & , -KI_5 \\ \nu I_3 & ,(1-\nu)I_1+\nu I_4, -KI_6 \end{bmatrix} \begin{bmatrix} p_{I1} \\ p_{I2} \\ p_{I3} \quad \text{known} \end{bmatrix} \tag{26}$$

$I_i = I_i(\underline{x},\underline{x}_I,\Delta\underline{x}_I)$; see sec. "discretisation"; $i = 1, \ldots, 6$

(27)

Here we also define $w_{Jt}(t= 1,2)(J= 1, \ldots, N)$ as $w_t(\underline{x}_J)$; s_{Jt} is simular.

$$\begin{bmatrix} w_{J1} \\ w_{J2} \end{bmatrix} = \begin{bmatrix} s_{J1} \\ s_{J2} \end{bmatrix} +$$

$$+ \frac{1}{\pi G} \sum_{I=1}^{N} \begin{bmatrix} (1-\nu)I_1+\nu I_2, \nu I_3 & , -KI_5 \\ \nu I_3 & ,(1-\nu)I_1+\nu I_4, -KI_6 \end{bmatrix} \begin{bmatrix} p_{I1} \\ p_{I2} \\ p_{I3} \quad \text{known} \end{bmatrix} \tag{28}$$

$I_i = I_i(\underline{x}_J,\underline{x}_I,\underline{x}_I)$; $i=1, \ldots 6$; $I=1, \ldots, N$; $J=1, \ldots, N$

<u>The discretisation of an evolution</u>

A continuous evolution was defined as a continuous sequence of shifts. We will discretise it not only in space, but also in time. Point of departure is (11):

$$\begin{aligned} \underline{\dot{w}}(\underline{x})\tau &= \underline{\dot{s}}\tau + \{\underline{u}_{t,w}(\underline{x},t) - \underline{u}_{t,r}(\underline{x},t)\} - \{\underline{u}_{t,w}(\underline{x},t-\tau) - \underline{u}_{t,r}(\underline{x},t-\tau)\} \\ &= \underline{\dot{s}}\tau + \{\underline{u}_t - \underline{\bar{u}}_t\}(\underline{x},t) - \{\underline{u}_t - \underline{\bar{u}}_t\}(\underline{x},t-\tau) \end{aligned} \tag{11}$$

Hence

$$\begin{bmatrix} \dot{w}_1(\underline{x})\tau \\ \dot{w}_2(\underline{x})\tau \end{bmatrix} = \begin{bmatrix} \dot{s}_1\tau-(u_1-\bar{u}_1)(\underline{x},t-\tau) \\ \dot{s}_2\tau-(u_2-\bar{u}_2)(\underline{x},t-\tau) \end{bmatrix} \begin{pmatrix} \text{known} \\ \text{term} \end{pmatrix} +$$

$$+ \frac{1}{\pi G} \sum_{I=1}^{N} \begin{bmatrix} (1-\nu)I_1+\nu I_2, \nu I_3 & , -KI_5 \\ \nu I_3 \quad ,(1-\nu)I_1+\nu I_4, & -KI_6 \end{bmatrix} \begin{bmatrix} p_{I1} & \\ p_{I2} & \\ p_{I3} & \text{known} \end{bmatrix} \tag{29}$$

$$I_i = I_i(\underline{x}_J, \underline{x}_I, \Delta\underline{x}_I)$$

Defining $w_{Jt}(t = 1,2)$ according to (27), we easily arrive at the analogue of (29), which we will not write out.

The discretisation for steady-state rolling

Point of departure is equation (13):

$$\begin{aligned} \underline{\dot{w}}(\underline{x})\tau &= \underline{\dot{s}}\tau+\{\underline{u}_{t,w}(\underline{x})-\underline{u}_{t,r}(\underline{x})\}-\{\underline{u}_{t,w}(\underline{x}+\underline{V}\tau)-\underline{u}_{t,r}(\underline{x}+\underline{V}\tau)\} \\ &= \underline{\dot{s}}\tau+\{\underline{u}(\underline{x})-\underline{\bar{u}}(\underline{x})\}-\{\underline{u}(\underline{x}+\underline{V}\tau)-\underline{\bar{u}}(\underline{x}+\underline{V}\tau)\} \end{aligned} \tag{13}$$

We introduce $w_{Jt}, \dot{s}_{Jt}$ according to (27). Then we have

$$\begin{bmatrix} w_{J1} \\ w_{J2} \end{bmatrix} = \begin{bmatrix} \dot{s}_{J1} \\ \dot{s}_{J2} \end{bmatrix} \tau +$$

$$+ \frac{1}{\pi G} \sum_{I=1}^{N} \left\{ \begin{bmatrix} (1-\nu)I_1+\nu I_2, \nu I_3 & , -KI_5 \\ \nu I_3 \quad ,(1-\nu)I_1+\nu I_2, & -KI_6 \end{bmatrix} \Big| (\underline{x}_J, \underline{x}_I, \Delta\underline{x}_I) + \right. \tag{30}$$

$$\left. - \begin{bmatrix} (1-\nu)I_1+\nu I_2, \nu I_3 & , -KI_5 \\ \nu I_3 \quad ,(1-\nu)I_1+\nu I_4, & -KI_6 \end{bmatrix} \Big| (\underline{x}_J+\underline{V}\tau, \underline{x}_I, \Delta\underline{x}_I) \right\} \begin{bmatrix} p_{I1} \\ p_{I2} \\ p_{I3} \end{bmatrix}$$

ALGORITHMS FOR THE NORMAL AND THE FRICTIONAL PROBLEM

We will now describe two numerical processes, one for the normal problem, and one for the frictional problem.

An algorithm for the normal contact problem

The oldest contact theory is due to Hertz (1881), see [1] pg.193. This theory applies when the bodies are not conformal, such as two spheres, and when the contact is frictionless. In order to obtain the distance function L, Hertz approximated the bodies by elliptic paraboloids, so that L becomes

$$L(x_1,x_2) = L_{11}x_1^2 + 2L_{12}x_1x_2 + L_{22}x_2^2 - L_{00}$$

$$L_{ij}\text{: constants} \tag{31}$$

$$L(x_1,x_2) = 0\text{: an ellipse with centre in the origin}$$

Next, the bodies are approximated by half-spaces, to perform the elastic analysis. Hertz showed that the contact area is an ellipse, and that the pressure distribution is semi-ellipsoidal, see fig. 7, where the Hertz theory is reproduced by the numerical process described below.
When the bodies are conformal in one or more directions, or when the distance function L cannot satisfy (31) for any other reason, other means have to be adopted to solve the frictionless contact problem. In fig. 8 are shown non-Hertzian contact areas for which the distance function is

not conformal, but has a discontinuous second derivative.
The first to give an algorithm for a very wide class of normal contact problems were Fridman and Chernina (USSR, 1967) [7]. Making use of a variational formulation of the contact problem they reduced it to a so-called quadratic program, which is a minimisation problem with a quadratic function to be minimised, linear equality constraints, and non-negative variables. This representation is very basic to the normal contact problem, and indeed necessary to prove existence-uniqueness of normal contact which was done in 1964 by Fichera [2]. Independent of Fridman and Chernina, Conry and Seirig (1971; USA) [8] and Kalker and Van Randen (1972; The Netherlands) [9] came up with optimization solutions of the normal contact problem. Kalker and Van Randen used a standard quadratic programming technique to solve the normal contact problem.

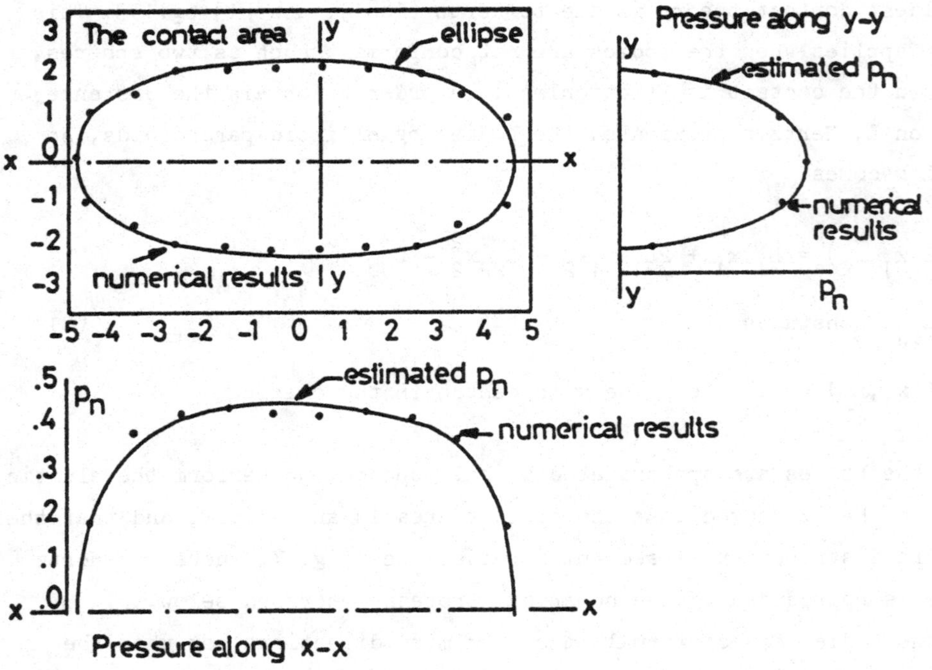

Fig. 7 - Numerical results suggestive of the Hertz theory.

Many authors followed, so that nowadays almost every institute which occupies itself with contact problems has a normal contact computer code. Here we will present a very simple and effective algorithm which was conceived by Johnson in the late sixties, early seventies; it was rediscovered by Ahmadi, Keer and Mura in 1981 [11] while I myself removed a restriction and gave a new proof of it (1982) [12]. But in my opinion it is no better than Fridman & Chernina, or Kalker & Van Randen. The algorithm is incorporated in the program Contact. It runs as follows:

Problem. Solve the normal contact problem when the tangential traction is given

Algorithm N

1. We start with a traction distribution p_{Ii}, $I = 1,\ldots,N$, $i = 1,2,3$.
   We initialize the contact area $K = \{I \mid p_{I3} > 0\}$, and the outside of the contact area $B = \{1,2,\ldots,N\} \setminus K$. In B $p_{I3} \le 0$.
2. We set $d_I \stackrel{def}{=} d(\underline{x}_I)$ = deformed distance $= L_I + (u_{I3} - \bar{u}_{I3}) = 0$ when $I \in K$; d: see (17), $u_{I3} - \bar{u}_{I3}$: see (39).
   We set $p_{I3} = 0$ when $I \in B$.
   These are N linear equations for the N normal pressures p_{I3}.
3. Remove from K those I for which the solution p'_{I3} of the equations of Step 2 are negative:

   $p'_{I3} < 0 \Rightarrow$ I goes from K to B.
4. If K has changed during Step 3, GO TO Step 2.
5. Now K has not changed during Step 3. Let d'_I be the deformed distance corresponding to the solution of the equations of Step 2. If $d'_I < 0$ for some $I \in B$ we have penetration, we add I to the contact area K.
6. If K has changed during Step 5, GO TO Step 2.
7. We have found a solution with positive p_{I3} in the contact, while penetration also does not occur $\Rightarrow$ we have found the solution.
   EXIT.

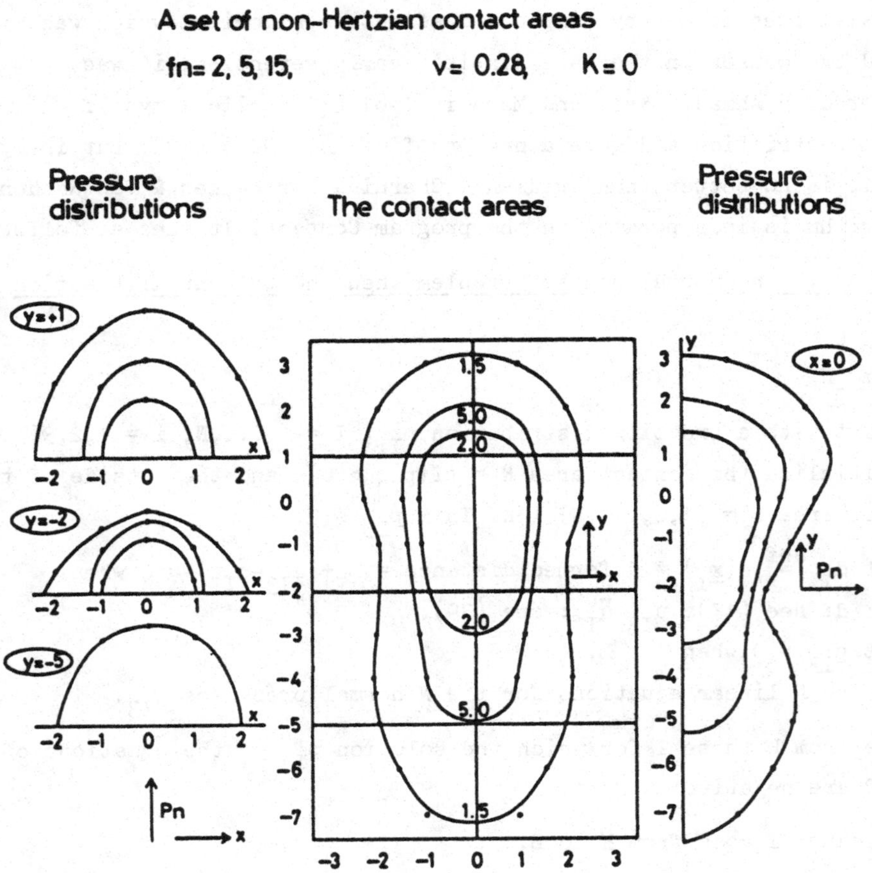

Fig. 8 - Three non-Hertzian contact areas.

In addition to the algorithms of which the above is an example there are algorithms which presuppose specialized forms of the normal pressure. Such a class of algorithms was initiated by Line Contact Theory (1971, [13]). In its original form, this theory presupposed that the contact area was slender and that its tips were not blunt. Very often under these circumstances the normal pressure distribution over the narrow direction of the contact area was semi-elliptical. Yet the theory did not become very popular owing to the requirement that the tips of the contact area should not be blunt. This restriction was removed by Reusner (1977; [14]) who assumed that the pressure distribution in the narrow direction was semi-elliptical. He has two parameters per element in the long direction, viz. the contact width and the maximal height of the elliptical pressure distribution. As data he uses the maximum penetration of L per element and the curvature L_{xx} of the undeformed distance in the x-direction. Reusner's method may be speeded up by not taking an elliptical but a parabolic pressure distribution, since then closed-form expressions of the displacement similar to (22)-(23) may be obtained.
We can compare algorithms of the type N and the Reusner type algorithms in the following manner:

Algorithms type "N"	Algorithms type "Reusner"
1. Many elements	Few elements
2. Influence matrix calculated only once	Influence matrix calculated many times
3. Complicated (2D) discretisation	Simple (1D) discretisation

It turns out that if the influence matrix of Algorithm "N" is precomputed and the parabolic traction distribution is adopted in Reusner's routine, the algorithms are about equally fast. The results of "parabolic" Reusner need some post-processing to improve their accuracy.

An algorithm for the frictional contact problem

The first algorithm for a wide class of frictional problems dealt with steady-state rolling. It was due to Kalker (1967) [15], and it was essentially restricted to elliptical contact areas. In addition, for reasons which never became quite clear, the program failed when the contact ellipse was longer in the rolling direction than it was wide; it was always assumed that the rolling direction coincided with a principal axis of the ellipse. The method employed by this algorithm is now completely antiquated.

In a paper dating from 1971 [16] Kalker introduced a variational principle for the problem of friction. This principle appeared to be quite powerful in two-dimensional frictional contact in which there are only the normal and one tangential direction. Interesting results on steady-state and non-steady-state 2D rolling were obtained. The principle could be adapted to a 3D situation, and indeed a program was generated by Kalker and Goedings (1972) [17], but although it did not suffer from the restrictions of Kalker's thesis program [15], it was too slow to be of much practical use.

An improved variational principle, due to Duvaut and Lions (1972) [3] led to a much faster steady-state program (Duvorol) which was, in addition, fully reliable (1979) [6]. The steady-state rolling algorithm to be described presently is about 5 times faster than Duvorol, but it is not always fully reliable. It is incorporated in the program Contact (1982) [12]. When the new algorithm fails, recourse must be had to the Duvorol algorithm, which is likewise incorporated in Contact.

The algorithm runs as follows:

Problem F. Solve the frictional contact problem when the normal pressure, the creepage, and the previous traction distribution are given

Algorithm F. Start with an arbitrary "present" traction distribution.

1. Initialize the areas of adhesion $A = \{I \| p_{I1}, p_{I2}| < fp_{I3}\}$ and slip $S = \{I \| p_{I1}, p_{I2}| \geq fp_{I3}\}$.

2. In the area of slip set $p_{It} := (fp_{I3}/|p_{I2}|)p_{It}$, $t = 1,2$; $I \in S$.
   p_{It} has now the Coulomb value.

3. $\underline{w}_J$ is the slip in $\underline{x}_J$, $J = 1,\dots,N$, see (41),(42),(43)
   $\underline{r}_I = (p_{I2}, -p_{I1}) \neq 0$, a vector orthogonal to $\underline{p}_{It} = (p_{I1}, p_{I2})$
   Δy is the increment of the variable y.
   Then the following equations hold:
   In area of adhesion A: $\underline{w}_J + \Delta\underline{w}_J = 0$
   The new slip vanishes in A.
   In area of slip S: $(\underline{w}_J + \Delta\underline{w}_J)^T(\underline{r}_J + \Delta\underline{r}_J) = 0$
   If this equation is satisfied, the slip is parallel to the traction.
   In area of slip S: $(\underline{p}_{It} + \Delta\underline{p}_{It})^T(\underline{p}_{It} + \Delta\underline{p}_{It}) = f^2 p_{I3}^2$
   This equation ensures that $|\underline{p}_{It}| = fp_{I3}$.
   These are 2N non-linear equations in the 2N unknowns p_{It} (t=1,2).
   They are solved by linearisation, i.e. neglect of Δ^2terms.

4. When $I \in A$ and $|\underline{p}_{It} + \Delta\underline{p}_{It}| > fp_{I3} \Rightarrow$ I is placed in slip area S.

5. If: - the area of slip has changed in Step 4.
   - or the equations have not yet converged: GO TO Step 2.

6. Now $|\underline{p}_{It}| \leq fp_{I3}, \underline{w}_I = \underline{0}$ in A, $\underline{w}_I // \underline{p}_{It}$ in S.
   Determine $\underline{p}_{It}\underline{w}_I$, $I \in S'$. If this form is positive then the slip has the wrong sense, and I is removed from the area of slip S.

7. If the area of slip has changed in Step 6: GO TO Step 2.

8. All conditions of the frictional problem are satisfied. EXIT.

An algorithm closely akin to the above has been proved in [12].
In fig. 9 are shown the areas of slip and adhesion for a circular Hertzian contact area and a hemispherical normal pressure distribution. The difference parameter K, see (20), vanishes, so that the tangential pressure does not affect the contact formation, which therefore remains Hertzian. Also shown in fig. 9 are the tangential tractions with their

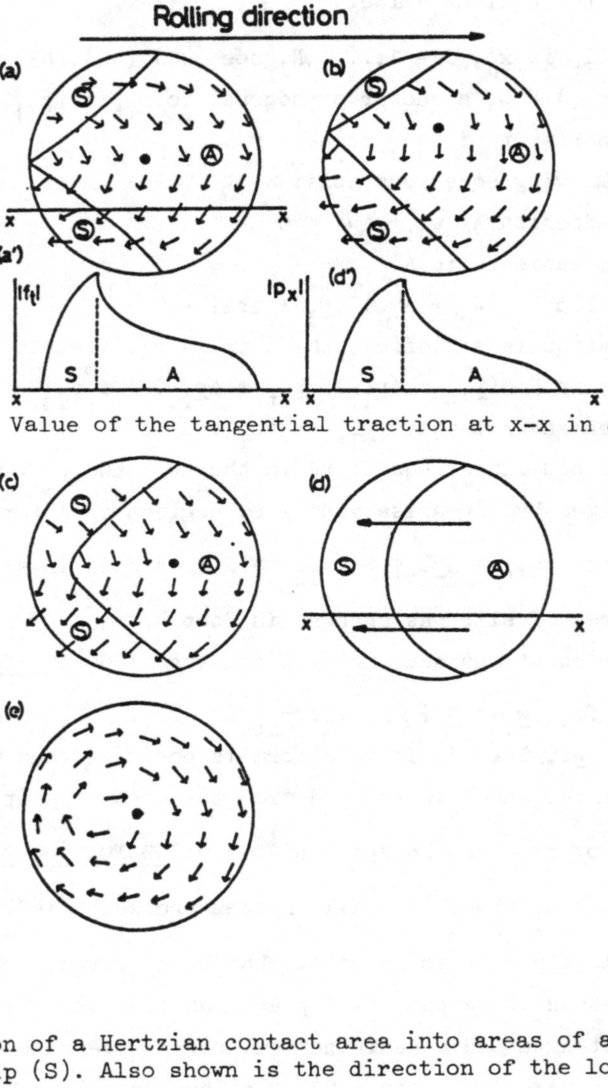

Fig. 9 - Division of a Hertzian contact area into areas of adhesion (A) and slip (S). Also shown is the direction of the local tangential traction.
a,a': pure spin, small; b: long. creep + spin, small; c: lat. creep + spin, small; d, d': long. creep, small.

Rolling with pure longitudinal creepage

$\nu=0.28$, $K=0.$ $u_x=0.3$ $fn=15$, $f=1$

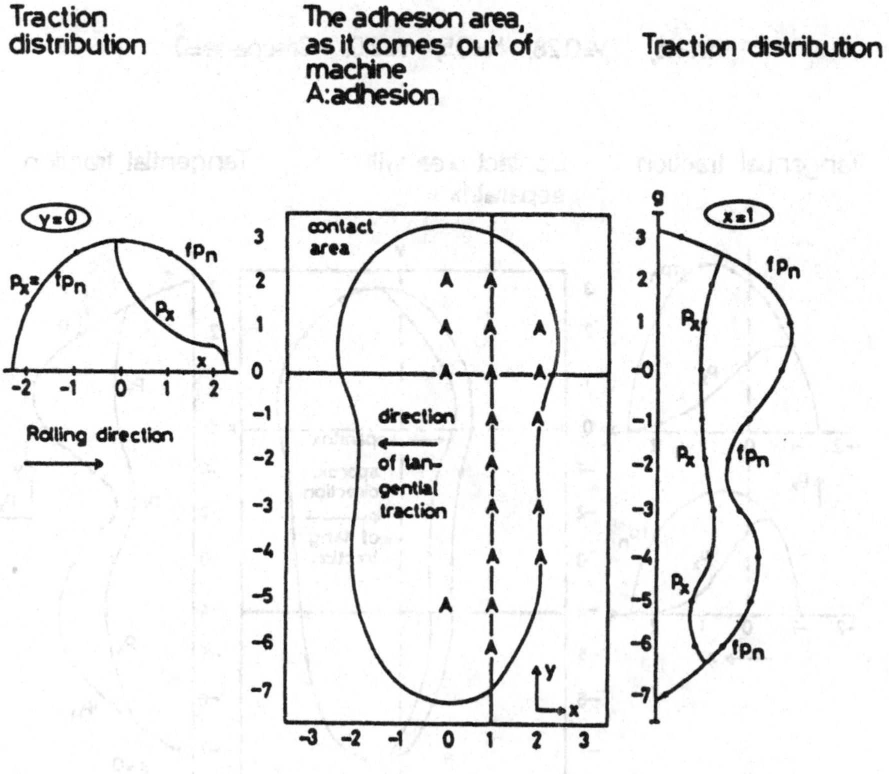

Fig. 10 - Rolling of identical bodies with creepage.

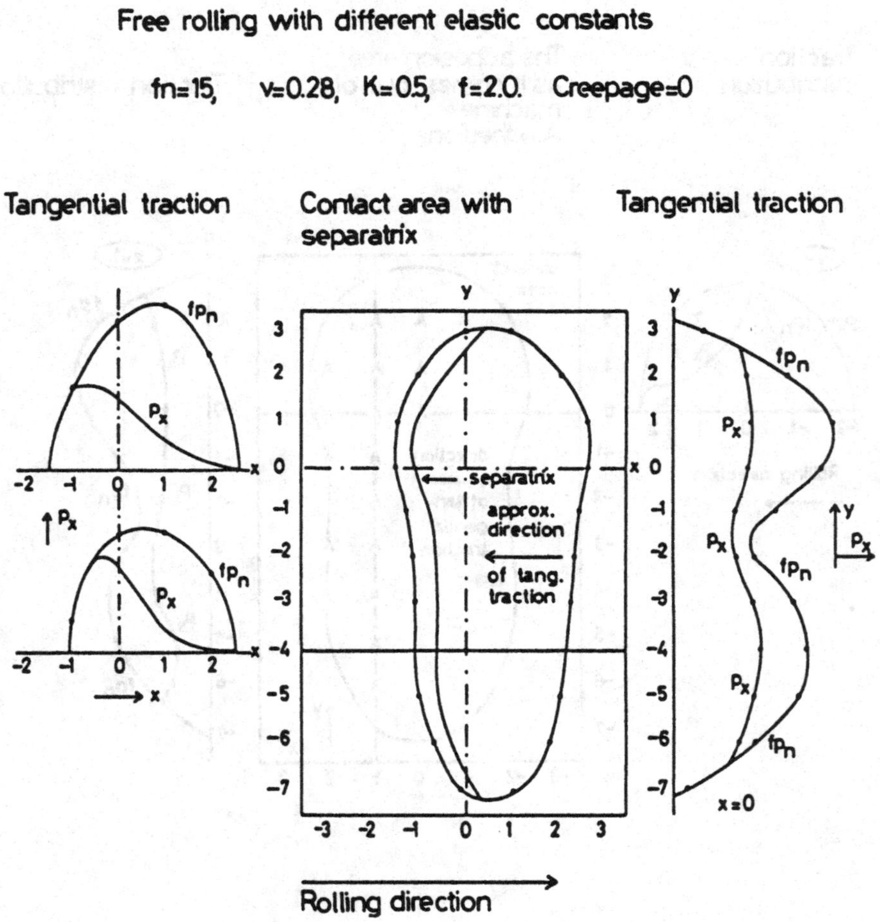

Fig. 11 - Rolling of non-identical bodies (creepage = 0).

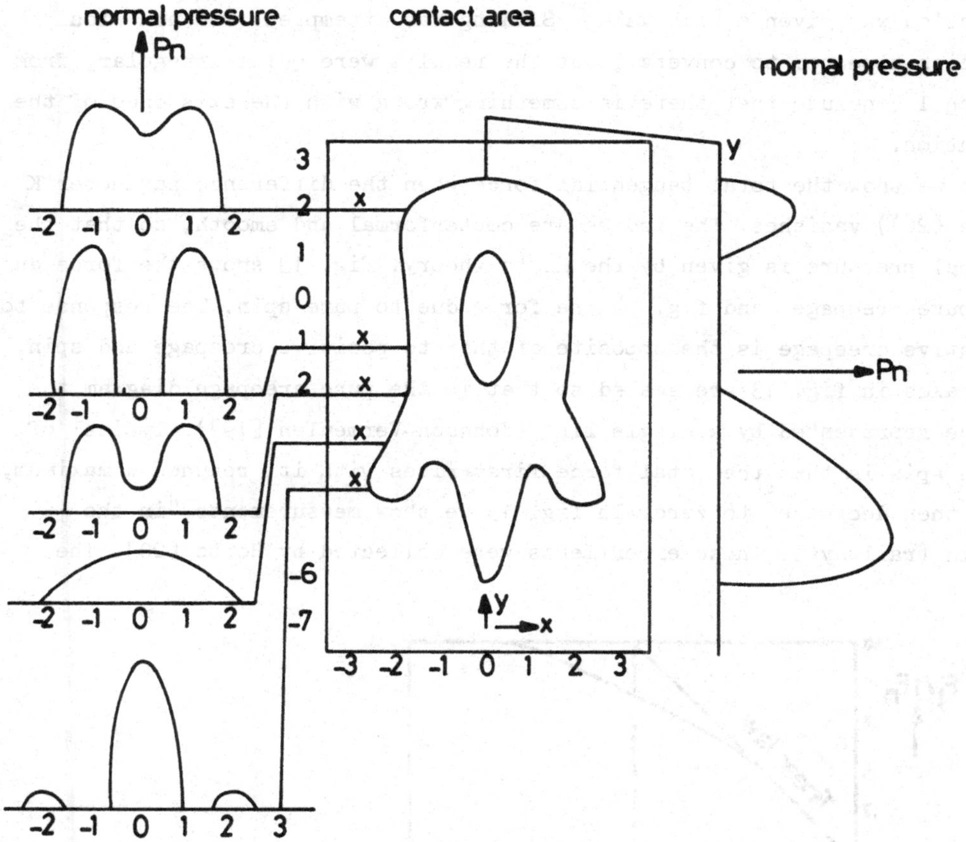

Fig. 12 - Failure of the Panagiotopoulos process.

directions which act at the contact area. The areas of adhesion and slip coincide extremely well with the photographic evidence of Ollerton and Pigott. Non-Hertzian cases are shown in figs. 10 and 11.
A failure of the Panagiotopoulos process is shown in fig. 12.
Fig. 12 was produced as follows. The contact area was discretised, the difference parameter K was set unequal to zero, and the coefficient of friction was given a high value. Sliding was attempted. The numerical process appeared to converge, but the results were quite irregular, from which I conclude that there is something wrong with the existence of the solution.
Next we show the total tangential force when the difference parameter K (see (20)) vanishes. The bodies are couterformal and smooth, so that the normal pressure is given by the Hertz theory. Fig. 13 shows the force due to pure creepage, and fig. 14 the force due to pure spin. The response to negative creepage is the opposite of that to positive creepage and spin. The axes in fig. 13 are scaled so that in the pure creepage diagram the force represented by a single line (Johnson-Vermeulen [19]). Typical of pure spin is that the total force first rises with it, reaches a maximum, and then decreases to zero. In fig. 15 we show measurements "in the field (railways). These experiments were collected by Hobbs [20]. The

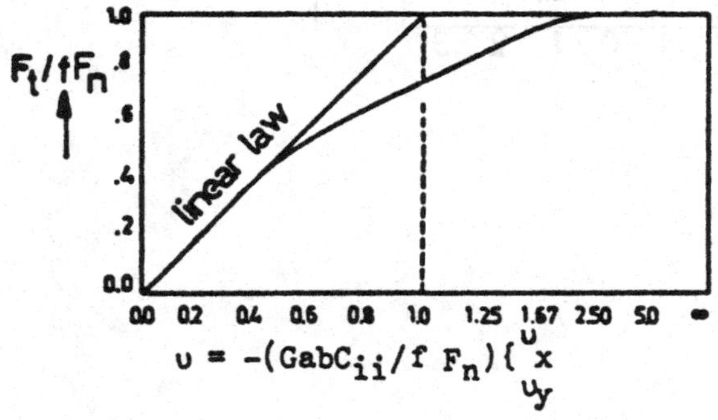

Fig. 13 - The tangential force in the no-spin case, calculated with the programs CONTACT.

difference with the theory is commonly attributed to measurement errors and by contamination of the contacting bodies. This is borne out by the experiments of Brickle [21] done under laboratory conditions. A sample of these experiments, together with the appropriate results of the program Duvorol, is shown in fig. 16.

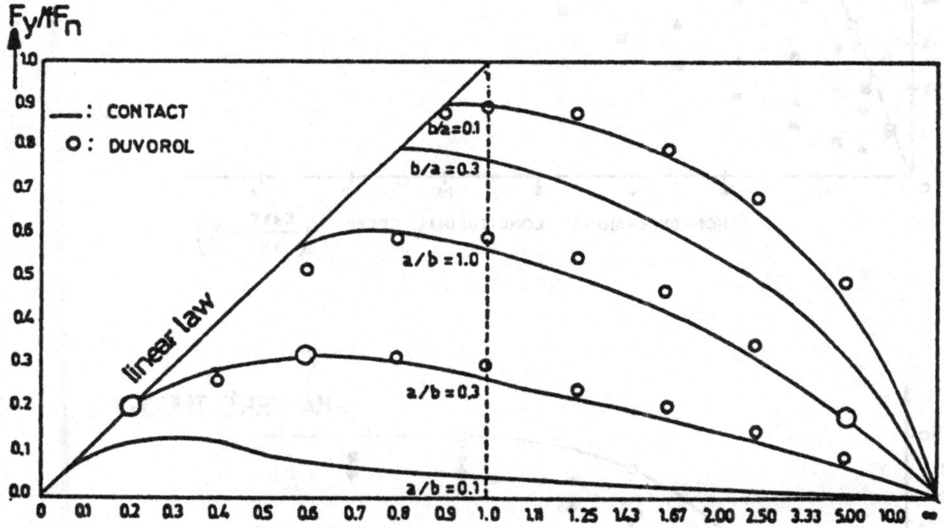

$\psi = -(G\sqrt{ab}^3\ (C_{23}/fF_n)\phi \rightarrow$

Fig. 14 - The tangential (lateral) force due to pure spin, calculated by the programs CONTACT, (Hertzian) DUVOROL.
Hertzian contact; a: semi-axis of contact ellipse in rolling direction, b: semi-axis of contact ellipse in lateral direction.

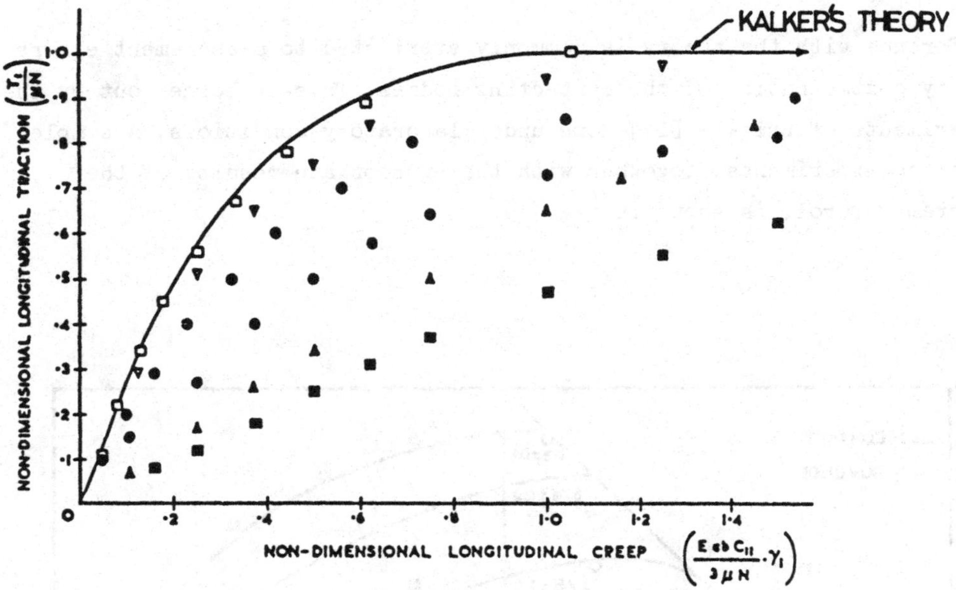

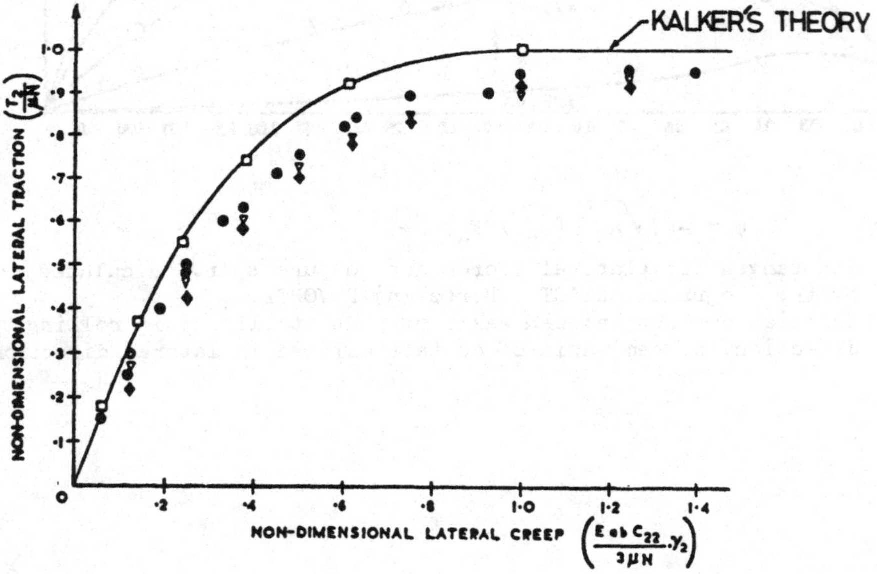

Fig. 15 - Experiments "in the field" compared with theory.

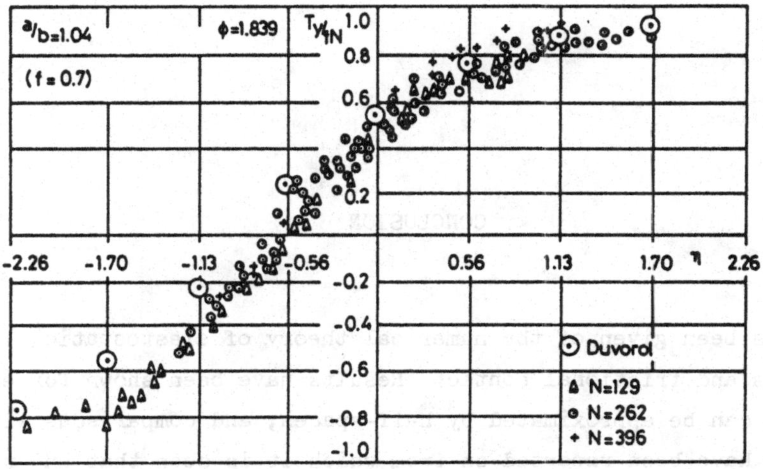

Fig. 16 - Comparison of the results of Duvorol with Brickle's measurements (constant spin, variable lateral creepage η).

CONCLUSION

A survey has been given of the numerical theory of elastostatic, frictionless and frictional contact. Results have been shown for elastic bodies that can be approximated by half-spaces, and comparisons with experiments have been reported on from which it is seen that under laboratory conditions good agreement has been found.

LITERATURE

[1] A.E.H. Love, A treatise on the mathematical theory of elasticity, 4th Ed. Cambridge UP (1926).

[2] G. Fichera, Problemi elastostatici con vincoli unilaterali: il problema di Signorini con ambigue condizioni al contorno, Mem. Ac. N. Lincei 8, 7 (1964) 116-140.

[3] G. Duvaut, J.-L. Lions, Les inéquations en mécanique et en physique, Dunod, Paris, 1972.

[4] P.D. Panagiotopoulos, A non linear programming approach to the unilateral contact and friction-boundary value problem in the theory of elasticity, Ing. Arch. 44 (1975) 421-432.

[5] K.L. Johnson, Tangential tractions and microslip in rolling contact, In: Proc. Symp. Rolling Contact Phenomena, Elsevier (1962) 6-28.

[6] J.J. Kalker, The computation of three-dimensional rolling contact with dry friction, Int. J. Num. Meth. Eng. 14 (1979) 1293-1307.

[7] V.M. Fridman, V.S. Chernina, Iteration methods applied to the solution of contact peoblems between bodies, Mekh. Tverd. Tela AN SSSR, 1 (1967) 116-120.

[8] T.F. Conry, A. Seirig, A mathematical programming method for design of elastic bodies in contact, J. Appl. Mech. 38 (1971) 387-392.

[9] J.J. Kalker, Y. van Randen, A minimum principle for frictionless elastic contact with application to non-Hertzian half-space contact problems, J. Eng. Math. 6 (1972) 192-206.

[10] J.J. Kalker, The contact between wheel and rail, In: Int. Cent. Transp. Stud. Vol.IV Proc. Series, Oct. 25-30/1982, p. 275-312.

[11] N. Ahmadi, L.M. Keer, T. Mura, Non-Hertzian stress analysis - normal and sliding contact, Rept. Dept. Civil Engineering, Northwestern University, USA, 1981.

[12] J.J. Kalker, Two algorithms for the contact problem in elastostatics, In: Proc.Int.Symp. Contact Mechanics and Wear of Rail/Wheel systems, ed. Gladwell, (1983).

[13] J.J. Kalker, On elastic line contact, J. Appl. Mech.39 (1972) 1125-1132.

[14] H. Reusner, Druckflächenbelastung und Oberflächenverschiebung im Wälzkontakt von Rotationskörpern. Thesis Karlsruhe, SKF Schweinfurt 1977, (German).

[15] J.J. Kalker, On the rolling contact of two elastic bodies in the presence of dry friction, Thesis Delft 1967.

[16] J.J. Kalker, A minimum principle for the law of dry friction with application to elastic cylinders in rolling contact, J.Appl.Mech. 38 (1971) 875-887.

[17] J.J. Kalker, H. Goedings, A program for three-dimensional steady-state rolling, Internal Report (1972), Delft U of T.

[18] M. Abramovitz, I.A. Stegun, Handbook of Mathematical Functions, Dover (1965).

[19] K.L. Johnson, P.J. Vermeulen, Contact of non-spherical bodies transmitting tangential forces, J.Appl.Mech. (1964), p. 338-340.

[20] A.E.H. Hobbs, A survey of creep, Brit. Rail Res. Dept. Dyn 52 (1967)(Derby, U.K.).

[21] B.V. Brickle, The steady state forces and moments on a railway wheel set including flange contact conditions. Loughborough Chr. Doct. Thesis (1973).

[22] N. Kikuchi, Oral Private Communication, July 1982.

NUMERICAL APPROXIMATION AND ERROR ESTIMATES FOR ELASTIC-PLASTIC TORSION PROBLEMS IN MULTIPLY CONNECTED DOMAINS

L.D. Marini
Istituto di Analisi Numerica del C.N.R.
Pavia

Summary A numerical approximation with conforming finite elements is presented for elastic-plastic torsion problems in multiply connected domains. The problem is formulated as a unilateral problem, of the obstacle type, with the obstacle depending on the solution.

1. INTRODUCTION

Before writing the mathematical formulation let us explain the physics of the problem under consideration. Consider first the case of a simply connected domain. Denote by Ω a bounded simply connected domain of $\mathbb{R}^2$ and by D the cylinder:

$$D = \{(x_1,x_2) \in \Omega, \quad 0<x_3<\ell\}.$$

The bar D, made up of a homogeneous elastic-perfectly plastic material, is supposed to be clamped at the bottom (x_3=0) and subjected to a prescribed torsional rotation θ at the top ($x_3=\ell$). No external forces are assumed to act on the lateral boundary of D. In the infinitesimal theory, the two nonvanishing components σ_{13} and σ_{23} of the stress tensor can be written as ∇u, where u (the stress potential) is the solution of the following variational inequality:

$$\left|\begin{array}{l} \text{Find } u \text{ such that:} \\ u \in \bar{K}, \\ \int_\Omega \nabla u \cdot \nabla(v-u)dx \geq \int_\Omega \mu(v-u)dx \qquad \forall v \in \bar{K}, \end{array}\right. \tag{1.1}$$

where:

$$\bar{K} = \{v \in H_0^1(\Omega): |\nabla v| \leq 1 \text{ a.e. in } \Omega\}, \tag{1.2}$$

and $\mu = 2\lambda\theta$, λ being the shear modulus. It has been proved by BREZIS & SIBONI [1], that problem (1.1),(1.2) is equivalent to the following obstacle problem: Find u such that:

$$\left|\begin{array}{l} u \in K, \\ \int_\Omega \nabla u \cdot \nabla(v-u)dx \geq \int_\Omega \mu(v-u)dx \qquad \forall v \in K, \end{array}\right. \tag{1.3}$$

where:

$$K = \{v \in H_0^1(\Omega): 0 \leq v \leq d(x) \text{ a.e. in } \Omega\}, \tag{1.4}$$

and $d(x)=\text{dist}(x,\partial\Omega)$. This formulation is numerically simpler and suitable for deriving approximations with conforming finite elements. Indeed, the constraint on the function has a local character and can be verified node by node. Hence, when using an iterative method for solving (1.3)-(1.4), the projection on K, at each step, is easily performed. On the other hand, the projection on $\bar{K}$ is more difficult, unless special finite elements are used, having the gradient as an independent variable (*).

(*) In this framework results can be found in FALK-MERCIER[2], where a discretization using mixed finite element methods is studied.

When Ω is a multiply connected domain having a finite number of holes Ω_k, $k=1,\dots,n$, the stress potential u is the solution of the variational inequality (1.1), with the convex set defined by:

$$\bar{K} = \{v \in H_0^1(Q): |\nabla v| \le 1 \text{ a.e. in } \Omega, |\nabla v|=0 \text{ in } \Omega_k, \forall k\}, \tag{1.5}$$

where Q is the simply connected domain:

$$Q = \Omega \cup (\bigcup_{k=1}^{n} \bar{\Omega}_k). \tag{1.6}$$

In order to transform problem (1.1),(1.5) into an obstacle problem, appropriate boundary conditions have to be found for u in Ω. These are u=0 on $\partial\Omega$, while on $\partial\Omega_k$ condition $|\nabla u| = 0$ implies u to be equal to some constant c_k, the value of c_k being unknown. One way for overcoming this problem is to associate with any $c=(c_1,\dots,c_n)$ a convex set K^c depending on c, and then to find conditions to evaluate the right value of c. The way for finding the proper conditions to evaluate the unknown constants (TING [3]) is shown in Sect. 2 below. There are other examples of problems which can be written as variational inequalities with the convex set depending on unknown parameters to be determined. For instance, some free boundary problems arising from the stationary filtration of a fluid through porous media (see e.g. BAIOCCHI [4]). In general the error analysis is quite difficult. The mathematical interest of the problem under consideration is that for it the error analysis can be carried out successfully. In Sect. 2 we recall the formulation of the problem as a two-obstacle problem, and we show how it can be reduced to a minimization problem in $\mathbb{R}^n$ (n being the number of holes). In Sect. 3 we present a numerical approximation by means of continuous piecewise linear finite elements and we derive optimal error bounds for both u and c. In Sect. 4 we present some numerical tests and suggest an algorithm very effective and low costly for solving the discrete minimum problem in the case of a domain Ω with one hole.

2. STATEMENT OF THE PROBLEM. PRELIMINARY RESULTS

Let Ω be a bounded multiply connected domain in $\mathbb{R}^2$, having a finite number of "holes" Ω_k, $k=1,\dots,n$, with respective boundaries $\Gamma_k = \partial\Omega_k$, $k=1,\dots,n$. Let Q be the simply connected domain whose boundary Γ is the exterior boundary of Ω: $Q=\Omega \cup \Omega^0$, $\Omega^0 = \bigcup_{k=1}^{n} \bar{\Omega}_k$. Γ and Γ_k are supposed to be smooth enough to ensure the regularity needed for the solution (see e.g. BREZIS & STAMPACCHIA [5], CAFFARELLI & FRIEDMAN [6], GERHARDT [7], GRISVARD [8]).

Let K be the convex set:

$$K = \{v \in H_0^1(Q): |\nabla v| \leq 1 \text{ a.e. in } \Omega,\ \nabla v = 0 \text{ in } \Omega_k\ \forall\, k\}. \tag{2.1}$$

We define:

$$a(u,v) = \int_Q \nabla u \cdot \nabla v \, dx = (u,v)_{1,Q} \qquad u,v \in H_0^1(Q), \tag{2.2}$$

$$J(v) = \frac{1}{2}a(v,v) - \int_Q \mu v \, dx \qquad v \in H_0^1(Q). \tag{2.3}$$

Consider the following problem:

$$\begin{cases} \text{Find } u^* \in K \text{ such that:} \\ J(u^*) \leq J(v) \quad \forall\, v \in K, \end{cases} \tag{2.4}$$

or the equivalent variational formulation:

$$\begin{cases} \text{Find } u^* \in K \text{ such that:} \\ a(u^*, v-u^*) \geq \int_Q \mu(v-u^*)dx \quad \forall\, v \in K. \end{cases} \tag{2.5}$$

It is well known [5] that problem (2.4) (or (2.5)) has a unique solution u^* in $H_0^1(Q)$ which takes a constant value c_k^* over each region Ω_k, $k=1,\dots,n$.

Let us define:

$$c^* = (c_1^*,\ldots,c_n^*)\ ;\ c_k^* = u^*_{|\Omega_k} \qquad k=1,\ldots,n, \tag{2.6}$$

$$d_k(x) = \mathrm{dist}(x,\Omega_k) \qquad (x \in \bar{Q}),\ k=1,\ \ldots,\ n, \tag{2.7}$$

$$d_e(x) = \mathrm{dist}(x,\Gamma) \qquad (x \in \bar{Q}), \tag{2.8}$$

and, for any vector $c \in \mathbb{R}^n$:

$$\psi_1(x) = \max_k\ (c_k - d_k(x)), \tag{2.9}$$

$$\psi_2(x) = \min\ (d_e(x),\ \min_k\ (c_k + d_k(x)), \tag{2.10}$$

$$K^c = \{v \in H_0^1(Q):\ \psi_1 \le v \le \psi_2\}. \tag{2.11}$$

Then we have that, [1,3] if u^* is the (unique) solution of (2.4) (or (2.5)) with K as defined in (2.1), then u^* is the unique solution of the problem:

$$\begin{cases} \text{Find } u^* \in K^{c^*} \text{ such that:} \\ J(u^*) \le J(v) \quad \forall\ v \in K^{c^*}, \end{cases} \tag{2.12}$$

with J(v) as in (2.3) and K^{c^*} as in (2.11) for $c=c^*$.

By the maximum principle, $c_k^* \ge 0\ \forall\ k$; moreover, since $|\nabla u| \le 1$, c^* belongs to the set $\mathcal{C}$ of the vectors $c = \{c_k\}$ of $\mathbb{R}^n$ satisfying:

$$0 \le c_k \le \mathrm{dist}(\Gamma_k,\Gamma) \qquad k=1,\ldots,n, \tag{2.13}$$

$$|c_k - c_j| \le \mathrm{dist}(\Gamma_k,\Gamma_j) \qquad k,j=1,\ldots,n. \tag{2.14}$$

Notice that (2.13) and (2.14) are equivalent to $\psi_2 \ge \psi_1$, so that K^c is not empty iff $c \in \mathcal{C}$.

<u>REMARK 2.1</u> The Fact that $u^* \in K$ implies $u^* \in W^{1,\infty}(\Omega)$; moreover, we also assume $u^* \in H^2(\Omega)$. This regularity can be derived under suitable smoothness assumptions on the boundaries Γ_k and Γ [3,5,7]. Other results concerning regularity and various properties of the solution of elastic-plastic torsion problems can be found for instance in

CAFFARELLI & FRIEDMAN [6], CAFFARELLI & RIVIERE [9], CAFFARELLI, FRIEDMAN & POZZI [10], FRIEDMAN & POZZI [11].

REMARK 2.2 In the two-obstacle problem (2.12), the definition of the obstacle depends on the constants c_k^* which are the unknown values of u^* in Ω_k. The evaluation of c^* can be done by noticing that u^* is such that its corresponding energy $J(u^*)$ is minimum. More precisely, for any $c \in \mathcal{C}$, denote by u_c the unique solution of the problem:

$$\begin{cases} u_c \in K^c \text{ such that:} \\ J(u_c) \leq J(v) \quad \forall\, v \in K^c, \end{cases} \tag{2.15}$$

where K^c is defined in (2.11). Note that, following [1] it can be proved that u_c satisfies $|\nabla u_c| \leq 1$ a.e. in Ω. Moreover, from results of [5,6] we also have $u_c \in H^2(\Omega)$. Consider the function $F: \mathcal{C} \longrightarrow \mathbb{R}$, defined by:

$$F(c) = J(u_c) \qquad \forall\, c \in \mathcal{C}. \tag{2.16}$$

Since F is strictly convex in $\mathcal{C}$ (see LEMMA 2.1 below), there exists a unique $\bar{c} \in \mathcal{C}$ such that $F(\bar{c}) \leq F(c)$ $\forall\, c \in \mathcal{C}$, that is, $J(u_{\bar{c}}) \leq J(u_c)$ $\forall\, c \in \mathcal{C}$. The uniqueness of u^* implies $u_{\bar{c}} = u^*$ and $\bar{c} = c^*$; hence the evaluation of c^* reduces to the following minimum problem (in $\mathbb{R}^n$):

$$\begin{cases} \text{Find } c^* \in \mathcal{C} \text{ such that:} \\ F(c^*) \leq F(c) \quad \forall\, c \in \mathcal{C} \end{cases} \tag{2.17}$$

LEMMA 2.1 The function $F(c)$ is strictly convex on $\mathcal{C}$, in the sense that, for any fixed pair $(c^1, c^2) \in \mathcal{C}$ and for any $\lambda \in [0,1]$, we have:

$$F((1-\lambda)c^1+\lambda c^2) \leq (1-\lambda)F(c^1) + \lambda F(c^2) - \gamma\lambda(1-\lambda), \tag{2.18}$$

with $\gamma = \gamma_0 \| c^1 - c^2 \|^2_{\mathbb{R}^n}$, γ_0 being a positive constant independent of λ, c^1, c^2.

Proof: Let $\alpha = (1-\lambda)c^1 + \lambda c^2$; from the definitions (2.16), (2.3) of F and J we have:

$$F(\alpha)=F((1-\lambda)c^1+\lambda c^2)=J(u_\alpha)\leq J(v)=\frac{1}{2}|v|^2_{1,Q}-\int_Q \mu v dx \quad \forall v \in K^\alpha. \tag{2.19}$$

Setting $v = (1-\lambda)u_{c^1} + \lambda u_{c^2}$, with u_{c^1} (resp. u_{c^2}) solution of problem (2.15) in K^{c^1} (resp. in K^{c^2}), it is immediate to see that $v \in K^\alpha$; moreover,

$$|v|^2_{1,Q}= (1-\lambda)^2|u_{c^1}|^2_{1,Q}+ \lambda^2|u_{c^2}|^2_{1,Q}+ 2\lambda(1-\lambda)(u_{c^1},u_{c^2})_{1,Q}. \tag{2.20}$$

Noticing that:

$$2(u_{c^1}, u_{c^2})_{1,Q} = |u_{c^1}|^2_{1,Q}+ |u_{c^2}|^2_{1,Q}- |u_{c^1} - u_{c^2}|^2_{1,Q}, \tag{2.21}$$

(2.20) becomes:

$$\begin{aligned}|v|^2_{1,Q}&= (1-\lambda)^2|u_{c^1}|^2_{1,Q}+ \lambda^2|u_{c^2}|^2_{1,Q}+ \lambda(1-\lambda)|u_{c^1}|^2_{1,Q}+\\ &+ \lambda(1-\lambda)|u_{c^2}|^2_{1,Q}- \lambda(1-\lambda)|u_{c^1}-u_{c^2}|^2_{1,Q} =\\ &= (1-\lambda)|u_{c^1}|^2_{1,Q}+ \lambda|u_{c^2}|^2_{1,Q}- \lambda(1-\lambda)|u_{c^1}-u_{c^2}|^2_{1,Q} .\end{aligned} \tag{2.22}$$

Substituting (2.22) in (2.19) we get:

$$F((1-\lambda)c^1+\lambda c^2) \leq (1-\lambda)F(c^1) + \lambda F(c^2) - \frac{\lambda(1-\lambda)}{2}|u_{c^1}-u_{c^2}|^2_{1,Q} . \tag{2.23}$$

Using Poincaré's inequality in (2.23) we have:

$$|u_{c^1}-u_{c^2}|^2_{1,Q} \geq \frac{1}{\beta}\|u_{c^1}-u_{c^2}\|^2_{0,Q} \geq \frac{1}{\beta}\|u_{c^1}-u_{c^2}\|^2_{0,\Omega^0} \geq$$

$$\geq \frac{1}{\beta} \min_k(\text{meas}\Omega_k)\|c^1-c^2\|^2 = M\|c^1-c^2\|^2 . \tag{2.24}$$

Substitution in (2.23) leads to:

$$F((1-\lambda)c^1+\lambda c^2) \leq (1-\lambda)F(c^1) + \lambda F(c^2) - \frac{M}{2}\|c^1-c^2\|^2\lambda(1-\lambda), \tag{2.25}$$

and (2.18) comes from (2.25) with $\gamma_0 = M/2$.

■

An immediate corollary of Lemma 2.1 is the following

<u>LEMMA 2.2</u> If c^* is the (unique) solution of (2.17), then for any $c \in \mathcal{C}$ we have:

$$F(c) \geq F(c^*) + \frac{M}{4}\|c-c^*\|^2 , \tag{2.26}$$

with M defined by (2.24).

<u>Proof</u>: Let $\bar{c} = \frac{c^*+c}{2}$; by applying (2.25) with $\lambda = 1/2$ to the pair (c^*,c) and since c^* is the minimum for F we deduce:

$$F(c^*) \leq F(\bar{c}) \leq \frac{1}{2}F(c^*) + \frac{1}{2}F(c) - \frac{M}{8}\|c-c^*\|^2 , \tag{2.27}$$

and, therefore,

$$F(c) \geq F(c^*) + \frac{M}{4}\|c-c^*\|^2 . \tag{2.28}$$

■

3. NUMERICAL APPROXIMATION OF PROBLEM (2.12)

Let us assume, for the sake of simplicity, that Q is convex and that the holes Ω_k are also convex. Let $\{\mathcal{T}_h\}_h$ be a regular family of decompositions of Q into triangles T (see e.g. CIARLET [12]). For any $\mathcal{T}_h$, let Q_h be the polygonal domain so obtained: $Q_h = \Omega_h \cup \Omega_h^0$, $\Omega_h^0 = \bigcup_k \Omega_k^h$. Note that in general Ω_h is not contained in Ω, unless the holes Ω_k are polygons. We consider continuous, piecewise linear elements defined by function values at the vertices of the triangles. With the same notations used in the previous Section, define:

$$V_h = \{v \in C^0(\bar{Q}): v_{|T} \in P_1(T) \quad \forall T \in \mathcal{T}_h, \ v=0 \text{ at nodes on } \partial Q\}, \quad (3.1)$$

$$\phi^I = \text{a piecewise linear interpolate of a continuous function } \phi, \quad (3.2)$$

$$c \in \mathcal{C} \rightarrow K_h^c = \{v \in V_h: \psi_1 \leq v \leq \psi_2 \text{ at nodes of } \mathcal{T}_h\}, \quad (3.3)$$

$$a_h(u,v) = \sum_{T \in \mathcal{T}_h} \int_T \nabla u \cdot \nabla v dx = \int_{Q_h} \nabla u \cdot \nabla v dx \qquad u,v \in V_h, \quad (3.4)$$

$$J_h(v) = a_h(v,v) - \int_{Q_h} \mu v dx \qquad v \in V_h, \quad (3.5)$$

$$\|v\|_{V_h}^2 = a_h(v,v) = |v|_{1,Q_h}^2 \qquad v \in V_h. \quad (3.6)$$

When necessary, functions $v \in V_h$ can be extended to the whole set Q by attributing them the zero value on $Q \setminus Q_h$. Note also that $v \in K_h^c$ implies $\nabla v=0$ in Ω_k, $\forall$ k.

The approximation of problem (2.15) is then:

$$\left| \begin{array}{l} \text{Find } u_h^c \in K_h^c \text{ such that:} \\ J_h(u_h^c) \leq J_h(v) \quad \forall\, v \in K_h^c, \end{array} \right. \quad (3.7)$$

or equivalently:

$$\begin{cases} \text{Find } u_h^c \in K_h^c \text{ such that:} \\ a_h(u_h^c, v-u_h^c) \geq \int_{Q_h} \mu(v-u_h^c)dx \quad \forall\, v \in K_h^c . \end{cases} \tag{3.8}$$

The uniqueness of the solution u_h^c is guaranteed from the coercivity of a_h on V_h. As for the continuous problem we introduce the function $F_h : \mathcal{C} \longrightarrow \mathbb{R}$ defined by:

$$F_h(c) = J_h(u_h^c) \qquad c \in \mathcal{C}. \tag{3.9}$$

<u>LEMMA 3.1</u> F_h is strictly convex in $\mathcal{C}$, in the sense that, for any fixed pair $(c^1, c^2) \in \mathcal{C}$ and for any $\lambda \in [0,1]$ we have:

$$F_h((1-\lambda)c^1 + \lambda c^2) \leq (1-\lambda)F_h(c^1)+\lambda F_h(c^2) - \gamma'\lambda(1-\lambda), \tag{3.10}$$

with $\gamma' = \gamma_0' \| c^1-c^2 \|^2_{\mathbb{R}^n}$, γ_0' being a positive constant independent of λ, c^1, c^2. The proof follows the scheme of LEMMA 2.1.

■

As a consequence, the problem:

$$\begin{cases} \text{Find } c_h^* \in \mathcal{C} \text{ such that:} \\ F_h(c_h^*) \leq F_h(c) \quad \forall\, c \in \mathcal{C} \end{cases} \tag{3.11}$$

has a unique solution c_h^*. The discrete analogue of LEMMA 2.2 can now be immediately proved:

<u>LEMMA 3.2</u> If c_h^* is the solution of (3.11), then $\forall\, c \in \mathcal{C}$ we have:

$$F_h(c) \geq F_h(c_h^*) + \frac{M'}{4} \| c-c_h^* \|^2 , \tag{3.12}$$

with M' a positive constant.

■

Let u_h^* be the solution of (3.7) for $c=c_h^*$. It follows from definition (3.9) that u_h^* verifies:

$$J_h(u_h^*) \leq J_h(u_h^c) \quad \forall\, c \in \mathcal{C} ; \tag{3.13}$$

we wish to prove that:

$$\| c^*-c_h^* \| = O(h) , \tag{3.14}$$

$$\| u^*-u_h^* \|_{V_h} = O(h) , \tag{3.15}$$

with u^* and c^* solutions of (2.12) and (2.17), respectively. To this end we need some preliminary lemmata.

<u>LEMMA 3.3</u> For c fixed in $\mathcal{C}$, denote by u^c the solution of (2.15) and by u_h^c the solution of (3.7). Then,

$$\| u^c-u_h^c \|_{V_h} \leq C\, h , \tag{3.16}$$

with C a constant independent of h (*).

<u>Proof</u> Let us write u, u_h instead of u^c, u_h^c, and let u^I be the interpolate of u in K_h^c. Then,

$$\| u-u_h \|_{V_h}^2 \leq a_h(u-u_h,u-u_h) = a_h(u-u_h,u-u^I)+a_h(u-u_h,u^I-u_h). \tag{3.17}$$

The first term in (3.17) is bounded by:

$$a_h(u-u_h,u-u^I) \leq \| u-u_h \|_{V_h} \| u-u^I \|_{V_h} . \tag{3.18}$$

Since u does not belong to $H^2(\Omega_h)$ ($\Omega_h \not\subset \Omega$) we cannot use directly the

(*) Hereafter we denote by C any constant independent of h.

usual error estimates. However, with standard techniques (see e.g. LIONS & MAGENES, [13]) we can define a function $\tilde{u} \in H^2(Q) \cap W^{1,\infty}(Q)$ such that:

$$\begin{aligned} &\tilde{u} = u \quad \text{in } \Omega , \\ &\|\tilde{u}\|_{H^2(Q)\cap W^{1,\infty}(Q)} \leq C\|u\|_{H^2(\Omega)\cap W^{1,\infty}(\Omega)} . \end{aligned} \tag{3.19}$$

Inserting $\tilde{u}$ in (3.18) we have:

$$\|u-u^I\|_{V_h} = |u-u^I|_{1,\Omega_h} \leq |u-\tilde{u}|_{1,\Omega_h} + |\tilde{u}-u^I|_{1,\Omega_h} . \tag{3.20}$$

Since $u^I=\tilde{u}^I$ in Ω_h, for the second term in (3.20) we get (see e.g. CIARLET & RAVIART [14], STRANG & FIX [15], CIARLET [12]):

$$|\tilde{u}-u^I|_{1,\Omega_h} \leq C\, h|\tilde{u}|_{2,\Omega_h} \leq C\, h|u|_{2,\Omega} . \tag{3.21}$$

On the other hand, by definition of $\tilde{u}$ the first term in (3.20) reduces to:

$$|u-\tilde{u}|_{1,\Omega_h} = \int_{S_h} |\nabla\tilde{u}|^2 dx, \tag{3.22}$$

where:

$$S_h = \Omega_h \setminus \Omega , \quad \text{meas}(S_h) = O(h^2). \tag{3.23}$$

Since $\tilde{u} \in W^{1,\infty}(Q)$, the Swartz-Hölder inequality in (3.22) gives:

$$|u-\tilde{u}|_{1,\Omega_h} \leq \|(\nabla\tilde{u})^2\|_{L^\infty(S_h)} \|1\|_{L^1(S_h)} = C\, h^2 . \tag{3.24}$$

Substituting (3.24) and (3.21) in (3.20) yields:

$$\|u-u^I\|_{V_h} \leq C\, h , \tag{3.25}$$

so that (3.18) becomes:

$$a_h(u-u_h,u-u^I) \le C\,h\,\|u-u_h\|_{V_h}. \tag{3.26}$$

For the second term in (3.17), the variational inequality (3.8) with $v=u^I$ implies:

$$\begin{aligned} a_h(u-u_h,u^I-u_h) &= a_h(u,u^I-u_h)-a_h(u_h,u^I-u_h) \le \\ &\le a_h(u,u^I-u_h) - \int_{Q_h} \mu(u^I-u_h)dx . \end{aligned} \tag{3.27}$$

After integrating by parts and introducing the variable $w=-\Delta u-\mu$ we find:

$$\begin{aligned} a_h(u,u^I-u_h)-\int_{Q_h} \mu(u^I-u_h)dx &= \\ = \int_\Omega w(u^I-u_h)dx + \int_{\partial\Omega^0} \frac{\partial u}{\partial \nu}(u^I-u_h)d\ell - \int_{S_h} \mu(u^I-u_h)dx . \end{aligned} \tag{3.28}$$

Recall that, by defining $w^+ = \sup\{0,w\}$, $w^- = \sup\{0,-w\}$, the variational inequality (2.15) implies (see BREZIS & STAMPACCHIA [5], BREZZI, HAGER & RAVIART [16]):

$$w^+(\psi_1-u) = 0 \quad , \quad w^-(\psi_2-u) = 0 \quad \text{a.e. in } \Omega . \tag{3.29}$$

Then:

$$\int_\Omega w(u^I-u_h)dx = \int_\Omega w^+(u^I-u_h)dx - \int_\Omega w^-(u^I-u_h)dx . \tag{3.30}$$

By adding and subtracting u, ψ_1, ψ_1^I we get:

$$\int_\Omega w^+(u^I-u_h)dx = \int_\Omega w^+(u^I-u+u-\psi_1+\psi_1-\psi_1^I+\psi_1^I-u_h)dx \leq$$
$$\leq \int_\Omega w^+[(\psi_1-u)-(\psi_1^I-u^I)]\,dx , \tag{3.31}$$

since $w^+(u-\psi_1)=0$ and $w^+(\psi_1^I-u_h) \leq 0$ a.e. in Ω. (In fact, $u_h \geq \psi_1^I$ at nodes of $\mathcal{T}_h$ implies $u_h \geq \psi_1^I$ a.e.). The last integral reduces to an integral on $\mathcal{F}_h \cap \Omega$, where $\mathcal{F}_h$ is a region of triangles which intersect the free-boundary, since (3.29) implies $w^+=0$ where $u > \psi_1$ and $u = \psi_1$ elsewhere. For h sufficiently small $\mathcal{F}_h$ is contained in a set where ψ_1 is regular [6]. Then we get from (3.31)

$$\int_\Omega w^+(u^I-u_h)dx \leq \|w^+\|_{0,\Omega}\|(\psi_1-u)-(\psi_1^I-u^I)\|_{0,\mathcal{F}_h\cap\Omega} \leq$$
$$\leq C\,h^2|\psi_1-u|_{2,\mathcal{F}_h\cap\Omega} , \tag{3.32}$$

possibly using $\widetilde{\psi_1-u}$ (see (3.19)) as in the proof of (3.25).

A similar relation holds with ψ_1 and w^+ replaced by ψ_2 and w^- respectively. Hence:

$$\int_\Omega w(u^I-u_h)dx = O(h^2). \tag{3.33}$$

The second term in (3.28) is easily bounded by:

$$\int_{\partial\Omega^0} \frac{\partial u}{\partial\nu}(u^I-u_h)d\ell \leq \text{meas}(\partial\Omega^0)\sup_{\partial\Omega^0}\left|\frac{\partial u}{\partial\nu}\right|\sup_{\partial\Omega^0}|u^I-u_h| , \tag{3.34}$$

where u^I and u_h are piecewise linear polynomials having the same value c on $\partial\Omega_h^0$. Moreover, $|\nabla u|\leq 1$ implies $|\nabla u^I|\leq 1$, and $\psi_1\leq u_h\leq\psi_2$ implies also $|\nabla u_h|\leq 1$ on each triangle intersecting $\partial\Omega^0$. Hence, both u^I and u_h are (on each of these triangles) of the type $\beta y+c$, with $|\beta| \leq 1$. Therefore, $\sup_{\partial\Omega^0}|u^I-u_h|=O(h^2)$ since $\text{dist}(\partial\Omega^0, \partial\Omega_h^0) = O(h^2)$, and (3.34) becomes:

$$\int_{\partial\Omega^0} \frac{\partial u}{\partial \nu}(u^I - u_h)\,d\ell \le C\,h^2 . \qquad (3.35)$$

With similar arguments and recalling (3.23) we also have that

$$\int_{S_h} \mu(u^I - u_h)dx \le \mu\ \mathrm{meas}(S_h) \sup_{S_h} |u^I - u_h| = O(h^4). \qquad (3.36)$$

Inserting (3.33), (3.35) and (3.36) in (3.28) we get:

$$a_h(u, u^I - u_h) - \int_{Q_h} \mu(u^I - u_h)dx = O(h^2), \qquad (3.37)$$

which gives in (3.27):

$$a_h(u - u_h, u^I - u_h) = O(h^2). \qquad (3.38)$$

Combining (3.26) and (3.38), (3.17) becomes:

$$\|u - u_h\|^2_{V_h} \le C\,h\|u - u_h\|_{V_h} + O(h^2). \qquad (3.39)$$

The estimates (3.16) is finally obtained by completing the square in (3.39).

∎

<u>LEMMA 3.4</u> Let F and F_h be defined in (2.16) and (3.9) respectively. Then we have:

$$\sup_{c\in\mathcal{C}} |F(c) - F_h(c)| = O(h^2) . \qquad (3.40)$$

<u>Proof</u>: For any $c \in \mathcal{C}$, let u be the solution of (2.15) and u_h be the solution of (3.7). (Recall that u_h and u^I are both equal to zero on $S_h = Q \setminus Q_h$). We can write:

$$F(c) - F_h(c) = \frac{1}{2}a_h(u,u) - \int_Q \mu u dx - \frac{1}{2}a_h(u_h, u_h) + \int_Q \mu u_h dx + \frac{1}{2}\int_{S_h} |\nabla u|^2 dx \le \qquad (3.41)$$

$$\le -\frac{1}{2}a_h(u-u_h,u-u_h)+a_h(u,u-u_h)-\int_Q \mu(u-u_h)dx + \tag{3.42}$$
$$+\frac{1}{2}\,\mathrm{Sup}|\nabla u|^2 \mathrm{meas}(S_h) \le$$

$$< \frac{1}{2}\|u-u_h\|^2_{V_h} + a_h(u,u-u^I) - \int_Q \mu(u-u^I)dx + \tag{3.43}$$
$$+ a_h(u,u^I-u_h) - \int_Q \mu(u^I-u_h)dx + O(h^2) =$$

$$= O(h^2) + a_h(u,u-u^I) - \int_Q \mu(u-u^I)dx\ , \tag{3.44}$$

where (3.44) is derived using (3.16) and the intermediate result (3.37) of LEMMA 3.3. For the remaining terms in (3.44) after integrating by parts and introducing the variable $w = -\Delta u-\mu$ we find:

$$a_h(u,u-u^I)-\int_Q \mu(u-u^I)dx = \int_\Omega w(u-u^I)dx + \int_{\partial\Omega^0} \frac{\partial u}{\partial\nu}(u-u^I)d\ell - \tag{3.45}$$
$$-\int_{S_h}\left[\mu(u-u^I) + \nabla u\cdot\nabla(u-u^I)\right]dx.$$

By inserting the function $\tilde{u}$ defined in (3.19), the first integral can be easily bounded with the usual error estimate $\|\tilde{u}-u^I\|_{0,\Omega_h} = O(h^2)$. The two remaining terms give immediately:

$$\int_{\partial\Omega^0} \frac{\partial u}{\partial\nu}(u-u^I)d\ell \le \mathrm{meas}(\partial\Omega^0)\ \sup_{\partial\Omega^0}\left|\frac{\partial u}{\partial\nu}\right| \sup_{\partial\Omega^0}|u-u^I| \le C\,h^2\ , \tag{3.46}$$

since $|\nabla u| \le 1$ a.e. in Ω and $u^I_{|\partial\Omega^0} = O(h^2)$, $u_{|\partial\Omega^0} = c$. Also we have:

$$\int_{S_h}\left[\mu(u-u^I)+\nabla u\cdot\nabla(u-u^I)\right]dx \le \mathrm{meas}(S_h)(\mu\ \sup_{S_h}|u|+\sup_{S_h}|\nabla u|^2)\le C\,h^2 \tag{3.47}$$

Hence we have from (3.44):

$$F(c) - F_h(c) = 0(h^2) . \tag{3.48}$$

On the other hand, using (3.16), (3.45) and (3.37) we have

$$F_h(c)-F(c) = \frac{1}{2} a_h(u-u_h,u-u_h)-[a_h(u,u-u_h) - \int_Q \mu(u-u_h)dx]+$$
$$+ \frac{1}{2} \int_{S_h} |\nabla u|^2 dx \leq \frac{1}{2}\|u-u_h\|^2_{V_h} -[a_h(u,u-u^I)-\int_Q \mu(u-u^I)dx]+ \tag{3.49}$$
$$+ \frac{1}{2} \sup_{S_h}|\nabla u|^2 \text{meas}(S_h) - [a_h(u,u^I-u_h)-\int_Q \mu(u^I-u_h)dx]=0(h^2) .$$

Combining (3.48) and (3.49) we obtain the desired result.

■

An immediate consequence of (3.40) is the following

LEMMA 3.5 If c^* and c_h^* are the solutions of (2.17) and (3.11) respectively, then:

$$\|c^* - c_h^*\| = 0(h). \tag{3.50}$$

Proof: By applying (2.26) with $c=c_h^*$ we find:

$$F(c_h^*) \geq F(c^*) + \frac{M}{4}\|c_h^* - c^*\|^2 . \tag{3.51}$$

On the other hand, (3.12) for $c=c^*$ gives:

$$F_h(c^*) \geq F_h(c_h^*) + \frac{M'}{4}\|c^* - c_h^*\|^2 . \tag{3.52}$$

By adding (3.51) and (3.52) we have:

$$\|c^*-c_h^*\|^2 \leq \frac{4}{M+M'} \left[|F(c_h^*)-F_h(c_h^*)| + |F_h(c^*) - F(c^*)|\right] \tag{3.53}$$

and (3.50) follows from (3.53) and (3.40).

■

<u>LEMMA 3.6</u> If $u^* \in K^{c^*}$ is the solution of (2.12) and $u_h^* \in K_h^{c_h^*}$ is the solution of (3.7) for $c=c_h^*$, the following estimate holds:

$$\|u^* - u_h^*\|_{V_h} \leq C\,h. \tag{3.54}$$

<u>Proof</u>: Denote by u the solution of (2.15) for $c=c_h^*$. We can write:

$$\|u^*-u_h^*\|_{V_h} \leq |u^*-u|_{1,Q} + \|u-u_h^*\|_{V_h} . \tag{3.55}$$

By the definition (2.3) of J(v) and the parallelogram rule, the first term in (3.55) gives:

$$\begin{aligned} |u^*-u|_{1,Q}^2 &= 2|u^*|_{1,Q}^2 + 2|u|_{1,Q}^2 - |u^*+u|_{1,Q}^2 = \\ &= 4J(u^*)+4J(u)+4\int_Q \mu(u^*+u)dx-4\left|\frac{u^*+u}{2}\right|_{1,Q}^2 = \\ &= 4F(c^*)+4F(c_h^*)-8\left[\frac{1}{2}\left|\frac{u^*+u}{2}\right|_{1,Q}^2-\int_Q \mu\left(\frac{u^*+u}{2}\right)dx\right] . \end{aligned} \tag{3.56}$$

Setting $v = \frac{u^*+u}{2}$, if follows that $v \in K^{\alpha}$, with $\alpha = \frac{c^*+c_h^*}{2}$. Then, if u_α is the solution of (2.15) for $c=\alpha$:

$$J(v) = \frac{1}{2}|v|_{1,Q}^2 - \int_Q \mu v dx \geq J(u_\alpha) = F\left(\frac{c^*+c_h^*}{2}\right). \tag{3.57}$$

Substitution in (3.56) gives:

$$|u^*-u|_{1,Q}^2 \leq 4F(c^*)+4F(c_h^*)-8F\left(\frac{c^*+c_h^*}{2}\right) \leq 4(F(c_h^*)-F(c^*)), \tag{3.58}$$

since $F\left(\frac{c^*+c_h^*}{2}\right) \geq F(c^*)$. On the other hand,

$$F(c_h^*)-F(c^*)=F(c_h^*)-F_h(c_h^*)+F_h(c_h^*)-F_h(c^*)+F_h(c^*)-F(c^*) \leq \leq F(c_h^*)-F_h(c_h^*)+F_h(c^*)-F(c^*) \ . \tag{3.59}$$

Note that $F_h(c_h^*)-F_h(c^*) \leq 0$ since c_h^* is the minimum of F_h. Then from (3.40) applied to (3.59) it follows that:

$$F(c_h^*)-F(c^*) \leq 2 \sup_c |F(c)-F_h(c)| = O(h^2), \tag{3.60}$$

and, from (3.58),

$$|u^*-u|_{1,Q} = O(h) \ . \tag{3.61}$$

Moreover, $\|u-u_h^*\|_{V_h} = O(h)$ as a result of LEMMA 3.3. Substitution in (3.55) completes the proof.

■

4. NUMERICAL RESULTS

The discretization proposed in Sect. 3 has been tested on two model problems, for the simplest case of multiply-connected domains having just one hole. The discrete problem (3.11) is a minimization problem for a convex function, F_h, on a closed bounded domain, $\mathcal{C}$. There are several methods for solving this kind of problems. Most of them use the gradient of the function to be minimized. Unfortunately in our case F_h may not to be differentiable, and anyway we do not know how to compute ∇F_h. What we are able to compute is just $F_h(c)$ for any given $c \in \mathcal{C}$.

In order to compute $F_h(c)$ for a given c, we have first to solve the discrete variational inequality (3.8) in K_h^c. This leads to a complementarity system wich has been solved with the over-relaxation method with projection (see e.g. CRYER [17], GLOWINSKI, LIONS & TREMOLIERES [18]). For each new $c \in \mathcal{C}$, the initial guess for the overrelaxation was the computed solution corresponding to the previous value of c. Once the solution u_h^c of (3.8) is found, the evaluation of $F_h(c):=J_h(u_h^c)$ is a simple matrix product.

In the two test problems presented here the minimization problem (3.11) reduces to a minimization problem over an interval. The golden section method has been chosen because it has two important advantages: it converges in a finite number of steps and, at each step, it demands the evaluation of F_h at only one new point c.

In the first problem the cross-section of the bar is the square $]-0.5,0.5[\times]-0.5,0.5[$ having a concentric circular hole with radius r=0.25. In this case the set $\mathcal{C}$ is the interval $[0,c_\infty]$, $c_\infty=0.5-r=0.25$. In the second problem the cross-section is the square $]-0.9,0.9[\times]-0.9,0.9[$ having a concentric square hole: $]-0.3,0.3[\times]-0.3,0.3[$. In this case we get $\mathcal{C}=[0,c_\infty]$, $c_\infty=0.6$. Because of simmetry of the solution, in both problems computations were carried out on a quarter of the domain. Many different values of the torsional rotation μ were tested, to check the behaviour of the contact regions: $\Omega^-=\{x\in\bar{\Omega}: u(x)=\psi_1(x)\}$, $\Omega^+=\{x\in\bar{\Omega}: u(x)=\psi_2(x)\}$. The results obtained are in agreement with the conclusions of CAFFARELLI & FRIEDMAN [6], CHIPOT [19]. For the first problem the lower obstacle is never reached, while the region $\Omega_h^+=\{P_i: u_h(P_i)=\psi_2(P_i)\}$ increases as μ increases (see figs. 4.1,-,4.4). Similar results are obtained for the second problem, as far as the behaviour of Ω_h^+ is concerned. Moreover, a small region Ω_h^- of contact with the lower obstacle appears in this case for 'big" values of μ. (See figs. 4.5,-,4.8, where the symbol "*" is used to plot the points of Ω_h^-). Notice that in this case the shape of the hole does not guarantee the regularity (of the exact solution of the continuous problem) we used for the error estimates of Sect. 3. Nevertheless, the computed discrete solution reproduces the theoretical behaviour discussed in [6].

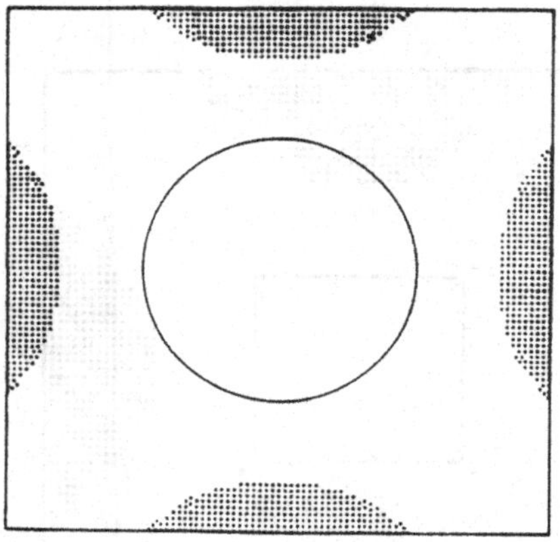

Fig. 4.1- μ=4. c*=.22119

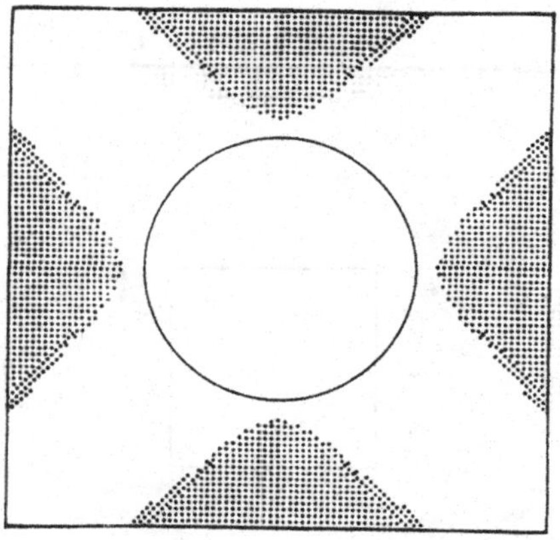

Fig. 4.2- μ=5. c*=.246513

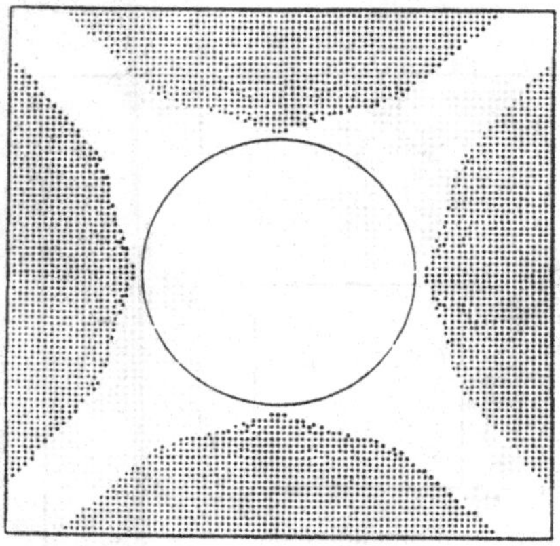

Fig. 4.3- μ=10. c*=.25

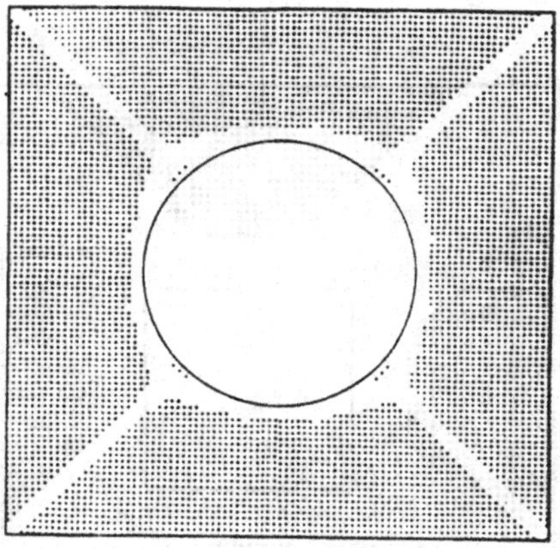

Fig. 4.4- μ=40. c*=.25

Square bar of edge 1. with a concentric circular cavity of radius 0.25

Mesh of 1296 triangles and 703 nodes

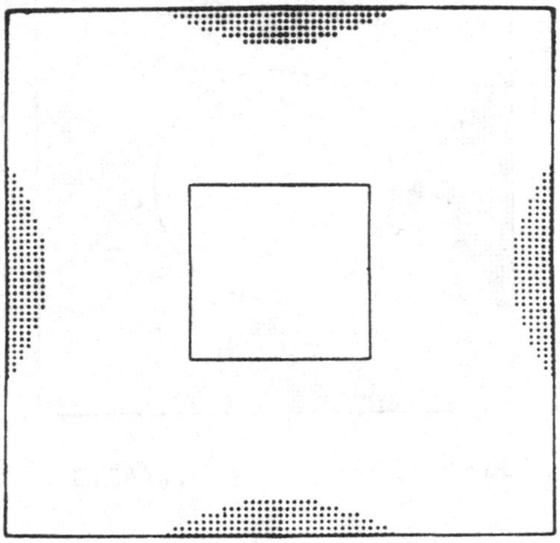

Fig. 4.5 - μ=2. c*=.406872

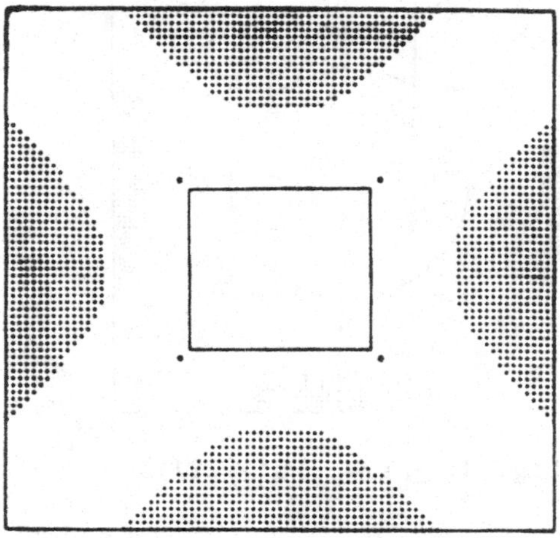

Fig. 4.6 - μ=3. c*=.518110

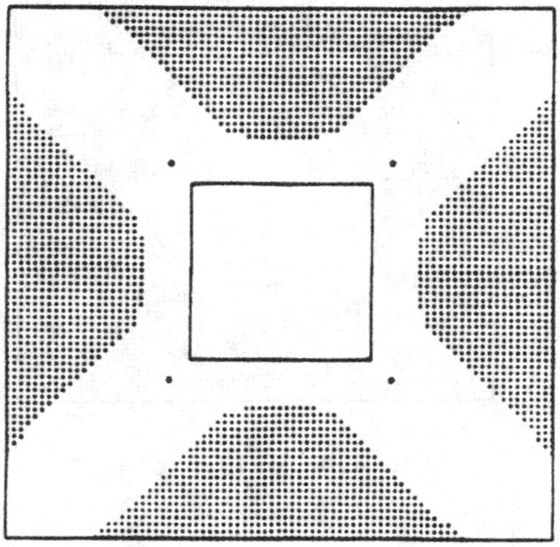

Fig. 4.7 - μ=4. c*=.565181

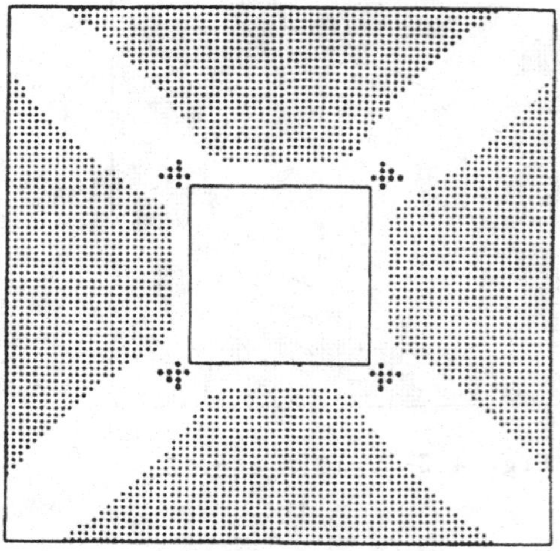

Fig. 4.8 - μ=6. c*=.596303

Square bar of edge 1.8 with a concentric square cavity of edge 0.6

Mesh of 1152 triangles and 625 nodes

REFERENCES

1. Brézis, H. and M. Sibony, Equivalence de deux inéquations variationnelles et applications, *Arch. Rational Mech. Anal.*, 41, 254, 1971.

2. Falk, R. and B. Mercier, Error estimates for elastoplastic problems, *R.A.I.R.O. Anal. numér.*, 11, 117, 1977.

3. Ting, T.W., Elastic-plastic torsion problem over multiply connected domains, *Ann. Scuola Norm. Sup. Pisa*, 4(4), 291, 1977.

4. Baiocchi, C., Su un problema di frontiera libera connesso a questioni di idraulica, *Ann. Mat. Pura Appl.*, 4(9), 107, 1972.

5. Brézis, H. and G. Stampacchia, Sur la régularité de la solution d'inéquations elliptiques, *Bull. Soc. Math. France*, 96, 153, 1968.

6. Caffarelli, L.A. and A. Friedman, The free boundary for elastic-plastic torsion problems, *Trans. Amer. Math. Soc.*, 252, 65, 1979.

7. Gerhardt, C., Regularity of solutions of nonlinear variational inequalities with a gradient bound as constraint, *Arch. Rational Mech. Anal.*, 58, 309, 1975.

8. Grisvard, P., Alternative de Fredholm relative au problème de Dirichlet dans un polygone ou un polyèdre, *Boll. U.M.I.*, 5(4), 132, 1972.

9. Caffarelli, L.A. and N.M. Riviere, The smoothness of the elastic-plastic free boundary of a twisted bar, *Proc. Amer. Math. Soc.*, 63, 56, 1977.

10. Caffarelli, L.A., Friedman, A. and G.A. Pozzi, Reflection methods in elastic-plastic torsion problems, *Indiana Univ. Math. J.*, 29, 205, 1980.

11. Friedman, A. and G.A. Pozzi, The free boundary for elastic-plastic torsion problems, *Trans. Amer. Math. Soc.*, 257, 411, 1980.

12. Ciarlet, P.G., *The finite element method for elliptic problems*, North Holland, Amsterdam, 1978.

13. Lions, J.L. and E. Magenes, *Non Homogeneous Boundary Value Problems and Applications*, 1, Springer, Berlin-Heidelberg-New York, 1972.

14. Ciarlet, P.G. and P.A. Raviart, General lagrange and hermite interpolation in $\mathbb{R}^n$ with applications to finite element methods, *Arch. Rational Mech. Anal.*, 46, 177, 1972.

15. Strang, G. and G. Fix, *An analysis of the finite element method*, Prentice-Hall, Englewood Cliffs (New Jersey), 1973.

16. Brezzi, F., Hager, W.W. and P.A. Raviart, Error estimates for the finite element solution of variational inequalities. Part I, *Numer. Math.*, 28, 431, 1977.

17. Cryer, C.W., The solution of a quadratic programming problem using systematic overrelaxation, *Siam J. Control*, 9, 385, 1971.

18. Glowinski, R., Lions, J.L. and R. Tremolières, *Analyse Numérique des Inequations Variationnelles*, 1, Dunod, Paris, 1976.

19. Chipot, M., Some results about an elastic-plastic torsion problem, *Nonlinear Anal. Theory Methods Appl.*, 3, 261, 1979.

THE AXISYMMETRIC BOUSSINESQ PROBLEM FOR SOLIDS WITH SURFACE ENERGY

D. Maugis
CNRS – LCPC
Paris

1 - Introduction

Besides Young's modulus E and Poisson ratio ν, isotropic elastic solids have a surface energy γ. E and ν reflect the behaviour of intermolecular forces for small displacements of atoms around their equilibrium position and 2γ is the work needed to cut these bonds along an imaginary plane of unit area and to reversibly separate the two parts of the solid. Surface energy thus characterises the nature of bonds ensuring the cohesion of the solid through this imaginary plane. Accordingly, metals and covalents have high surface energy (from 1000 to 3000 $mJ.m^{-2}$), ionic crystals (100 to 500 $mJ.m^{-2}$) and molecular crystals (γ < 100 $mJ.m^{-2}$) have lower surface energy. The first to have coupled surface energy and elasticity was Griffith[1] : to extend the area of a crack by dA the work $2\gamma dA$ is needed ; it is taken from the elastic field and/or the potential energy of the system. Later Irwin[2] introduced the strain energy G released when the crack area varies by dA, and stated the Griffith criterion for a crack in (stable or unstable) equilibrium as $G \Rightarrow 2\gamma$. The singularity of stresses near a crack tip was pointed out by Sneddon[3] ; Irwin[4,5] introducing the stress intensity factor K controlling the intensity of the stress fields showed the relation between the energy approach and that in terms of stress fields.

Attractive forces exist also between bodies in contact , and the work per unit area needed to cut these bonds is the Dupré energy of adhesion $w = \gamma_1 + \gamma_2 - \gamma_{12}$, where γ_1 and γ_2 are the surface energy of solids 1 and 2, and γ_{12} the interfacial energy. These forces ensure adhesion between solids in the same way that they ensure cohesion at the interior of a single solid. Thus a rigid punch on an elastic half-space can sustain tensile stresses, can have a finite area of contact under zero load, and a force (adherence force) must be applied to separate it from the supporting half-space. Kendall[6] and Johnson et al[7] (JKR) were the first to calculate the adherence of flat punches or spheres, via an energy balance between elastic energy U_E, potential energy U_P and total surface energy $U_S = -\pi a^2 w$ (where a is the radius of the circular contact area). In analogy with the Griffith approach, Maugis and Barquins[8] considered the reduction of contact area as a crack propagating at the interface, introduced the energy release rate G and stated the equilibrium of the punches as $G = w$; they showed that the adherence force at fixed load can be different from that at fixed grips. To compute the elastic energy stored under a rigid sphere loaded by P, JKR started from the Hertz contact under a load P_1, introduced adhesion over the area πa^2 and unloaded the sphere to

P at constant radius of contact a. In this rigid body displacement from P_1 to P, singular stresses appear at the edge of the contact area and Savkoor and Briggs[9] showed that a stress intensity factor K_1 can be introduced like in fracture mechanics.

Are these solutions compatible with the principles of the theory of elasticity ? If so, why have they not been worked out before 1971 ? The answer to these questions can be found in the early work of Boussinesq[10]. In the general problem of a convex frictionless punch indenting an elastic half-space under a normal load P, the area of contact is generally unknown. Since the work of Boussinesq[10] the problem has been solved by adding the condition that the normal component of stress must be zero at the limit of the contact area, this condition ensuring a tangential connection between the punch and the elastic half-space, and prescribing the value δ_o of the depth of penetration of the punch. As noted by Boussinesq, a difference in the depth of penetration corresponds to the superposition of a rigid displacement $\delta_o - \delta$ of the punch and gives rise to singular stresses and to the displacement discontinuities typical of a flat punch ; singular compressive stresses are to be ruled out because they lead to a negative discontinuity of displacement and there fore to a penetration of the half space into the punch ; on the other hand singular tensile stresses seemed to be impossible, unless the punch and the half-space are allowed to sustain normal tensile forces ; they were also ruled out, so that the necessity of zero stress at the edge seemed well-established for a century.

Sneddon[11] derived a rigorous solution of the axisymmetric Boussinesq problem for frictionless punches covering the classical results of Hertz[12], Boussinesq[10] and Love[13] as particular cases. In this solution an arbitrary rigid body displacement $\chi(1)$ appears, which must vanish in order the stresses at the edge of the contact area be finite. Barquins and Maugis[14] showed that if $\chi(1) \neq 0$, a stress singularity and a discontinuity of displacement appear at the edge of the contact ; moreover, using a power expansion of Sneddon's formulae, they found that stresses and displacement are those of fracture mechanics with a stress intensity factor K_1 proportional to $\chi(1)$, and thus with an energy release rate proportional to $\chi^2(1)$. Thus, the supplementary condition needed to solve the problem is the Griffith criterion $G = w$. If $w = 0$ all the classical results are obtained. Thus, the problem of the adherence of punches reveals to be a particular case of the Boussinesq problem, which is itself a problem of fracture mechanics for exterior interface cracks. For example the problems arising for a flat punch with infinite friction (sometimes called adhesive contact), studied by Mossakovskii[15] and Spence[16], are similar to those found in interface cracks by Williams[17], Erdogan[18,19], England[20] and others : i.e. the same Dundurs[21] β parameter is introduced, and the stresses oscillate rapidly at the crack tip. This unrealistic behaviour can be removed by allowing a partial slip at the edge of the punch (Spence[22,23]) or at the tip of the crack (Comninou[24,25]).

In this paper, we examine the Boussinesq problem of frictionless punches capable of sustaining tensile forces. The proper denomination is frictionless adhesive contact, but unfortunately the word adhesive has been already used for contact with infinite friction. Also the locutions smooth and rough contact sometimes used to designate frictionless contacts and contacts with finite friction should be avoided, because real smooth solids can have finite coefficient of friction.

2 - The energy approach

The energy approach was employed by Maugis and Barquins[26] using the classical Legendre transformation. Let us consider the system made of two

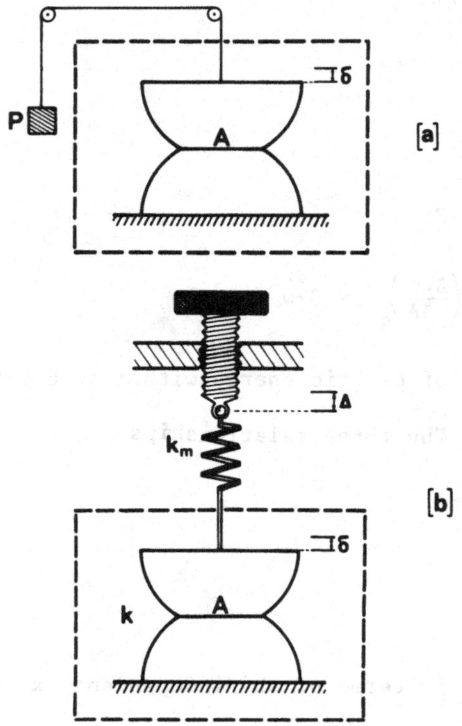

Fig. 1.

elastic solids in contact through the area A. This system is allowed to exchange work and heat, but not matter, with the exterior. A force P (compressive or tensile) can be applied to either of the two elastic bodies, either by a dead load as in fig. 1 a, or by imposing a displacement Δ to a spring of stiffness k_m as in fig. 1 b.

The energy U of the system is assumed to be a function of the elastic displacement δ, of the area of contact A, and of the entropy S ; it can be decomposed in elastic energy U_E and interface energy $U_S = -wA$. The first differential of the energy can be written in the form

$$dU = TdS + Pd\delta + (G-w)dA \qquad (1)$$

with

$$\left(\frac{\partial U}{\partial S}\right)_{\delta,A} = T$$

$$\left(\frac{\partial U}{\partial \delta}\right)_{A,S} = \left(\frac{\partial U_E}{\partial \delta}\right)_{A,S} = P$$

$$\left(\frac{\partial U}{\partial A}\right)_{S,\delta} = \left(\frac{\partial U_E}{\partial A}\right)_{S,\delta} + \left(\frac{\partial U_S}{\partial A}\right)_{S,\delta} = G-w$$

G which describes the variation of elastic energy with A at constant δ, is the strain energy release rate. The three relationships

$$T = T\ (\delta,A,S) \qquad (2a)$$

$$P = P\ (\delta,A,S) \qquad (2b)$$

$$G = G\ (\delta,A,S) \qquad (2c)$$

expressing intensive parameters in terms of the independent extensive parameters, are the equations of state of the system. Knowledge of these three equations of state is equivalent to knowledge of the fundamental equation $U = U(S,\delta,A)$, and gives a thermodynamically complete description of the system.

A system is in equilibrium if virtual perturbations of the extensive variables let its energy constant. However, equilibrium is often studied in the presence of constraints such as constant pressure, constant volume, etc... In this case, the Legendre transformation of the energy U is used to exchange any variable X_j with its derivative $P_j = (\frac{\partial U}{\partial X_j})$. The equilibrium of the system at constant P_j corresponds to the extremum of the function $\Psi = U - P_j X_j$. Ψ which is the Legendre transform of the energy U, is called thermodynamic potential. In the present case we wish to study equilibrium at constant temperature, by allowing perturbations of the area of contact at constant load P, or at constant displacement δ (fixed load or fixed grips conditions). Of interest are thus the Helmoholtz free energy $F=U-TS$, and the Gibbs free energy $g = U - TS - P\delta$, whose differentials are

$$dF = Pd\delta + (G-w)dA - SdT \tag{3}$$

$$dg = \delta dP + (G-w)dA - SdT \tag{4}$$

Noting that $-P\delta$ is the potential energy U_P of the load, these expressions show that

$$G = \left(\frac{\partial U_E}{\partial A}\right)_{\delta,S} = \left(\frac{\partial U_E}{\partial A} + \frac{\partial U_P}{\partial A}\right)_{P,S} \tag{5}$$

Equilibrium at fixed temperature ($dT = 0$) and fixed load ($dP = 0$) corresponds to an extremum of g, and equilibrium at fixed temperature and fixed grips ($d\delta = 0$) to an extremum of F. In either case, equilibrium is characterised by

$$G = w \tag{6}$$

The same result would be obtained at fixed Δ with a spring of finite stiffness k_m.

Eq.(6) is the Griffith criterion which links two of the three variables δ, P, A of the equations of state (2), so that the equilibrium curves $\delta(A)$, A(P), P(δ), are function of w. Analogy with fracture mechanics arises because the edge of the contact area can be considered as an interface crack tip in mode I (opening mode) that recedes or advances as the area of contact increases or decreases.

If $G \neq w$ the contact area will spontaneously change so as to decrease the thermodynamic potential. If $G < w$, eq. (3) and (4) show that A must increase, and the crack recedes. Conversely, if $G > w$ the area of contact must decrease to give $dg < 0$ or $dF < 0$, and the crack extends. GdA is the mechanical energy released when the crack extends by dA. The breaking of interfacial bonds requires an amount of energy wdA, and the excess (G-w)dA in the absence of internal dissipation is entirely changed into kinetic energy. G is often called the crack extension force but, properly speaking, the crack extension force is G-w, which is zero at equilibrium.

The equilibrium can be stable, unstable or neutral. A state of equilibrium of a thermodynamic system under a given constraint is stable if the corresponding thermodynamic potential is minimum, i.e., if its second derivative is positive. Thus, from eq.(3) and (4) stability is defined by

$$\left(\frac{\partial G}{\partial A}\right)_\delta > 0 \quad \text{at fixed grips,}$$

$$\left(\frac{\partial G}{\partial A}\right)_P > 0 \quad \text{at fixed load,}$$

or, more generally, by $(\partial G/\partial A)_\Delta$ if the machine has a finite stiffness. It can be shown that the stability range monotonically increases with the stiffness, from the fixed load case $k_m = 0$ to the fixed grips case $k_m = \infty$. If a state of stable equilibrium is perturbed in such a way that A decreases, then G decreases, so that one has $G < w$: the crack recedes to its equilibrium position. It can only advance if the load P or the displacement δ is slowly varied, bringing back G to the value w : one is dealing with controlled rupture of an adhesive joint. In this case one has

$$dG = \left(\frac{\partial G}{\partial A}\right)_P dA + \left(\frac{\partial G}{\partial P}\right)_A dP = 0, \tag{7}$$

or

$$dG = \left(\frac{\partial G}{\partial A}\right)_\delta dA + \left(\frac{\partial G}{\delta \delta}\right)_A d\delta = 0 \tag{8}$$

Starting from stable equilibrium with the two bodies compressed ($P > 0$, $\delta > 0$), let us quasistatically decrease the load : one generally encounters a progressive reduction in the contact area, i.e., a controlled rupture at $dG = 0$, until a negative load P_c is reached where $\left(\frac{\partial G}{\delta A}\right)_P < 0$; the equilibrium becomes unstable and the crack spontaneously extends toward complete separation at constant P_c, with G-w increasing as A decreases. The load P_c corresponding to the limit of stability is the adherence force in an experiment at fixed load. At fixed grips the adherence force could be different.

Let us return to eq.(3). It can be rewritten in the form

$$dU_E = Pd\delta + GdA, \tag{9}$$

which shows that evaluation of elastic energy in systems with surface energy needs special care. Besides external forces, there are forces due to molecular attraction that cause elastic deformation and elastic energic energy storage.

3 - The elasticity approach

Let us consider, fig. 2, a rigid frictionless axisymmetric punch with a profile given by f(r) (with f(0) = 0), in contact with an elastic half-space over an area of radius a.

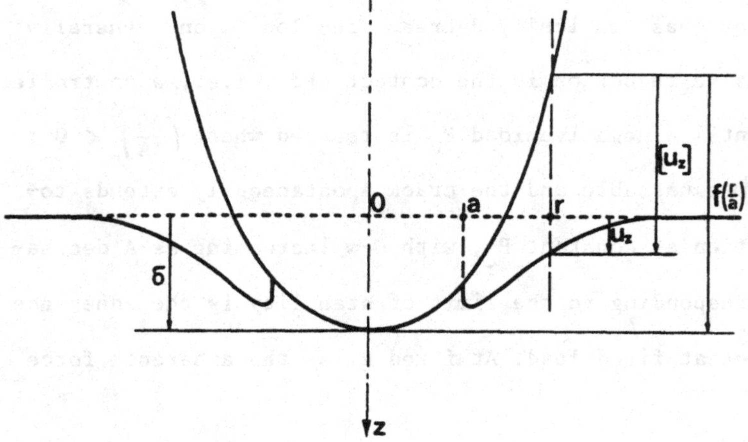

Fig. 2

Letting $\rho = r/a$, and denoting by δ the depth of penetration, the mixed boundary conditions are :

$$u_z(\rho,0) = \delta - f(\rho) \qquad 0 \leqslant \rho \leqslant 1 \tag{10}$$

$$\left.\begin{aligned} \sigma_z(\rho,0) &= 0 \qquad (11) \\ \tau_{rz}(\rho,0) &= 0 \qquad (12) \end{aligned}\right\} \quad \rho > 1$$

From Sneddon's[11] solution, the distribution of pressure under the punch , the displacement of the surface, and the load P which must be applied in order the depth of penetration be δ, are given by

$$P = \frac{\pi a E}{1-\nu^2} \int_0^1 \chi(t)dt = \frac{2aE}{1-\nu^2}\left(\delta - \int_0^1 \frac{\rho f(\rho)}{(1-\rho^2)^{1/2}} d\rho\right), \tag{13}$$

$$\sigma_z(\rho,0) = -\frac{E}{2(1-\nu^2)a}\left[\frac{\chi(1)}{(1-\rho^2)^{1/2}} - \int_\rho^1 \frac{\chi'(t)}{(t^2-\rho^2)^{1/2}} dt\right], \rho < 1, \tag{14}$$

$$u_z(\rho,0) = \int_0^1 \frac{\chi(t)}{(\rho^2-t^2)^{1/2}} dt , \quad \rho > 1, \tag{15}$$

where $\chi(t)$ is defined by

$$\chi(t) = \frac{2}{\pi}\left(\delta - t \int_0^t \frac{f'(\rho)}{(t^2-\rho^2)^{1/2}} d\rho\right). \tag{16}$$

The depth of penetration is thus

$$\delta = \int_0^1 \frac{f'(\rho)d\rho}{(1-\rho^2)^{1/2}} + \frac{\pi}{2}\chi(1),$$

where $\chi(1)$ is the arbitrary rigid body displacement which was cancelled to satisfy the Boussinesq condition for punches with continuous profile. Similar equations were deduced by Shield[27] using Betti's theorem.

In fact, the hypothesis $\chi(1) = 0$ is not necessary ; on the contrary, the rigid body displacement $\chi(1)$ describes the attractive action of molecular forces and gives rise to singular tensile stresses and to a displacement discontinuity at the edge of the contact area as predicted by Boussinesq[10]. As shown by Barquins and Maugis[14], the stress σ_z and the displacement discontinuity

$$[u_z] = f(r) - \delta + u_z(r,o) \tag{18}$$

at a distance x from the edge of the contact area ($r = a \pm x$) can be written

$$\sigma_z(a-x,o) \simeq \frac{K_1}{(2\pi x)^{1/2}} \tag{19}$$

$$[u_z] \simeq \frac{4(1-\nu^2)}{E} K_1 \left(\frac{x}{2\pi}\right)^{1/2} \tag{20}$$

with

$$K_1 = -\frac{E}{2(1-\nu^2)} \left(\frac{\pi}{a}\right)^{1/2} \chi(1) \tag{21}$$

Eq. (19) and (20) are those of fracture mechanics for mode I, and K_1 is the stress intensity factor. Eq. (20) corresponds to plane deformation as for any tridimensional crack (Kassir and Sih[28], Sih and Liebowitz[29]). The strain energy release rate is thus

$$G = \frac{1}{2} \frac{1-\nu^2}{E} K_1^2 \tag{22a}$$

$$= \frac{\pi E}{8(1-\nu^2)} \frac{\chi^2(1)}{a} , \tag{22b}$$

where the factor 1/2 is due to the fact that the punch is not deformable and does not store elastic energy. GdA is the work done by singular stresses to close the crack between a + da and a. Eq. (13) and (22) are the two equations of state (2b) and (2c) of the system. Eq. (13) expresses δ as a function of the two variables P, a, and allows us to plot the penetration curves versus radius of contact at different fixed loads, but it does not allow to compute the equilibrium curves $\delta(a)$, $\delta(P)$, $a(P)$. One equation is still missing, giving the values of $\chi(1)$ at equilibrium as a function of the surface properties of the solids. This equation is the Griffith equation (6). Of course, for zero surface energy (w = 0), this equilibrium equation reduces to $\chi(1) = 0$, and one returns to Boussinesq's conditions and to classical results.

Let P_1 be the fictitious load that would give for $\chi(1) = 0$, the same radius of contact a observed under the load P for $x(1) \neq 0$

$$P_1 = \frac{2Ea}{1-\nu^2} \left[\int_o^1 \frac{f'(x)dx}{(1-x^2)^{1/2}} - \int_o^1 \frac{xf(x)dx}{(1-x^2)^{1/2}} \right] \tag{23}$$

It can be shown be shown easily that

$$P_1 - P = - \frac{\pi Ea}{1-\nu^2} \chi(1) \tag{24}$$

so that eq. (17), (14) and (15) become :

$$\delta = \int_0^1 \frac{f'(x)dx}{(1-x^2)^{1/2}} - \frac{1-\nu^2}{E}\,\frac{P_1-P}{2a}\,, \tag{25}$$

$$\sigma_z(\rho,0) = \frac{E}{2a(1-\nu^2)} \int_\rho^1 \frac{\chi'(t)dt}{(t^2-\rho^2)^{1/2}} + \frac{P_1-P}{2\pi a^2}\,\frac{1}{(1-\rho^2)^{1/2}},\ \rho<1, \tag{26}$$

$$u_z(\rho,0) = -\int_0^1 \chi'(t)\sin^{-1}\frac{t}{\rho}\,dt - \frac{1-\nu^2}{\pi E}\,\frac{P_1-P}{a}\,\sin^{-1}\frac{1}{\rho}\,,\ \rho>1. \tag{27}$$

They show that the stresses and displacements for a punch with surface energy having a contact radius a under the load P are obtained by summing the classical stresses and displacements for that punch under the load P_1 giving that radius a, with the stresses and displacements of a flat punch of radius a under the tensile load P_1-P. In particular from eq. (22) and (24) one has

$$G = \frac{1-\nu^2}{E}\,\frac{(P_1-P)^2}{8\pi a^3} \tag{28}$$

for any axisymmetric punch.

Note that the Maxwell relations (obtained from the equality of the mixed partial derivatives of g and F) give

$$\left(\frac{\partial G}{\partial P}\right)_A = -\left(\frac{\partial \delta}{\partial A}\right)_P = -\frac{1-\nu^2}{E}\,\frac{P_1-P}{4\pi a^3}$$

$$\left(\frac{\partial G}{\partial \delta}\right)_A = \left(\frac{\partial P}{\partial A}\right)_\delta = -\frac{P_1-P}{2\pi a^2}$$

so that the extrema of the curves δ(a,P) occur on the classical curves δ(a) or P(a) obtained for w = 0.

4 - The flat punch :

The load-displacement relation given by Boussinesq[10] in the case of a compressive load is

$$\delta = \frac{1-\nu^2}{E} \frac{P}{2a}$$

Using the same relation for a tensile load, and imposing energy balance between the elastic energy $U_E = \frac{1}{2} P\delta$, the potential energy $U_P = -F\delta$, and the surface energy $-\pi a^2 w$. Kendall[6] obtained the adherence force :

$$P_c = -\left(\frac{8\pi E}{1-\nu^2} a^3 w\right)^{1/2} \tag{29}$$

The same result can be reached directly from by considering a profile of equation $f(\rho) = 0$ and by setting $P_1 = 0$ in eq (28). The equilibrium, again obtained by imposing G = w is always unstable, both at fixed load and fixed grips, so that the equilibrium load is the adherence force P_c, eq (29).

For a flat punch K_1 and G were first given by Kassir and Sih[30] by studying external cracks in an elastic solid.

The stress tensor for a flat punch was computed by Sneddon[31]. It is interesting to study because it is the singular part of the stress tensor for punches with surface energy. Using polar coordinates (ρ,Ψ) in a plane containing the z-axis :

$$z = -\rho\sin\Psi$$

$$r = a-\rho\cos\Psi$$

Barquins and Maugis[14] showed that at the edge of the contact area, the stress tensor is

$$\sigma_r = \frac{K_1}{\sqrt{2\pi\rho}} \cos\frac{\Psi}{2}\left(1 - \sin\frac{\Psi}{2}\sin\frac{3\Psi}{2}\right) \tag{30}$$

$$\sigma_\theta = \frac{K_1}{\sqrt{2\pi\rho}} 2\nu\cos\frac{\Psi}{2} \tag{31}$$

$$\sigma_z = \frac{K_1}{\sqrt{2\pi\rho}} \cos\frac{\Psi}{2}\left(1 + \sin\frac{\Psi}{2}\sin\frac{3\Psi}{2}\right) \tag{32}$$

$$\tau_{rz} = \frac{K_1}{\sqrt{2\pi\rho}} \cos\frac{\Psi}{2}\left(\sin\frac{\Psi}{2}\cos\frac{3\Psi}{2}\right) \tag{33}$$

This is the same as the classical result for mode I in plane strain, with $\sigma_\theta = \nu(\sigma_r + \sigma_z) + O(\rho)$. Displacements were also given by Sneddon[31]. In polar coordinates, at the edge of the contact area, one has

$$u_z(\rho,\Psi) - \delta = \frac{K_1}{2\mu}\sqrt{\frac{\rho}{2\pi}}\sin\frac{\Psi}{2}\cdot(3-4\nu-\cos\Psi), \tag{34}$$

which has the same form as in fracture mechanics, but

$$u_r(\rho,\Psi) - u_r(0,\Psi) = \frac{K_1}{\mu}\sqrt{\frac{\rho}{2\pi}}\cos\frac{\Psi}{2}\left(1-2\nu + \frac{1-2\nu}{2\nu}\sin^2\Psi\right) \tag{35}$$

which differs from the corresponding expression in fracture mechanics in that $(1-2\nu)/2\nu$ is replaced by 1. This probably arises from the hypothesis of frictionless contact which implies a radial displacement under the punch.

The case of the flat punch clearly shows the connection between energy balance, mechanics of contact and fracture mechanics.

5 - The spherical punch

Johnson et al.[7] (JKR) used energy balance to calculate adherence force, with special care in computing elastic energy. They first loaded the sphere, with w assumed zero, until the radius a is reached at the hertzian load

$$P_1 = \frac{a^3K}{R} \tag{36}$$

with

$$\frac{1}{K} = \frac{3}{4}\,\frac{1-\nu^2}{E}$$

then unloaded from P_1 to P at constant radius of contact (flat punch displacement). By minimizing the energy, they obtained

$$a^3 = \frac{PR}{K}\left\{1 + \frac{3\pi wR}{P} + \left[\frac{6\pi wR}{P} + \left(\frac{3\pi wR}{P}\right)^2\right]^{1/2}\right\} \tag{37}$$

The term in brackets is the correction to Hertz's theory. The adherence force is the minimum load P for which the squate root is real, i.e.

$$P_c = -\frac{3}{2}\pi wR \tag{38}$$

A more direct route is obtained with the elasticity approach. By using the approximation

$$f(x) = \frac{a^2}{2R}x^2 \tag{39}$$

we have from eq.(13) to (16) :

$$\chi(t) = \frac{2}{\pi}(\delta - \frac{a^2}{R} t^2) \tag{40}$$

$$\chi(1) = \frac{2}{\pi}(\delta - \frac{a^2}{R}) \tag{41}$$

$$P = \frac{2Ea}{1-\nu^2}(\delta - \frac{a^2}{3R}) \tag{42}$$

$$\sigma_z(r,o) = -\frac{3}{2}\frac{P_1}{\pi a^2}(1 - \frac{r^2}{a^2})^{1/2} + \frac{P_1-P}{2\pi a^2}(1 - \frac{r^2}{a^2})^{-1/2} \tag{43}$$

$$u_z(r,o) = \frac{a^2}{\pi R}\left[(\frac{r^2}{a^2}-1)^{1/2}+(2-\frac{r^2}{a^2})\sin^{-1}\frac{a}{r}\right] - \frac{1-\nu^2}{\pi E}\frac{P_1-P}{a}\sin^{-1}\frac{a}{r} \tag{44}$$

Eq. (42) was first given by Shield[27]. Eq. (23) for P_1 gives the classical Hertz result, eq (36), and the Griffith relation with G given by equ (28), leads to the JKR solution, equ. (37). The limit of stability at fixed load $(\partial G/\partial A)_P = 0$ gives the adherence force equ (38), whereas the limit of stability at fixed grips $(\partial G/\partial A)_\delta = 0$ gives $P_c = -\frac{5}{6}\pi wR$.

Equ (37) was verified by Johnson et al[7] and by Maugis and Barquins [8,26] and has become a standart method for evaluating w. Fig. 3 represents the equation of state, eq (42). Superimposed on the curves $\delta(P,a)$ independent of w, are the curve G = 0, which is the Hertz equation for solids with zero surface energy (w = 0), and the curve G = w which is the JKR solution. Quasistatic loading or unloading follows the equilibrium curve $\delta(a)$; but for an instantaneous unloading from

P to P', i.e. an instantaneous increase in G, one observes an instantaneous displacement at constant a (branch LM or LM') followed by a crack propagation at constant P towards a new equilibrium G = w if $P' > -\frac{3}{2}\pi wR$ (branch MN) or towards rupture if $P' < -\frac{3}{2}\pi wR$ (branch M'Q).

Conversely, if one applies an instantaneous loading from P' to P, G immediately decreases at constant a, but as K_1 cannot be negative,

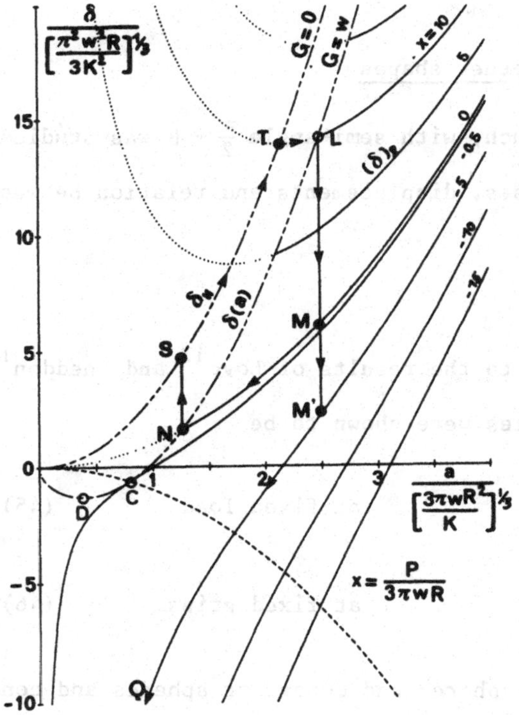

Fig. 3

the branch ST of the Hertz curve (G = 0) is followed (with a tangential connection between the sphere and the half-space), then the crack recedes at constant load P towards its new equilibrium G = w (branch TL). Such pathes have been effectively observed by Maugis and Barquins[8,26] and the agreement with the theory is quite satisfactory.

The stress tensor for such contacts, together with the surface profile for various values of $P/3\pi wR$, was computed by Barquins and Maugis[14] by superposing the stress tensor for a flat punch (Sneddon[30]) to the stress tensor for the sphere (Huber[32]).

6 - Axisymmetric punches of other shapes

The case of a conical punch, with semi-angle $\frac{\pi}{2} - \beta$ was studied by Maugis and Barquins[33]. Stresses, displacements and relation between load have been computed using

$$f(\rho) = (a\,tg\beta)\rho$$

for the profile. They reduce to the results of Love[13] and Sneddon[11] for w = 0. The adherence forces were shown to be

$$P_c = -\frac{54(1-\nu^2)w^2}{\pi E tg^3\beta} \qquad \text{at fixed load} \qquad (45)$$

$$P_c = -\frac{6(1-\nu^2)w^2}{\pi E tg^3\beta} \qquad \text{at fixed grips} \qquad (46)$$

Sectionally smooth ended spheres and cones, or spheres and cones with tips of different radii were also studied by Maugis and Barquins[34] Results for flat-ended spheres and cones reduce to those of Ejike[35] for w = 0.

7 - Experimental verifications on elastomers

Experiments on equilibrium contacts of spheres or flat-ended spheres of glass with rubber agree very well with the theory, see the curves a(P) on fig. 4. Theory can also be tested by studying the equation of state $\delta(P,a)$ in crack propagation at fixed load P as shown in fig.3 ; the agreement is also quite satisfactory (see ref. 8,26,34).

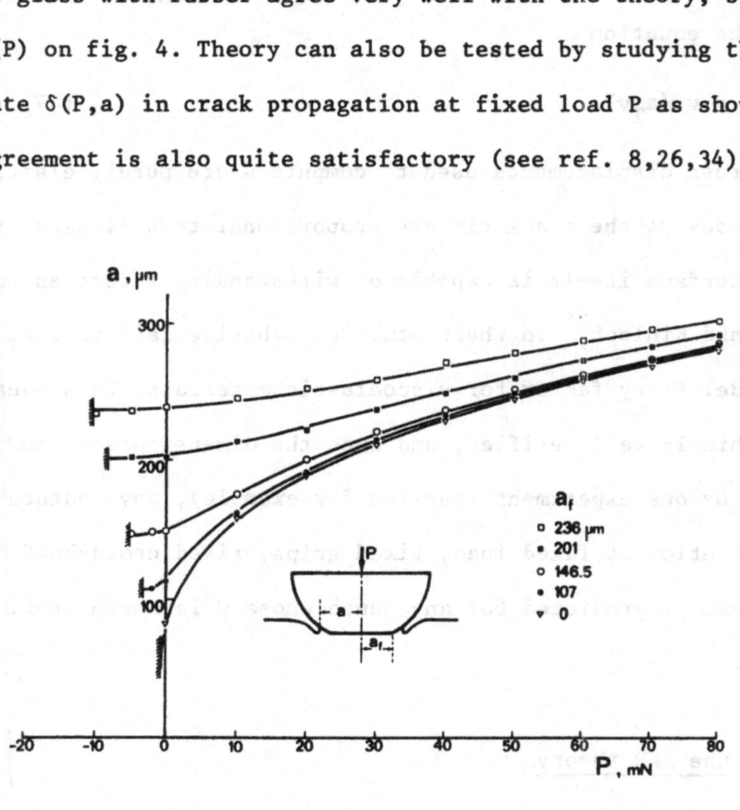

Fig. 4- Equilibrium radii of contact for flat-ended spheres Comparison with theory (R = 2.10 mm, K : 8.9 MPa, w = 50 mJ.m^{-2}). (from réf. |34|)

More impressive are the results concerning the kinetics of crack propagation. In a perfectly elastic solid, there is no dissipation and a crack undergoes a constant acceleration when subjected to a constant force G-w per unit length. In real solids, there is always some internal damping, whose influence is maximal at the crack tip, where strain rates are very high ; consequently the crack takes a speed v, depending on the temperature. To account for damping Maugis and Barquins[8] have proposed the equation

$$G - w = w\,\phi(a_T v) \tag{47}$$

assuming that gross displacements used to compute G are purely elastic, and that the losses at the crack tip are proportional to w (losses only arise if the interface itself is capable of withstanding stress as noted by Andrews and Kinloch[36] in their study of adhesive failure). a_T is the William-Landel-Ferry factor for viscoelastic materials. This equation is astonishingly well verified, and once the dimensionless function $\phi(a_T v)$ is known by one experiment (peeling for example), any feature such as crack kinetics at fixed load, fixed grips, fixed cross-head displacements, can be predicted for any punch whose G is known (ref. 8, 26, 34, 37, 38).

8 - Validity of the JKR theory

The JKR theory, as presented here, follows directly from the principles of the theory of elasticity. However, it was first given by means of an energy balance approach, a dangerous method if the stored elastic energy is not properly evaluated. It is this relative uncertainty in evaluating the elastic deformation due to surface forces that explains the ten years of discussions on the validity of the JKR theory. Considering this theory as an approximate thermodynamic method, Derjaguin, Muller and Toporov[39] proposed a different theory (DMT theory)in which they take into account the molecular attractions in a ring-shaped zone round the contact area. Considering the reverse problem of a deformable sphere on a rigid plane, they assumed that under the influence of surface forces, the sphere is deformed near the contact region according to the Hertz solution ("hemispherical" stress distribution in the area of contact and tangential connnection to the plane), but with an enlarged radius of contact. They found an adherence force $P_c = -2\pi wR$. Their paper was criticized by Tabor[40], and a long dispute followed[41-45] on the shape of the deformed zone (vertical tangent or smooth profile) and on the action of intermolecular forces outside the contact area. From equ (18) and (44) the power expansion near the edge of the contact, at $r = a(1+\varepsilon)$ is

$$[u_z] = -\chi(1)\sqrt{2\varepsilon} + \left[5\chi(1) + 32\frac{a^2}{\pi R}\right]\frac{(2\varepsilon)^{3/2}}{24}$$

where the leading term is in accordance with equ (20). This expression can be given in the alternative form

$$[u_z] = \frac{8\sqrt{2}}{3}\frac{a^2}{\pi R}\left(\frac{r}{a} - 1\right)^{3/2} + \left(\frac{4}{3\pi}\frac{a^2}{R} - \frac{1-\nu^2}{\pi E a}P\right)\cos^{-1}\frac{a}{r}$$

$$= h + h_1$$

as given by Tabor[40] (and corrected by Deryagin et al.[43]) for P = 0. Taking at P = 0 (which implies $P_1 = 6\pi wR$ and $a^3 = 6\pi wR^2/K$) the "neck" of the contact as the flat punch displacement

$$\delta = \frac{1-\nu^2}{2E}\;\frac{P_1-P}{a}$$

and comparing this neck to the equilibrium separation Z_o between atomes, Tabor[40] and Muller et al[45] introduced the parameter

$$\lambda = \frac{\delta}{Z_o} = \left[\frac{6\pi^2 w^2 R(1-\nu^2)^2}{E^2 Z_o{}^3}\right]^{1/3}$$

and arrived at the conclusion that the JKR theory holds for large values of λ and that for very small radii (colloid size) the DMT approach must be used. In fact the discussion here limited to P = 0, can be generalised to any load. The discontinuity of displacement at equilibrium, derived from eq.(20), (22a) and (6)

$$[u_z] = \left(\frac{16(1-\nu^2)wx}{\pi E}\right)^{1/2} ,$$

must be larger than Z_o at a distance say $3Z_o$, i.e.

$$\frac{w}{EZ_o} \gg 1$$

If not, attractive interactions appears between the lips of the Griffith crack, and one must return to the Barenblatt analysis of crack tips. This point was also examined by Greenwood and Johnson[46], and by Savkoor[47]. The situation is not yet very clear, and the subject of exact deformation of elastic solids under the action of surface forces is still under examination (Hughes and White[48,49], Muller et al.[50]).

9 - Conclusion

Contact of elastic solids, adherence of elastic solids, and interface exterior cracks appear to be three facets of the same subject. Progress in the field must now include contact with adhesion and finite friction (the case of infinite friction has been studied by Savkoor[47]), and tangential loading. A comparison between Mindlin[51] solution and Comninou solution (partial crack closure) at interface cracks, or between bubble-like cracks (Comninou[55], Janach[56]) and Schallamach waves[57,58] would be of interest.

REFERENCES -

1 - Griffith, A.A., The phenomena of rupture and flow in solids, *Phil. Trans. Roy. Soc. A*, 221, 1920.

2 - Irwin, G.R., Kies, J.A., Critical energy rate analysis of fracture strength of large welded structures, *Welding Journal (Res. Suppl.)*, 33, 193, 1954.

3 - Sneddon, I.N., The distribution of stress in the neighborhood of a crack in an elastic solid, *Proc. Roy. Soc. A*, 187, 229, 1946.

4 - Irwin, G.R., Analysis of stresses and strains near the end of a crack traversing a plate, *J. Appl. Mech*, 24, 361, 1957.

5 - Irwin G.R., Fracture, *in Encyclopedia of Physics*, vol VI, Flügge, Springer Verlag, 1958, p. 551.

6 - Kendall K, The adhesion and surface energie of elastic solids, *J. Phys. D : Appl. Phys.* 4, 1186, 1971.

7 - Johnson, K.L. Kendall, K, Roberts, A.D., Surface energy and the contact of elastic solids, *Proc. Roy. Soc. A,* 324, 301, 1971.

8 - Maugis, D, Barquins, M, Fracture mechanics and the adherence of viscoelastics bodies, *J. Phys. D : Appl. Phys.* 11, 1989, 1978.

9 - Savkoor, A.R., Briggs, G.A.C. The effect of tangential force in the contact of elastic solids in adhesion, *Proc. Roy. Soc. A,* 356, 103, 1977.

10 - Boussinesq, J, *Application des potentiels à l'étude de l'équilibre et du mouvement dessolides élastiques,* Gauthiers Villars, Paris (Blanchard, Paris 1969) p. 208.

11 - Sneddon, I.N. The relation between load and penetration in the axisymmetric Boussinesq problem for a punch of arbitrary profile, *Int. J. Engng Sc.* 3, 47, 1965

12 - Hertz, H, Uber die Berührung fester Elastischer Korper, *J. für die reine und Angewandte Mathematik,* 92, 156, 1881

13 - Love, E.E.H., Boussinesq's problem for a rigid cone, *Quat. J. Math (Oxford),* 10, 161, 1939.

14 - Barquins, M, Maugis, D, Adhesive contact of axisymmetric punches on an elastic half-space : the modified Hertz-Huber's stress tensor for contacting sphres. *J. Meca. Theor. Appl.* 1, 131, 1982.

15 - Mossakovski, V.I, Compression of elastic bodies under condition of adhesion (axisymmetric case), *P M M,* 27, 418, 1963.

16 - Spence, D.A, Self similar solutions to adhesive contact problems with incremental loading, *Proc. Roy. Soc.* A, 305, 55, 1968.

17 - Williams, M.L, The stresses around a fault or a crack in dissimilar media, *Bulletin of the Seismological Soc. Am,* 49, 199, 1959.

18 - Erdogan, F, Stress distribution in an nonhomogeneous elastic plate with cracks, *J. Appl. M ech.* 30, 232, 1963.

19 - Erdogan, F, Stress distribution in bonded dissimilar materials with cracks, *J. Appl. Mech.,* 32, 403, 1965.

20 - England, A.H., A crack between dissimular media, *J. Appl. Mech.,* 32, 400, 1965.

21 - Dundurs, J. Discussion, *J. Appl. Mech.,* 36, 650, 1969.

22 - Spence, D.A, Similarity considerations for contact between dissimilar elastic bodies, in *The mechanics of the contact between deformable bodies,* de Pater A.R. and Kalker J.J, Eds, Delft University Press, Delft 1975, p. 67.

23 - Spence, D.A., The Hertz contact problem with finite friction, *J. Elasticity,* 5, 297, 1975.

24 - Comninou M, The interface crack, *J. Appl. Mech.* 44, 631, 1977.

25 - Comninou, M, Interface crack with friction in the contact zone, *J. Appl. M ech,* 44, 780, 1977.

26 - Maugis, D., Barquins, M., Fracture mechanics and adherence of viscoélastic solids, in *Adhesion and adsorption of polymers,* part A, Lee, L.H., Ed, Plenum Publ. Corporation, New York, 1980, p. 203-277.

27 - Shield, R.T., Load-displacement relations for elastic bodies, *Z. Agnew Math. Phys.* 18, 682, 1967.

28 - Kassir, M.K, Sih, G.C., Three dimensional stress distribution around an elliptical crack under arbitrary loading, *J. Appl. Mech,* 33, 601, 1966.

29 - Sih, G.C, Liebowitz, H, Mathematical theory of brittle fracture, in *Fracture, an advanced treatise,* vol. 2, Liebowitz, H., ed, Academic Press, New York, 1968, p 67-190.

30 - Kassir, M.K, Sih, G.C, External elliptic crack in elastic solid, *Int. J. Fracture Mech,* 4, 347, 1968.

31 - Sneddon, I.N, Boussinesq's problem for a flat-ended cylinder *Proc. Cambridge Phil. Soc.,* 42, 29, 1946.

32 - Huber, M.T, Zur theory der Berührung fester elastischer Korper, *Ann. Physik,* 14, 153, 1904.

33 - Maugis, D, Barquins, M, Adhesive contact of a conical punch on an elastic half-space, *J. Phys. Lettres* 42, L95, 1981.

34 - Maugis, D, Barquins, M, Adhesive contact of sectionally smooth-ended punches on elastic half-spaces : theory and experiment, *J. Phys. D : Appl. Phys.,* 16, 1843, 1983.

35 - Ejike, U.B.C.O, The stress on an elastic half space due to sectionnally smooth ended punch, *J. Elasticity,* 11, 395, 1981.

36 - Andrews, E.H. Kinloch, A.J. Mechanics of adhesive failure I, *Proc. Roy. Soc.A,* 332, 385, 1973.

37 - Barquins, M, Maugis, D, Tackiness of elastomers, *J. Adhesion,* 13, 53, 1981.

38 - Barquins, M, Influence of the stiffness of testing machine on the adherence of elastomers, *J. Appl. Polym. Sci.,* 28, 2647, 1983.

39 - Derjaguin, B.V, Muller, V.M, Toporov, Yu. P, Effect of contact deformations on the adhesion of particles, *J. Colloid Interface Sci.* 53, 314, 1975.

40 - Tabor, D, Surface forces and surface interactions, *J. Colloid Interface Sci.*, 58, 2, 1977.

41 - Derjaguin, B.V, Muller, V.M, Toporov, Yu. P, On the role of molecular forces in contact deformations (critical remarks concerning Dr. Tabor's report), *J. Colloid Interface Sci.*, 67, 378, 1978.

42 - Tabor, D, On the rôle of molecular forces in contact deformation. *J. Colloid Interface Sci.*, 67, 380, 1978.

43 - Derjaguin, B., Muller, V, Toporov, Yu, On different approaches to the contact mechanics, *J. Colloid Interface Sci.*, 73, 293, 1980.

44 - Tabor, D, Rôle of molecular forces in contact deformations, *J. Colloid Interface Sci.*, 73, 294, 1980

45 - Muller, V.M., YUSHENKO, V.S, Derjaguin, B.V, On the influence of molecular forces on the deformation of an elastic sphere and its sticking to a rigid plane, *J. Colloid Interface Sci.*, 77, 91, 1980.

46 - Greenwood, J.A, Johnson, K.L, The mechanics of adhesion of viscoelastic solids, *Phil. Màg A*, 43, 697, 1981.

47 - Savkoor, A.R, The mechanics and physics of adhesion of elastic solids, in *Microscopic aspects of adhesion and lubrication*, Georges, J.M. Ed,Elsevier, Amsterdam, 1982, p. 279.

48 - Hughes, B.D, White, L.R, "Soft" contact problems in linear elasticity, *Quat. J. Mech. Appl. Math*, 32, 445, 1979.

49 - Hughes, B.D, White, L.R, Implications of elastic deformation on the

direct measurement of surface forces, *J.C.S. Faraday I*, 76, 963, 1980.

50 - Muller, V.M, Yushenko, V.S, Derjaguin B.V, General theoretical consideration on the influence of surface forces on contact deformations and the reciprocal adhesion of elastic spherical particles, *J. Colloid Interface Sci.*, 92, 92, 1983.

51 - Mindlin, R.D, Compliance of elastic bodies in contact, *J. Appl. Mech.* 16, 259, 1949.

52 - Comninou, M., Exterior interface cracks, *Int. J. Engng. Sci.*, 18, 501, 1980.

53 - Janach, W, Separation bubble at the tip of a shear crack under normal Pressure, *Int. J. Fracture*, 14, R 235, 1978.

54 - Schallamach, A, How does rubber slide ? , *Wear*, 17, 301, 1971.

55 - Barquins, M, Energy dissipation in Schallawach waves, *Wear*, 91, 103, 1983.

STANDARD INELASTIC SHOCKS AND THE DYNAMICS OF UNILATERAL CONSTRAINTS

J.J. Moreau
Institut de Mathématiques
Université des Sciences et Techniques du Languedoc
Montpellier

1. INTRODUCTION

This paper is devoted to mechanical systems with a finite number of degrees of freedom ; let $q^1,\dots,q^n$ denote (possibly local) coordinates in the configuration manifold $\mathcal{Q}$. In addition to the constraints, bilateral and frictionless, which have permitted such a finite-dimensional parametrization of $\mathcal{Q}$, we assume the system submitted to a finite family of <u>unilateral constraints</u> whose geometrical effect is expressed by ν inequalities

$$f_\alpha(q) \leqslant 0 \tag{1.1}$$

defining a closed region L of $\mathcal{Q}$. As every greek index in the sequel, α takes its values in the set $\{1,2,\ldots,\nu\}$. The ν functions f_α are supposed C^1, <u>with nonzero gradients</u>, at least in some neighborhood of the respective surfaces $f_\alpha = 0$; for the sake of simplicity, we assume them independent of time.

The typical instance of such a setting is provided by a system of perfectly rigid bodies which may enter into contact, and detach from each other, but can never interpenetrate. In view of this example, we shall refer to a configuration in which equality $f_\alpha = 0$ holds, by saying that <u>contact</u> α <u>takes place</u>. But the formalism applies as well to the unilateral constraints realized by means of inextensible strings ; equality $f_\alpha = 0$ in that case expresses that the corresponding string is taut.

Kinematically, for every motion $t \to q(t)$, the right-velocity $\dot{q}^+$, if it exists at the considered instant, is an element of the n-dimensional linear space $E(q)$, the tangent space to $\mathcal{Q}$ at the point $q(t)$; its components in this space are $\dot{q}^{i+}$, the right-derivatives of the real functions $t \to q^i(t)$. If the moving point $q(t)$ remains in L for every t and if the contact $f_\alpha = 0$ takes place at the considered instant, one immediately finds

$$\sum_i \dot{q}^{i+} \frac{\partial f_\alpha}{\partial q^i} \leqslant 0 . \tag{1.2}$$

Therefore, if we put, for every $q \in L$,

$$J(q) = \{\alpha \in \{1,\ldots,\nu\} : f_\alpha(q) = 0\} \tag{1.3}$$

the right-velocity vector necessarily belongs to the convex polyhedral cone defined in $E(q)$ as

$$V(q) = \{v \in E(q) : \forall\, \alpha \in J(q), \sum_i v^i \frac{\partial f_\alpha}{\partial q^i} \leqslant 0\} \; ; \qquad (1.4)$$

in particular, this is the whole of $E(q)$ if $J(q) = \phi$. This set is usually called the <u>tangent cone</u> at the point q to the region L .

Symmetrically, the left-velocity $\dot{q}^-$, if it exists, belongs to the cone $-V(q)$. If the proper velocity $\dot{q} = \dot{q}^+ = \dot{q}^-$ exists at the considered instant, it belongs to the linear subspace $V(q) \cap -V(q)$ of $E(q)$

We are to study the dynamics of the system, submitted from another part to some given forces, under the hypothesis made explicit in Sect. 2 that the unilateral constraints are <u>frictionless</u>.

Some practical instances have long been discussed ; this usually exhibits finite successions of time intervals : when t ranges over the interior of each of these intervals, $J(q(t))$ remains a constant subset of $\{1,2,\ldots,\nu\}$. Let us call this a <u>motion of finite sort</u>. Counter-examples can be produced where no motion of this sort satisfies the equations of Dynamics [1] . Practising mechanists may consider such instances as pathological and restrict themselves to the search of motions of finite sort ; even so, they will have to face two crucial questions :

<u>Question 1</u>. Starting from an instant t_o , with $q(t_o) = q_o \in L$, $\dot{q}^+(t_o) = \dot{q}_o^+ \in V(q_o)$, determine which of the contacts $f_\alpha = 0$, $\alpha \in J(q_o)$, persist during a subsequent interval.

<u>Question 2</u>. If some interval during which $f_\alpha(q(t)) > 0$ ends at an ins-

tant t_1 such that $f_\alpha(q(t_1)) = 0$, a shock is expected to occur ; determine the right-velocity $\dot{q}^+(t_1)$.

E. Delassus showed[2] that Question 1 had been incorrectly addressed by his eminent predecessors. One usually begins with the tentative assumption that all the contacts $\alpha \in J(q_o)$ such that (1.2) holds as an equality persist ; this amounts to treat the corresponding constraints $f_\alpha = 0$ as bilateral and the associated reactions are then calculated from the equations of Dynamics. If the calculation yields for one or more of these reactions a sign incompatible with the unilaterality of the corresponding constraint, the tentative assumption must be rejected and smaller subsets of $J(q_o)$ have to be tried in the same way. By very simple counter-examples, Delassus demonstrated that the contacts which cease are not necessarily those corresponding to conflicting signs in the first calculation. Even the existence of some satisfactory subset of $J(q_o)$ is a priori to prove, as well as its uniqueness. Delassus'arguments toward a correct solution seem today difficult to read ; a much clearer account of his ideas can be found in [3] (also annexed to the last edition of [4]).

The present author [5,6] developed a more expedient approach to the same question, using convex optimization. The result may be viewed as the extension of the Gauss-Appell "principle of least constraint" to unilateral situations. (The same idea is applied, with an infinite number of degrees of freedom, to the unilaterality of the incompressibility constraint in a liquid in [7,8,9], papers discussing the inception of cavitation when capillary effect may be neglected).

About Question 2, it is classical that adapting the no-friction hypothesis to the dynamics of <u>percussions</u> does not yield enough information to determine $\dot{q}^+(t_1)$. The shock is classically called <u>elastic</u> if it preserves energy ; adding this assumption is known to determine $\dot{q}^+(t_1)$ only in the case of a single contact.[10,11] However, it is widely recognized today that mechanical models have not necessarily to be deterministic. M. Schatzman[12,13] has effectively studied the dynamics of a system of finite freedom with frictionless unilateral constraints and elastic shocks, under the convenience assumption that L consists of a convex subset of $\mathbb{R}^n$; her approach is based on regularization techniques. For the case $\nu = 1$, see also C. Buttazzo and D. Percivale,[14,15] who apply the concept of Γ-convergence to the approximation of solutions. (On the other hand, as an example of problem with infinite freedom, numerous papers have in recent years been devoted to the vibrating string in the presence of an obstacle, initially considered by L. Amerio and G. Prouse[16] ; see e.g.[17,18] Another example is provided by the longitudinal dynamics of a rectilinear bar, an end of which hits an obstacle ; see C. Do[19]).

In contrast, the present paper rests on the recently introduced concept of a <u>standard inelastic shock</u>,[20] essentially dissipative, whose properties are discussed in Sect. 5 below. This results (Sect. 8) in a synthetic formulation of the evolution problem, which embodies in particular the jump conditions, in the event of a shock of the said sort, and the equations of dynamics for possible phases of smooth motion (considered in Sect. 3). The formulation may also be turned into a form in which the <u>sweeping process</u>[21], plays the central role. Sect. 7 summarizes some properties of this process which reveals itself as the basic example of evolution under unilateral and irreversible conditions, with possible jumps. In view of the jumps, the solutions of the process are defined as vector functions with locally bounded variation ; their time change is thus expressed in terms of vector measures, for which fundamental inequalities are given. The importance of right-continuity in this connection is explained.

However, the availability of an elaborate theory for the sweeping process does not readily solve all questions raised by the present dynamical problem : the existence of solutions, possible uniqueness, approximation procedures are still under investigation.

Assuming standard inelastic shock in calculating the motion of some elementary systems yields conclusions in agreement with common observation (a very simple example is given in Sect. 5) ; physical situations undoubtedly exist, where this concept gives a reasonably accurate description of reality. But, to the author's opinion, the main interest of the concept lies in the internal mathematical consistency of the resulting evolution problem. People facing technological applications may question such an attitude ; let us suggest the following answer.

In many domains of applied science, one is dealing with physical laws which, although nonlinear, are smooth enough to admit some linear approximation, after what various corrections, arising from experimental data, may be effected in order to reach a better agreement with reality. In contrast, when facing such situations as unilateral mechanics (or also dry friction or plasticity) one has to treat highly nonlinear, in fact non differentiable, relations. No linear approximation may be used as a first approach ; fortunately the tools of convex analysis are able to provide, as in the present paper, a mathematical framework as consistent and almost as simple as linear analysis ; related to it, a numerical machinery has been previously devised on the purpose of optimization techniques. The solidity of the theoretical core so constructed minimizes the risk of numerical and logical unconsistency when empirical corrections are afterwards added.

In support to the assertion that the quest for mathematical harmony is more than academic decorum, let us recall how successful such an attitude has been in theoretical physics, during the past century.

2. UNILATERAL REACTIONS

During a time interval of smooth motion, all actions experienced by the system are expressed in terms of forces. In the framework of analytical dynamics that we adopt in this paper, if the configuration of the system is q, forces are represented as elements of $E'(q)$, the cotangent space at the point q to the configuration manifold $\mathcal{Q}$. For every possible velocity $v \in E(q)$ of the system through the said configuration, the power of a force $f \in E'(q)$ is, by definition, the "scalar product" $\langle v,f \rangle$, the bilinear form which places the linear spaces $E(q)$ and $E'(q)$ in duality.

In particular, the mechanical realization of the condition $f_\alpha(q) \leqslant 0$ involves some <u>force of constraint</u>, or "reaction", $R_\alpha \in E'(q)$ about which we shall make the following usual assumptions :

1° The reaction R_α vanishes unless $f_\alpha(q) = 0$, i.e.

$$\alpha \notin J(q) \Rightarrow R_\alpha = 0 . \tag{2.1}$$

2° The possible "contact" $f_\alpha(q) = 0$ is frictionless, i.e. the power $\langle v,R_\alpha \rangle$ is zero for every $v \in E(q)$ such that $\langle v,\nabla f_\alpha(q) \rangle = 0$, where $\nabla f_\alpha(q) \in E'(q)$ denotes the gradient of f_α at the point q (non-zero, by hypothesis). This is known to be equivalent to

$$\exists\ \lambda_\alpha \in \mathbb{R} \quad \text{such that} \quad R_\alpha = -\lambda_\alpha \nabla f_\alpha(q) \ . \tag{2.2}$$

3° The direction of R_α is such that the above power is $\geqslant 0$ for every v directed toward the permitted region $f_\alpha \leqslant 0$, i.e.

$$\lambda_\alpha \geqslant 0 \ . \tag{2.3}$$

The latter involves in particular that <u>no adhesion</u> occurs at any proper contact ; for the case of the unilateral constraint realized by means of some inextensible string, it involves that <u>the string exhibits no stiffness</u>.

Conversely, we shall suppose that every value of $R_\alpha \in E'(q)$ satisfying conditions (2.1), (2.2), (2.3) is feasible ; this means that the physical realization of the considered unilateral constraint suffers <u>no strength limitation</u>.

Therefore, a value $R \in E'(q)$ is feasible as the sum of the reactions of the ν unilateral constraints if and only if

$$-R \in N(q) \ , \tag{2.4}$$

where $N(q)$ denotes the <u>convex cone generated in</u> $E'(q)$ <u>by the elements</u> $\nabla f_\alpha(q)$, $\alpha \in J(q)$ (by convention reduced to the zero of $E'(q)$ if $J(q) = \phi$) .

In view of elementary Convex Analysis, $N(q)$ is a closed convex polyhedral cone, equal to the <u>polar cone</u>, relative to the scalar product $<.,.>$, of the closed convex polyhedral cone $V(q)$ defined in (1.4), i.e.

$$N(q) = \{r \in E'(q) : \forall\, v \in V(q)\ ,\ <v,r> \leqslant 0\} \tag{2.5}$$

and symmetrically, with N and V exchanged.

Classically, $N(q)$ is called the <u>outward normal cone</u> to the region L of $\mathcal{Q}$ at the point q.

REMARK. Let us discuss more precisely than in Sect. 1 the meaning of $V(q_1)$, the so-called tangent cone at some point q_1 of L. For <u>each</u> $\alpha \in J(q_1)$, condition $<v, \nabla f_\alpha(q_1)> \leqslant 0$ is indeed necessary and sufficient for the existence of a motion $t \to q(t)$ starting from q_1 with v as initial right-velocity and verifying $f_\alpha(q(t)) \leqslant 0$ in the sequel. But the following counter-example show that, some element v being chosen in $V(q_1)$, it may prove impossible to construct a motion satisfying <u>all</u> conditions $f_\alpha(q(t)) \leqslant 0$ together.

Take $n = 3$, with three inequalities

$$f_1(q) \equiv -q^1 \leqslant 0 \qquad (2.6$$

$$f_2(q) \equiv q^1 - q^2 q^3 \leqslant 0 \tag{2.7}$$

$$f_3(q) \equiv -q^2 - q^3 \leqslant 0 \tag{2.8}$$

For $q_1 = (0,0,0)$, one has $J(q_1) = \{1,2,3\}$ and

$$V(q_1) = \{v = (v^1,v^2,v^3) \;:\; v^1 = 0 \;,\; v^2 + v^3 \geqslant 0\} \qquad (2.9)$$

Every motion $t \to q(t)$ starting from q_1 at time 0, with right-velocity $v = (0,2,-1) \in V(q_1)$ yields for $t > 0$

$$q^2(t) = 2t + o(t) \quad , \quad q^3(t) = -t + o(t) \; .$$

Hence $q^2(t)q^3(t) = -2t^2 + o(t^2)$ is negative in some right-neighborhood of zero ; this contradicts inequalities (2.6),(2.7) .

The following additional regularity assumption is known to secure equivalence between $v \in V(q)$ and the existence of a motion in L , starting from q with v as initial right-velocity (cf. [22] ; in Optimization Theory, this is called a "qualification" condition) :

$$\text{interior } V(q) \neq \phi \qquad (2.10)$$

By classical Convex Analysis, this in turn is found equivalent to the existence of a compact base for the polar cone $N(q)$ of $V(q)$, i.e. there exists in $E'(q)$ a hyperplane, not containing the origin, which intersects all the half-lines generated by the $\nabla f_\alpha(q)$, $\alpha \in J(q)$ (this is understood to hold, trivially, if $J(q) = \phi$) .

The above counter-example leads to question the generality of the so-called Principle of Fourier. This asserts that an element R of $E'(q)$ is a feasible value for the total reaction if and only if $\langle \delta q, R \rangle \geqslant 0$ for every "virtual displacement" (this is another word for the velocity of an imagined motion) starting from q and directed into the permitted region L. If condition (2.10) is satisfied, this property is indeed equivalent to (2.4). Examining the region L of $\mathbb{R}^3$ defined by (2.6),(2.7),(2.8), which does not verify condition (2.10) at the point $q_1 = (0,0,0)$, throws some light on the situation. The subset W of $V(q_1)$ defined by $v^1 = 0$, $v^2 \geqslant 0$, $v^3 \geqslant 0$ might be more legitimately interpreted as the tangent cone to L at q_1. Whether every element of its polar cone $W°$, larger than $N(q_1)$, is a feasible value of the total reaction of the supposedly frictionless unilateral constraints appears as a mechanically irrelevant question. The tangent planes at q_1 to the smooth boundaries defined by equations $f_1 = 0$, $f_2 = 0$, $f_3 = 0$ make zero angles. This allows for q, considered as a material point, to be "pinched" between these boundaries in the position q_1; so infinitely large values of the normal components of the boundary reactions may arise as a response to some moderate driving force acting on q. Under such circumstances, however small may be the friction coefficient between q and the boundaries, friction cannot be neglected.

Points at which (2.10) is not satisfied are not necessarily isolated : for instance, in $\mathbb{R}^3$, some boundaries may meet at zero angle all along a curve. The discussion of would-be frictionless bilateral constraints in classical Analytical Mechanics exhibits similar "pathological" situations ; see e.g.[11], Sect. 9.2.b.

3. DYNAMICAL EQUATIONS OF SHOCKLESS MOTION

In addition to the reactions of constraints, the system is supposed to experience some configuration-dependent forces, represented, in our setting of analytical dynamics, by giving the coefficients Q_i of the differential form "virtual work" of this system of forces, as continuous functions of $(q^1,\ldots,q^n)$; this amounts to define, on the manifold $\mathcal{Q}$, a continuous field of covectors, say $q \to Q(q) \in E'(q)$, possibly dependent also on time.

On the other hand, the expression $T(q,\dot{q})$ of the kinetic energy is given ; for simplicity's sake we restrict ourselves in this paper to the scleronomic case, implying that T does not contain t as independent variable and is a quadratic form relatively to $\dot{q} \in E(q)$.

Then the system of Lagrange equations for every smooth motion writes down as

$$P = Q + R \tag{3.1}$$

where P denotes, as classical, the element of $E'(q)$ whose components, relative to the chosen parametrization of $\mathcal{Q}$ are

$$P_i = \frac{d}{dt}\frac{\partial T}{\partial \dot{q}^i} - \frac{\partial T}{\partial q^i} . \tag{3.2}$$

Eliminating R through (2.4), we give (3.1) the form of a <u>second order differential inclusion</u>

$$Q(t,q(t)) - P(q(t),\dot{q}(t),\ddot{q}(t)) \in N(q(t)) . \tag{3.3}$$

By a solution of (3.3) over some, possibly unbounded, time interval I , we mean a differentiable motion $t \to q(t)$ such that the n derivatives $t \to \dot{q}^i(t)$ are absolutely continuous functions on every compact subinterval of I , with derivatives $t \to \ddot{q}^i(t)$ satisfying (3.3) up to the possible exception of a Lebesgue-negligible subset of I . If I possesses an origin t_o and contains it, this implies the existence of the right-derivatives $\dot{q}^{i+}(t_o)$ and makes the <u>initial conditions</u> $q(t_o) = q_o$ given in L , $\dot{q}^+(t_o) = \dot{q}_o$ given in $V(q_o)$, meaningful.

If we put the natural convention

$$N(q) = \emptyset \quad \text{for} \quad q \notin L , \tag{3.4}$$

<u>the requirement</u> $q(t) \in L$ <u>for every</u> t <u>in</u> I <u>is involved in</u> (3.3).

We shall prove now that every solution of (3.3) in the above sense, actually satisfies a somewhat <u>stronger</u> differential inclusion.

The assumptions made imply that the velocity $\dot{q} \in E(q)$ exists for every t in the interior of I ; since the motion $t \to q(t)$ takes place in the region L , it has been observed in Sect. 1 that $\dot{q} \in V(q) \cap -V(q)$.

In view of (2.5), we conclude that, for every t in the interior of I and such that (3.3) holds, one has

$$\langle \dot{q}, Q-P \rangle = 0 . \tag{3.5}$$

For every subset A of a linear space, we shall denote by ψ_A or $\psi(A,.)$ the <u>indicator function</u> of A, i.e.

$$\psi(A,x) = 0 \text{ if } x \in A , \quad +\infty \text{ if } x \notin A .$$

This function is convex (resp. lower semi-continuous) if and only if the set A is convex (resp. closed). For a pair of mutually polar convex cones, such as $V(q)$ and $N(q)$ above, the respective indicators $\psi(V,.)$ and $\psi(N,.)$ are known to constitute a pair of Fenchel <u>conjugate functions</u>. Now, (3.3) means that $\psi(N(q),Q-P) = 0$. Then, in view of (3.5) and of the fact that $\dot{q} \in V(q)$, one has, for almost every t in I, the equality

$$\psi(V(q),\dot{q}) + \psi(N(q),Q-P) - \langle \dot{q}, Q-P \rangle = 0 ,$$

expressing that $\dot{q}$ in $E(q)$ and $Q-P$ in $E'(q)$ are <u>conjugate points</u> relative to the above pair of conjugate functions. This in turn is known to be equivalent to

$$Q(t,q) - P(q,\dot{q},\ddot{q}) \in \partial\psi(V(q),\dot{q}) , \tag{3.6}$$

where the subdifferential $\partial\psi(V(q),\dot{q})$ classically equals the outward normal cone at the point $\dot{q}$ to the closed convex set $V(q)$. This normal cone is essentially a subset of $N(q)$; here again, convention (3.4) makes that $q(t) \in L$ is involved in (3.6) .

REMARK. The dynamics of supposedly shockless motions in the presence of scleronomic frictionless unilateral constraints, as developed here, exhibits the same reversibility in time as the traditional bilaterally constrained case. In fact the above reasoning could also yield symmetrically

$$Q(t,q) - P(q,\dot{q},\ddot{q}) \in -\partial\psi(-V(q),\dot{q}) \ . \tag{3.7}$$

On the other hand, (3.5) expresses that the total reaction $R \in E'(q)$ develops a zero power in the actual motion. The assumptions made imply that the function $t \to T(q(t),\dot{q}(t))$ is absolutely continuous on every compact subinterval of I , hence differentiable almost everywhere. A classical calculation, based on the fact that $T(q,\dot{q})$ is a quadratic form in its second argument, yields the "energy equation"

$$\frac{d}{dt} T(q(t),\dot{q}(t)) = \langle \dot{q}(t),Q(t,q(t)) \rangle \ . \tag{3.8}$$

It permits to establish a priori bounds of $\dot{q}$ for supposed solutions of the initial value problem. This equation is specially useful when the field of covectors $q \to Q(t,q)$ derives from a time-independent potential function $q \to \mathcal{W}(q)$, i.e. $Q(q) = -\nabla\mathcal{W}(q)$; then it comes that $T(q,\dot{q}) + \mathcal{W}(q)$ is a constant of the motion ; this is the familiar conservation property of the total energy.

4. SHOCK DYNAMICS

If an interval of smooth motion ends at some instant t_s such that the left-velocity $\dot{q}^-(t_s)$, more shortly denoted by $\dot{q}_s^-$, does not belong to $V(q_s)$, $q_s = q(t_s)$, a shock necessarily occurs. Classically, by integrating the Lagrange equation (3.1) over the "infinitely short" duration of this shock, one obtains

$$p_s^+ - p_s^- = \Pi \; . \tag{4.1}$$

Here p_s^+ and p_s^- respectively denote the right and left limits at time t_s of the <u>momentum</u>

$$p = \frac{\partial T}{\partial \dot{q}} \in E'(q) \; . \tag{4.2}$$

The <u>percussion of constraint</u> $\Pi \in E'(q_s)$ is introduced as the integral, over the shock duration, of the "infinitely large" reaction $R(t)$; as this reaction is supposed to satisfy (2.4) with $N(q) = N(q_s)$, a closed convex cone which does not vary during the shock, one has

$$-\Pi \in N(q_s) \tag{4.3}$$

(a discussion of this argument may be found in [11], Sect. 9.7.c, Remarque).

On the other hand, the kinematical condition

$$\dot{q}_s^+ \in V(q_s) \tag{4.4}$$

holds as before.

Even in the special case of a single contact, i.e. $J(q_s)$ consisting of a singleton, conditions (4.1), (4.3) and (4.4) are well known[10,11] to bring insufficient information to derive $\dot{q}_s^+$ from the data q_s and $\dot{q}_s^-$.

The parametrization (q^i) of the configuration manifold is supposed kinetically regular, in the sense that the quadratic form defined on each tangent space $E(q)$ by

$$v \rightarrow 2T(q,v) = \sum_{i,j} a_{ij}(q)\, v^i v^j$$

(with $a_{ij} = a_{ji}$) is positive definite. Let us equip the linear space $E(q)$ with a Euclidean structure by taking $2T(q,v)$ as the definition of the squared norm $\|v\|^2$. Equivalently, the Euclidean scalar product of two elements v and w of $E(q)$ is expressed by

$$v.w = \sum_{i,j} a_{ij}(q)\, v^i w^j \; . \tag{4.5}$$

For geometrical and notational simplicity, we shall perform the classical trick of using this Euclidean structure of the linear space $E(q)$ in order to identify it with its dual space $E'(q)$. From the standpoint of calculation, this means the following : to each choice of a parametrization (q^i) of the manifold Q corresponds a base in the tangent space $E(q)$, say (e_i), $i = 1,\dots,n$. The derivatives $\dot{q}^i$ of the functions $t \to q^i(t)$ representing some motion constitute, as before, the components relative to this base of the velocity vector $\dot{q}$. On the other hand, in view of (4.5), the expressions

$$p_i = \frac{\partial T}{\partial \dot{q}_i} = \sum_j a_{ij}(q)\,\dot{q}^j$$

equal the <u>covariant components</u> of the same element $\dot{q}$ of $E(q)$ relative to the said base, i.e. the respective scalar products $\dot{q}.e_i$. The identification trick amounts to declaring that the element $\dot{q}$ of $E(q)$ and the element p of $E'(q)$ constitute the same object.

Similarly, the partial derivatives $\partial f_\alpha / \partial q^i$ are interpreted as the covariant components of the gradient $\nabla f_\alpha(q)$, now considered as an element of the Euclidean linear space $E(q)$. We continue to denote by $N(q)$ the convex cone generated by the $\nabla f_\alpha(q)$, $\alpha \in J(q)$; this is now a closed convex polyhedral cone in $E(q)$, actually the polar cone of $V(q)$ since (2.5) still holds with $\langle v,r \rangle$ equal to the Euclidean scalar product $v.r$.

In view of the above identification, (4.1) takes on the form

$$\dot{q}_s^+ - \dot{q}_s^- = \Pi , \tag{4.6}$$

while (4.3) and (4.4) stay unchanged.

The shock is traditionally called <u>elastic</u> if it preserves energy, that is, in terms of the Euclidean norm of $E(q_s)$,

$$\frac{1}{2}\|\dot{q}_s^+\|^2 = \frac{1}{2}\|\dot{q}_s^-\|^2 . \tag{4.7}$$

In the special case where $J(q)$ reduces to a singleton, say $J(q) = \{1\}$, conditions (4.3), (4.4), (4.6) and (4.7) are found, by elementary geometry, equivalent to : <u>the vector</u> $\dot{q}_s^+$ <u>equals the mirror image of the vector</u> $\dot{q}_s^-$ <u>relative to the hyperplane of</u> $E(q_s)$ <u>with normal</u> $\nabla f_1(q_s)$.

On the other hand, for the same special case $J(q) = \{1\}$, the shock is called <u>soft</u> or <u>inelastic</u> if, instead of (4.7), one has

$$\dot{q}_s^+ . \nabla f_1(q_s) = 0 . \tag{4.8}$$

We propose, in the Section to come, a generalization of the latter.

5. STANDARD INELASTIC SHOCK

Let us first recall a few facts of elementary convex analysis in a Euclidean linear space E (also valid in an infinite-dimensional real Hilbert space). For every nonempty closed convex subset C of E , every point z of E possesses a unique proximal point in C , denoted here by prox(C,z) . Then (cf. [23] or, for more generality, [24]) the following non-linear generalization of the classical decomposition of E into the sum of orthogonal subspaces holds :

LEMMA OF THE TWO CONES. If V, N denote a pair of mutually polar closed convex cones in E and if x, y, z are three points of E , assertions i) and ii) below are equivalent :

i) $x = \text{prox}(V,z)$, $y = \text{prox}(N,z)$

ii) $z = x + y$, $x \in V$, $y \in N$, $x.y = 0$.

COROLLARY

$$x = \text{prox}(V,z) \iff z - x = \text{prox}(N,z) .$$

Using again the setting of Sect. 4, let us propose :[20]

DEFINITION. The shock at time t_s is said standard inelastic if the three following conditions, equivalent in view of (4.3), (4.4), (4.6), hold

$$\dot{q}_s^+ = \text{prox}\,(V(q_s), \dot{q}_s^-) \tag{5.1}$$

$$-\Pi = \text{prox}\,(N(q_s), \dot{q}_s^-) \tag{5.2}$$

$$\Pi . \dot{q}_s^+ = 0 \; . \tag{5.3}$$

Equivalence immediately results from the above Lemma, by taking $z = \dot{q}_s^-$, $x = \dot{q}_s^+$, $y = -\Pi$.

Condition (5.1) presents the reassuring aspect of an economy principle : among all the values of $\dot{q}_s^+$ kinematically compatible with the unilateral constraints, it imposes the nearest one to $\dot{q}_s^-$, in the sense of the kinetic metric of $E(q_s)$.

Symmetrically, (5.2) may be written as

$$\Pi = \text{prox}\,(-N(q_s), -\dot{q}_s^-) \; . \tag{5.4}$$

Using the equations of the dynamics of percussions under the form (4.6), one sees that $-\dot{q}_s^-$ equals the percussion <u>which should be applied to the system in order to obtain</u> $\dot{q}_s^+ = 0$. Then (5.4) expresses that, in the set of the values of Π permitted by the law (4.3) of frictionless unilaterality, the actual solution consists in the <u>nearest point to this stopping percussion</u>.

Concerning condition (5.3), one gives it in view of (4.6) the equivalent form

$$\frac{1}{2}\|\dot{q}_s^+\|^2 - \frac{1}{2}\|\dot{q}_s^-\|^2 = -\frac{1}{2}\|\dot{q}_s^- - \dot{q}_s^+\|^2 \tag{5.5}$$

This displays a loss of kinetic energy : <u>the process described by the above definition is essentially dissipative</u>. Observe that (5.3) holds in particular if $\dot{q}_s^+$ happens to be kinematically consistent with the permanence of all the contacts $f_\alpha = 0$, $\alpha \in J(q_s)$. Hence (5.5) may be viewed as a generalization of a classical theorem of Carnot, pertaining to the sudden introduction of persistent, bilateral, constraints.[11]

In that connection, when the cone $V(q_s)$ is given, the mapping $\dot{q}_s^- \to -\Pi$ defined by (5.2) appears as a relation between some "velocity" and some "force" of the form currently called <u>a standard dissipative process</u> [25,26] ; in fact this mapping equals the gradient of some convex function, namely $v \to (\text{dist}(v,V_s))^2/2$.

EXAMPLE. Let the system consist of a single particle moving in some vertical plane, with (q^1,q^2) as orthonormal Cartesian coordinates, the q^2 axis vertical and oriented upward. Fixed frictionless boundaries are assumed to impose

$$q^2 \geqslant 0 \quad , \quad q^1 \cos\theta + q^2 \sin\theta \leqslant 0$$

with θ given in $]-\frac{\pi}{2},+\frac{\pi}{2}[$.

A phase of motion :

$$t < 0 \, , \quad q^1(t) = wt \, , \quad q^2(t) = 0$$

($w > 0$ constant) ends with a shock at time $t_s = 0$, $q_s = (0,0)$, since the left-velocity $\dot{q}_s^- = (w,0)$ does not belong to

$$V(q_s) = \{v \in \mathbb{R}^2 : v^2 \geqslant 0 , \quad v^1 \cos\theta + v^2 \sin\theta \leqslant 0\} .$$

Here the kinetic norm coincides with the natural Euclidean norm of the plane. Using (5.1) to determine $\dot{q}_s^+$ leads to distinguish between two cases.

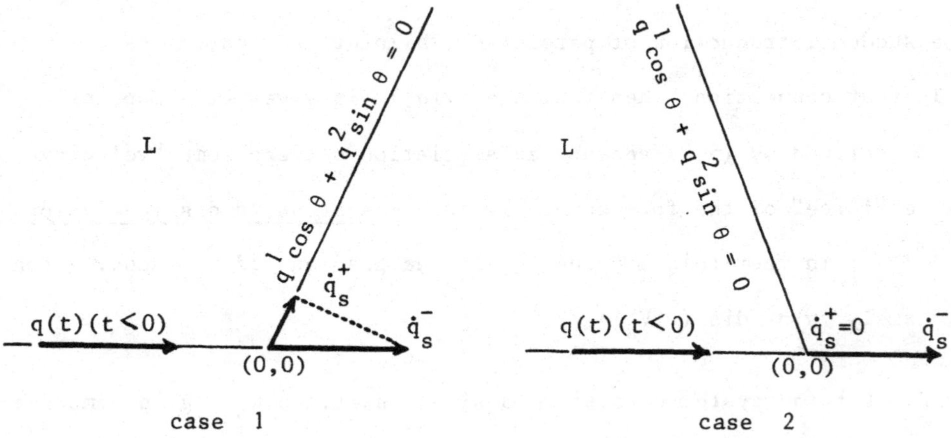

Case 1 : $-\frac{\pi}{2} < \theta < 0$. The subsequent motion takes place on the ascendent boundary line, with initial velocity :

$$\dot{q}_s^+ = (w \sin^2\theta , -w \sin\theta \cos\theta) .$$

Case 2 : $0 \leqslant \theta < \frac{\pi}{2}$. The subsequent motion is rest.

REMARK. Minimizing over $V(q_s)$ the distance to $\dot{q}_s^-$ is the same as minimizing the square of this distance ; hence the determination of $\dot{q}_s^+$ from (5.1) constitutes a problem of "quadratic programming". Let us only observe that the base (e_i) in $E(q_s)$ corresponding to some parametrization (q^i) of $\mathcal{Q}$ has no reason to be orthonormal relatively to the Euclidean structure precedingly defined ; in fact (4.5) shall be used to express the squared norm. Symmetrically, (5.2) makes the determination of Π a problem of quadratic programming ; however it should be kept in mind that the elements $\nabla f_\alpha(q_s)$, $\alpha \in J(q_s)$, generating the convex polyhedral cone $N(q_s)$ possess the respective partial derivatives $\frac{\partial f_\alpha}{\partial q^i}(q_s)$ as <u>covariant components</u> relative to the above base, i.e. such a derivative equals the scalar product of $\nabla f_\alpha(q_s)$ by the base vector e_i . If calculations are to be performed by using this sort of components, recall that, for a pair of vectors v , w with covariant components v_i , w_j , the scalar product is expressed by

$$v.w = \sum_{i,j} a^{ij} v_i w_j$$

where a^{ij} is the inverse matrix of a_{ij} .

6. DIFFERENTIAL MEASURES

Let I denote a real interval ; for brevity, we shall restrict the sequel to the case where I is open on the right, possibly unbounded, but is closed on the left, with origin t_o . A function f of I into a real Banach space X is said to have a <u>locally bounded variation</u> if it has a bounded variation, in the sense of the norm of X , over every compact subinterval of I ; notation $f \in \mathrm{lbv}(I,X)$.

This classically implies the existence of a measure df on I , with values in X , called the <u>differential measure</u> of f , with the characteristic property that, for every compact subinterval [a,b] of I,

$$\int_{[a,b]} df = f^+(b) - f^-(a) \, . \tag{6.1}$$

The existence of the right and left limits f^+ and f^- is secured by the bounded variation assumption ; by convention, $f^-(t_o)$ is taken equal to $f(t_o)$. In particular, by making a = b , (6.1) yields that every discontinuity point of f constitutes an atom of the vector measure df , with a mass equal to the jump of f .

Incidentally for every open subset Ω of I , the restriction of the measure df to Ω constitutes the derivative, in the sense of

Schwartz's distributions in Ω , of the vector measure f dt (dt : the Lebesgue measure in Ω) .

From (6.1) it ensues that, for every $t \in I$,

$$f^+(t) = f(t_o) + \int_{[t_o,t]} df \tag{6.2}$$

$$f^-(t) = f(t_o) + \int_{[t_o,t[} df . \tag{6.3}$$

Hence, knowing the measure df only allows one to reconstruct f^+ and f^- but not the actual function f ; this shows that the familiar pattern, according to which the "differential" df determines f up to an additive constant can subsist in the present context only under some assumption connecting f with its one-side limits. For a vast class of evolution problems, formulated in terms of differential measures, the good assumption turns out to be $f = f^+$, i.e. f <u>right-continuous</u> ; notation $f \in rclbv(I,X)$. This has the immediate advantage of making, for such an evolution problem, the <u>initial condition</u> $f(t_o) = f_o$ meaningful ; the following goes a little deeper in explaining the situation.

Let $B : X \times X \to \mathbb{R}$ denote a norm-continuous bilinear form. For every couple f , g of elements of lbv(I,X) the function $t \to B(f(t),g(t))$ belongs to $lbv(I,\mathbb{R})$ and the following calculation rules holds to express its differential measure :[27,28]

$$\begin{aligned} dB(f,g) &= B(df,g^+) + B(f^-,dg) \\ &= B(df,g^-) + B(f^+,dg) . \end{aligned} \tag{6.4}$$

Under some general definition that we shall not develop here, the symbols on the right-hand side represent real (signed) measures ; in usual instances, in particular when X is finite-dimensional or is a Hilbert space, a differential measure, such as df above, possesses a density relative to some nonnegative real measure $d\mu$, i.e. there exists a function $f' \subset \mathcal{L}^1_{loc}(I,d\mu;X)$ such that $df = f'd\mu$. Then the said definition is found to yield

$$B(df,g^+) = B(f',g^+)d\mu \; , \tag{6.5}$$

a meaningful expression since g^+ , as every element of $lbv(I,X)$, is locally bounded and universally measurable.

Let us make now the additional assumptions that the bilinear form B is symmetric and that the quadratic form $x \to B(x,x)$ it generates is nonnegative. Then it is proved that, for every $f \in lbv(I,X)$, the following inequalities hold, in the sense of the ordering of real (signed) measures

$$2B(f^-,df) \leqslant dB(f,f) \leqslant 2B(f^+,df) \; . \tag{6.6}$$

The inequality on the right explains the importance of the assumption $f = f^+$, i.e. $f \in rclbv(I,X)$, when "energy inequalities" for the considered evolution problems are to be derived ; see Sect. 7 below.

7. THE SWEEPING PROCESS

Here is a basic evolution problem of the unilateral sort, intensively studied by the author, with the theory of elastoplastic systems as primary motivation.[21,25] This problem will be shown in the sequel to present a close connection with the subject of this paper.

Let be given a moving set $t \to C(t)$, or multifunction from the interval I into some real Hilbert space H , with closed convex values. A moving point $u : I \to H$ is called a solution of the sweeping process by C if it satisfies the differential inclusion

$$-\dot{u}(t) \in \partial\psi(C(t),u(t)) . \qquad (7.1)$$

At the first level of the theory, the time-derivative $\dot{u}(t)$, i.e. the velocity vector of the moving point, is defined in the elementary way and (7.1) is supposed to hold for every t in I . If $C(t)$ possesses a non-empty interior, a particularly suggestive interpretation of (7.1) may be given : as long as $u(t)$ lies in this interior, the set $\partial\psi(C(t),u(t))$, namely the outward normal cone to $C(t)$ at this point, reduces to the zero of H ; then (7.1) implies that the point u stays at rest. It is only when caught up with by the boundary of $C(t)$ that u takes up a motion, in some inward normal direction so as to go on belonging to the moving set ; in fact the right-hand side of (7.1) would be empty if $u(t) \notin C(t)$.

The existence of differentiable or absolutely continuous solutions to (7.1) clearly requires some smoothness assumptions concerning the motion of C ; they can be formulated, for instance, in terms of the Hausdorff distance between successive positions of the moving set in H . At a more elaborate level, some "unilateral" rating for the evolution of C is put forward, sensitive only to retraction.[29]

If, on the contrary, the multifunction $t \to C(t)$ is discontinuous, jumps are expected to occur in the motion of u , so that condition (7.1) has to be given some extended meaning :

DEFINITION. We shall say that $u : I \to H$ is a solution of the sweeping process by C in the sense of differential measures if $u \in \mathrm{rclbv}(I,H)$ and if there exists (non uniquely) a nonnegative real measure $d\mu$, with a vector function $u'_\mu \in \mathcal{L}^1_{loc}(I,d\mu;H)$, such that $du = u'_\mu \, d\mu$ and that

$$-u'_\mu(t) \in \partial\psi(C(t),u(t)) \qquad (7.2)$$

holds for every t in I .

As usual, we denote with $\mathcal{L}$ the non-separated topological linear spaces consisting of everywhere defined functions, while L refers to the corresponding separated spaces whose elements are equivalence classes of such functions.

Observe that (7.2) implies $u(t) \in C(t)$ for every $t \in I$: otherwise the right-hand side would be empty. Instead of requiring of (7.2) to be satisfied everywhere, one could equivalently except a possible $d\mu$-negligible subset S of I, provided the condition $u(t) \in C(t)$ is additionally prescribed everywhere in I ; in fact the latter ensures that the right-hand side contains at least the zero of H, hence the latitude of correcting u'_μ in S, so as to finally satisfy (7.2) for every t.

A prominent feature is that, if u complies with the above Definition, the same holds when replacing $d\mu$ by any other nonnegative real measure $d\nu$, relatively to which the vector measure du possesses a "density" $u'_\nu \in \mathcal{L}^1_{loc}(I,d\nu;H)$. This easily results from the fact that every set $\partial\psi(C(t),u(t))$ is a cone : use the sum $d\mu + d\nu = d\beta$ as a "base" measure, relatively to which $d\mu$ and $d\nu$ both possess nonnegative density functions μ'_β and ν'_β ; observe that every subset of I throughout which $\nu'_\beta = 0$ is $d\nu$-negligible.

In particular, with the vector measure du it is associated the non-negative real measure $|du|$, called its <u>absolute value</u>,[30] and du possesses a density relatively to it (this is from elementary measure theory if H is finite-dimensional, but also holds for an arbitrary Hilbert space : such a space is said to possess the Radon-Nikodym property). So $d\mu$ in the above Definition may equivalently be taken equal to $|du|$.

Let us now give an illustration of the importance of the right-continuity involved in the requirement $u \in$ rclbv. Let u and $\bar{u}$ denote two solutions of the sweeping process by C, satisfying (7.2) with $du = u'_\mu \, d\mu$ and $d\bar{u} = \bar{u}'_{\bar{\mu}} \, d\bar{\mu}$ respectively. The nonnegative measures $d\mu$

and $d\bar{\mu}$ possess nonnegative densities relatively to $d\beta = d\mu + d\bar{\mu}$; hence, using again the fact that every $\partial\psi$ is a cone, one obtains

$$- u'_\beta(t) \in \partial\psi(C(t),u(t))$$

$$- \bar{u}'_\beta(t) \in \partial\psi(C(t),\bar{u}(t)) .$$

Classically, for every t , the multifunction $x \to \partial\psi(C(t),x)$ from H into itself is <u>monotone</u> ; therefore

$$(\bar{u}'_\beta(t) - u'_\beta(t)).(\bar{u}(t) - u(t)) \leqslant 0 .$$

The scalar product of H , denoted here by . , may be taken as the bilinear form B in the right-hand inequality (6.6). Due to the right-continuity of u and $\bar{u}$, this yields that the real measure $d\|\bar{u} - u\|^2$ is nonpositive. Consequently, <u>for every pair</u> $u , \bar{u}$ <u>of solutions of the sweeping process by</u> C , <u>the distance</u> $\|\bar{u} - u\|$ <u>is a nonincreasing function of</u> t ; this implies in particular that <u>at most one such solution may agree with some initial condition</u> $u(t_0) = u_0 \in C(t_0)$.

Concerning the <u>existence</u> of this solution, see [21] .

In the above definition of the sweeping process is involved a precise <u>jump condition</u> for every discontinuity point t_s of a solution u . In fact, at such a point, the vector measure du presents an atom of value

$$u(t_s) - u^-(t_s) = u'_\mu(t_s)\mu_s$$

where $\mu_s > 0$ denotes the value of the corresponding atom of $d\mu$. Then (7.2) implies

$$u^-(t_s) - u(t_s) \in \partial\psi(C(t_s),u(t_s)) \;.$$

Due to the classical characterization of proximal points in a Hilbert space, this is equivalent to

$$u(t_s) = \text{prox}\,(C(t_s),u^-(t_s)) \;. \tag{7.3}$$

This jump law is to be compared with a natural <u>time-discretization procedure</u> for approximating solutions. Let us start with the elementary formulation (7.1), involving the existence of the time derivative $\dot{u}$. An increasing sequence of points t_i is chosen in I and the vector

$$\frac{1}{t_{i+1}-t_i}(\hat{u}(t_{i+1}) - \hat{u}(t_i))$$

is adopted as an approximant of the derivative at point t_{i+1} of some approximate solution $\hat{u}$. Then replacing (7.1) by

$$\frac{1}{t_{i+1}-t_i}(\hat{u}(t_i) - \hat{u}(t_{i+1})) \in \partial\psi(C(t_{i+1}),\hat{u}(t_{i+1}))$$

constitutes a discretization scheme of the <u>implicit</u> sort. Because $\partial\psi$ is a cone, the positive factor $1/t_{i+1} - t_i$ may be dropped, hence equivalently

$$\hat{u}(t_{i+1}) = \text{prox}\,(C(t_{i+1}),\hat{u}(t_i)) \;, \tag{7.4}$$

a relation similar to the jump condition (7.3) . In fact the above approximation procedure amounts to replace the given moving set $t \to C(t)$ by another one which moves stepwise, say $t \to \hat{C}(t)$, with the constant value $\hat{C}(t) = C(t_i)$ for every $t \in [t_i, t_{i+1}[$. The corresponding solution $\hat{u}$ of the sweeping process, agreeing with the initial condition $\hat{u}(t_o) = u_o$ is a right-continuous step-function whose successive values are inductively constructed by means of (7.4). This procedure may be called the catching-up algorithm, since the moving point $\hat{u}(t)$, instead of being swept by $C(t)$, is left at rest during the time-interval $[t_i, t_{i+1}[$, at the end of which it has to catch up with $C(t_{i+1})$ by the shortest way.

In the case of a smoothly moving set C , one obtains the uniform convergence of $\hat{u}$ toward the exact solution when the subdivision (t_i) is uniformely refined.[25] In the discontinuous case, the uniform approximation of u by $\hat{u}$ can visibly be expected only if one includes in the sequence (t_i) the instants at which the jumps of u exceed in magnitude the accepted incertainty. This is not unrealistic in the present problem, since the discontinuities of u can only occur at points of discontinuity of the given multifunction $t \to C(t)$. In view of (7.3) , every jump of $C(t)$, expressed in terms of Hausdorff distance, exceeds the corresponding jump of $u(t)$. One proves in fact a property of uniform convergence relative to the set of the finite subsets of I , such as (t_i) above, partially ordered by inclusion, i.e. the uniform convergence of some directed net of step-functions.[21]

But, when tackling the approximation of discontinuous processes, another sort of convergence seems more promising : instead of measuring the distance between u and û by the uniform norm, one considers the Hausdorff distance between their graphs. This amounts to take into account jointly some uncertainty about the values of the investigated functions of t and some uncertainty about the values of t at which they are computed. In that context, a practical convergence theorem, relative to uniform refinements of the discretization, is established.[31]

8. A SYNTHETIC FORMULATION

Let us come back now to the problem formulated in Sect. 3 . In the general expression of the kinetic energy

$$T(q,\dot{q}) = \frac{1}{2} \sum_{i,j} a_{ij}(q)\, \dot{q}^i \dot{q}^j , \tag{8.1}$$

the functions $a_{ij}(q) = a_{ji}(q)$ are assumed smooth enough for the Lagrange equations to make sense. Actually the dependence of a_{ij} on q would in the sequel cause only technical complication, so for the sake of simplicity we shall restrict ourselves to the case where some parametrization of $\mathcal{Q}$ may be found, such that these $n(n+1)/2$ functions are constants. This amounts to say that the Riemannian metric constructed on $\mathcal{Q}$ by putting

$$ds^2 = \sum_{i,j} a_{ij}(q)\, dq^i\, dq^j$$

turns out to be locally Euclidean ; in other words, we let aside the possible effect of Riemann curvature.

Equivalently $\mathcal{Q}$ may, at least locally, be identified with an open subset of some Euclidean linear space E , with dimension n . In view of this Euclidean structure, the tangent space $E(q)$ at every point is identified with E itself, as well as the cotangent space $E'(q)$. The expression (8.1) of the kinetic energy now reduces to

$$T(q,\dot{q}) = \frac{1}{2} \|\dot{q}\|^2 \tag{8.2}$$

and the Lagrange equation (3.1), for every interval of smooth motion, becomes

$$\ddot{q} = Q(t,q) + R . \tag{8.3}$$

Using (2.4) as before in order to eliminate R we obtain

$$Q(t,q) - \ddot{q} \in N(q) . \tag{8.4}$$

As in Sect. 3, for every interval of smooth motion, this is found equivalent to the stronger differential inclusion

$$Q(t,q) - \ddot{q} \in \partial\psi(V(q),\dot{q}) . \tag{8.5}$$

Let us formulate now an evolution problem involving differential measures in the way explained in Sect. 6 and 7 ; here again I is a real interval, open on the right or unbounded, but containing t_o as origin.

PROBLEM P . <u>Given</u> $q_o \in L$ <u>and</u> $\dot{q}_o \in V(q_o)$, <u>to find</u> $u \in \mathrm{rclbv}(I,E)$, <u>with</u> $u(t_o) = \dot{q}_o$, <u>such that by putting</u>

$$q(t) = q_o + \int_{t_o}^{t} u(\tau)d\tau \tag{8.6}$$

one has

$$Q(t,q(t))dt - du \in \partial\psi(V(q(t)),u(t)) \tag{8.7}$$

in the following sense : there exists (non uniquely) a nonnegative real measure $d\mu$ on I relatively to which the Lebesgue measure dt and the differential measure du of u admit the respective densities $t'_\mu \in \mathcal{L}^1_{loc}(I,d\mu;R)$ and $u'_\mu \in \mathcal{L}^1_{loc}(I,d\mu;E)$ verifying for every $t \in I$

$$Q(t,q(t))t'_\mu(t) - u'_\mu(t) \in \partial\psi(V(q(t),u(t))) \ . \tag{8.8}$$

In the same way as in Sect. 7 for the sweeping process, the above formulation is found to be indifferent to changing the "base" measure $d\mu$

A possible solution u of this Problem is locally absolutely continuous over some open subinterval I' of I if and only if there exists $\dot{u} \in \mathcal{L}^1_{loc}(I',dt;E)$ such that, in restriction to I', one has $du = \dot{u}\,dt$; then the following equalities hold between the restrictions to I' of the considered measures

$$u'_\mu\, d\mu = \dot{u}\, dt = \dot{u}\, t'_\mu\, d\mu \ .$$

Consequently $u'_\mu = \dot{u}\, t'_\mu$ throughout I', with the possible exception of a $d\mu$-negligible, hence dt-negligible, subset. On the other hand the function t'_μ is nonnegative and the subset in which it vanishes is dt-negligible. Therefore (8.8) implies that

$$Q(t,q(t)) - \dot{u}(t) \in \partial\psi(V(q(t),u(t)) \tag{8.9}$$

holds Lebesgue-a.e. in I'. This amounts to say that the function $t \to q(t)$ satisfies over I' the dynamical condition (8.5) of shockless motion.

Suppose on the other hand that a solution u of Problem P is discontinuous at time $t_s \in I$. Then the vector measure du presents at this point an atom with value $u(t_s) - u^-(t_s)$; equivalently the real measure $d\mu$ presents an atom with mass $\mu_s > 0$ and

$$u'_\mu(t_s)\mu_s = u(t_s) - u^-(t_s) .$$

Since the Lebesgue measure dt has no atom, one obtains $t'_\mu(t_s) = 0$ and finally (8.8) implies

$$u^-(t_s) - u(t_s) \in \partial\psi(V(q(t_s)),u(t_s)) ,$$

equivalent to

$$u(t_s) = \text{prox}\,(V(q(t_s)),u^-(t_s)) . \tag{8.10}$$

In view of the right-continuity of u , this is nothing else than the definition property (5.1) of a standard inelastic shock ; in fact (8.6) elementarily implies that the right and left limits of u respectively equal the right and left derivatives of q .

Conversely, one checks in a similar way that a motion consisting of a sequence of intervals of smooth motion, connected by standard inelastic shocks, is a solution of Problem P . But it cannot be expected a priori that every solution of Problem P exhibits such a simple structure, i.e. is of finite sort, according to the terminology of Sect. 1 .

Let us establish now a power balance formula which generalizes both relations (3.8) and (5.3). For every $t \in I$, condition (8.8) expresses that $u(t)$ and the right-hand member form a pair of conjugate points relative to the pair of Fenchel conjugate functions equal to the respective indicators of $V(q(t))$ and $N(q(t))$; therefore these elements are orthogonal, i.e.

$$\forall\, t \in I : u(t).[Q(t,q(t))\, t'_{\mu}(t) - u'_{\mu}(t)] = 0 \,.$$

As an element of $lbv(I,E)$, the function u is locally bounded and universally measurable ; hence the above expression constitutes the density, relative to $d\mu$, of some real measure. Using the notations of Sect. 6 this yields the equality of real measures

$$u.du = u(t).Q(t,q(t))dt \,. \tag{8.11}$$

Therefrom we shall derive some inequalities, emphasizing in general the dissipative character of the mechanical process under study.

Let us call the speed of the system the real function

$$\sigma(t) = \| u(t) \| \quad ,$$

belonging to $lbv(I, \mathbb{R})$, since the norm mapping is Lipschitz from E into $\mathbb{R}$. Clearly $\sigma^+(t) = \| u^+(t) \| = \sigma(t)$ in view of the right-continuity, and $\sigma^-(t) = \| u^-(t) \|$. Observe that (8.10), established for every discontinuity point t_s , is also trivially true at continuity points since $u^- = u \in V(q)$ at such points. As the mapping "prox" is nonexpanding, it comes out that

$$\forall\, t \quad : \quad \sigma(t) \leqslant \sigma^-(t) \; . \tag{8.12}$$

Applying (6.6) successively with $X = \mathbb{R}$ and $X = E$ one obtains, in view of $u = u^+$ and using (8.11) and (8.12),

$$\sigma^- d\sigma \leqslant \frac{1}{2}\, d(\sigma^2) = \frac{1}{2}\, d\| u \|^2 \leqslant u.du \leqslant \sigma \| Q \| \; dt \leqslant \sigma^- \| Q \| \; dt \; . \tag{8.13}$$

If an upper bound M of $\| Q(t,q) \|$ is available for the considered problem, this allows one to derive an upper bound of $\sigma(\tau)$, namely

$$\sigma(\tau) \leqslant \| \dot{q}_o \| + (\tau - t_o)\, M \; . \tag{8.14}$$

In fact in the special case where σ^- does not vanish in the interval

$]t_o,\tau[$, (8.13) yields, for the restriction of the considered measures to this interval,

$$d\sigma \leqslant M\,dt \tag{8.15}$$

from which (8.14) would ensue by integration and by using (6.3) and (8.12). If the subset of $]t_o,\tau[$ where σ^- vanishes is nonempty, let us denote by τ_o the l.u.b. of this subset. Every left-neighborhood of τ_o in I contains points where $\sigma^- = 0$, hence $\sigma^-(\tau_o) = 0$ (recall that the function $t \to \sigma^-(t)$ is essentially left-continuous, see e.g.[27]) thus $\sigma(\tau_o) = 0$ by (8.12). Inequality (8.15) holds for the restriction of the considered measures to $]\tau_o,\tau[$, then, by integration

$$\int_{]\tau_o,\tau[} d\sigma = \sigma^-(\tau) - \sigma(\tau_o) \leqslant (\tau-\tau_o)M .$$

Using again (8.12) one concludes

$$\sigma(\tau) \leqslant \sigma^-(\tau) \leqslant (\tau-\tau_o)M \leqslant (\tau-t_o)M$$

which establishes (8.14).

We now discuss the connection of Problem P with the sweeping process presented in Sect. 7.

Let us introduce the new unknown function $v \in \mathrm{rclbv}(I,E)$, $v(t_o) = \dot{q}_o$, by

$$v(t) = u(t) - \int_{t_o}^{t} Q(\tau,q(\tau))d\tau \quad .$$

In view of (8.6), the function $t \to q(t)$ is related to v as being the solution of the smooth integro-differential equation

$$\frac{dq}{dt} - \int_{t_o}^{t} Q(\tau,q(\tau))dt = v(t\ ) \ ,$$

with initial condition $q(t_o) = q_o$. Let us express this by writing $q = Sv$; the nonlinear operator

$$S : \mathrm{rclbv}(I,E) \to W^{1,\infty}_{loc}(I,E)$$

enjoys reasonably good continuity properties. The right-hand side of (8.7) now writes down as

$$\partial\psi(V(q(t)),u(t)) = \partial\psi(C_v(t),v(t)) \ ,$$

with

$$C_v(t) = V(Sv(t)) - \int_{t_o}^{t} Q(\tau,Sv(\tau))d\tau \quad . \tag{8.16}$$

Under these notations, Problem P takes the equivalent form : <u>To find</u> $v \in \mathrm{rclbv}(I,E)$, $v(t_o) = \dot{q}_o$, <u>satisfying</u>

$$- dv \in \partial\psi(C_v(t),v(t)) \ . \tag{8.17}$$

In other words, v is a solution of the sweeping process by some moving convex set C which itself depends on v. Whether this allows one to prove solution existence by some fixed point argument is still an open question. But the above provides at least some insight into the structure of the problem and naturally suggests various approximation procedures.

For instance, one may apply to (8.17) the catching-up algorithm described in Sect. 7, with C_v adjusted at each step through the concomitant discretization of (8.16).

Another sort of approximation would proceed by "regularization" : replacing the indicator function $\psi(C_v,.)$ in (8.17) by a penalty function of the set C_v, namely

$$x \to \frac{1}{2\lambda}\,[\text{dist}\,(x,C_v)]^2 \quad ,$$

where λ is a "small" positive constant. Hence the differential inclusion (8.17) is replaced by the differential equation

$$-\frac{dv}{dt} = \frac{1}{\lambda}\,[v-\text{prox}(C_v,v)] \quad .$$

The single-valued Lipschitz mapping in the right-hand side may also be viewed as resulting from the regularization, in Yosida's style, of the monotone multifunction $\partial\psi(C_v,.)$ (see e.g.[32,33]).

Here, as well as in the catching-up discretization, a certain extent of violation of the geometrical condition $q \in L$ has to be accepted, with some adequate definition of V(q) for $q \notin L$. The regularization procedure used in[12,13] consists in exerting some elastic pull-back as soon as this condition is violated : this agrees with the concept of an energy-preserving shock. In contrast, the above amounts to apply some breaking action of the viscous type, with coefficient $1/\lambda$, as soon as the kinematical condition $\dot{q}^+ \in V(q)$ is violated : this is consistent with the dissipative character of inelastic shocks.

8. REFERENCES

1. Bressan, A., Incompatibilità dei teoremi di esistenza e di unicità del moto per un tipo molto comune e regolare di sistemi meccanici, Ann. Scuola Norm. Sup. Pisa, Ser. III, 14, 333, 1960.
2. Delassus, E., Mémoire sur la théorie des liaisons finies unilatérales, Ann. Sci. Ecole Norm. Sup., 34, 95, 1917.
3. Bouligand, G., Compléments et exercices sur la mécanique des solides, 2ème édition, Vuibert, Paris, 1945.
4. Bouligand, G., Mécanique rationnelle, 5ème édition, Vuibert, Paris, 1954.
5. Moreau, J.J., Les liaisons unilatérales et le principe de Gauss, C.R. Acad. Sci. Paris, 256, 871, 1963.
6. Moreau, J.J., Quadratic programming in mechanics : dynamics of one-sided constraints, SIAM J. Control, 4, 153, 1966 (Proceedings of the First International Congress on Programming and Control).
7. Moreau, J.J., Sur la naissance de la cavitation dans une conduite, C.R. Acad. Sci. Paris, 259, 3948, 1965.
8. Moreau, J.J., Principes extrémaux pour le problème de la naissance de la cavitation, J. de Mécanique, 5, 439, 1966.

9. Moreau, J.J., One-sided constraints in hydrodynamics, in Non linear Programming, Abadie, J., Ed., North-Holland Pub. Co., Amsterdam, 1967, 257.

10. Kilmister, C.W. and Reeve, J.E., Rational mechanics, Longmans, London, 1966.

11. Moreau, J.J., Mécanique classique, Vol 2, Masson, Paris, 1971.

12. Schatzman, M., Le système différentiel $(d^2u/dt^2) + \partial\varphi(u) \ni f$ avec conditions initiales, C.R. Acad. Sci. Paris, Série 1, 284, 603, 1977.

13. Schatzman, M., A class of non linear differential equations of 2nd order in time, J. Nonlinear Analysis, Theory, Methods and Appl., 2, 355, 1978.

14. Buttazzo, G. and Percivale, D., Sull'approssimazione del problema del rimbalzo unidimensionale, Scuola Norm. Sup. Pisa, E.T.S. Pisa, 1980.

15. Buttazzo, G. and Percivale, D., The bounce problem on n-dimensional Riemannian manifolds, Scuola Norm. Sup. Pisa, E.T.S. Pisa, 1981.

16. Amerio, L. and Prouse, G., Study of the motion of a string vibrating against an obstacle, Rend. Mat., 8, 563, 1975.

17. Schatzman, M., A hyperbolic problem of second order with unilateral constraints : the vibrating string with a concave obstacle, J. Math. Anal. Appl., 73, 138, 1980.

18. Cabannes, H. and Haraux, A., Mouvements presque-périodiques d'une corde vibrante en présence d'un obstacle fixe, rectiligne ou ponctual, Int. J. Non-linear Mechanics, 16, 449, 1981.

19. Do, C., On the dynamic deformation of a bar against an obstacle, in

Variational Methods in the Mechanics of Solids, S. Nemat-Nasser, Ed., Pergamon Press, 1980.

20. Moreau, J.J., Liaisons unilatérales sans frottement et chocs inélastiques, C.R. Acad. Sci. Paris, Série II, 296, 1473, 1983.

21. Moreau, J.J., Evolution problem associated with a moving convex set in a Hilbert space, J. Diff. Equ., 26, 347, 1977.

22. Abadie, J., On the Kuhn-Tucker theorem, in Non linear Programming, Abadie, J., Ed., North-Holland Pub. Co., Amsterdam, 1967, 17.

23. Moreau J.J., Décomposition orthogonale d'un espace hilbertien selon deux cones mutuellement polaires, C.R. Acad. Sci. Paris, 255, 238, 1962.

24. Moreau, J.J., Proximité et dualité dans un espace hilbertien, Bull. Soc. Math. France, 93, 273, 1965.

25. Moreau, J.J., On unilateral constraints, friction and plasticity, in New variational Techniques in Mathematical Physics, G. Capriz and G. Stampacchia, Eds., CIME II Ciclo 1973, Edizioni Cremonese, Roma, 1974, 173.

26. Halphen, B. and Nguyen, Q.S., Sur les matériaux standards généralisés, J. de Mécanique 14, 39, 1975.

27. Moreau, J.J., Sur les mesures différentielles de fonctions vectorielles, in Séminaire d'Analyse Convexe, Montpellier, 5, exp. 17, 1975.

28. Moreau, J.J., Sur les mesures différentielles et certains problèmes d'évolution, C.R. Acad. Sci. Paris, Sér. A-B, 282, 837, 1976.

29. Moreau, J.J., Multiapplications à rétraction finie, Ann. Scuola Norm.

Sup. Pisa, Cl. Sci. Ser. IV, 1, 169, 1974.

30. Bourbaki, N., Intégration, Chap. 6 (Eléments de Mathématique, fasc. XXV), Hermann, Paris, 1959.

31. Moreau, J.J., Approximation en graphe d'une évolution discontinue, RAIRO Analyse Numérique, 12, 75, 1978.

32. Brezis, H., Problèmes unilatéraux, J. Math. Pures Appl., 51, 1, 1972.

33. Brezis, H., Opérateurs Maximaux Monotones et Semi-groupes de Contractions dans les Espaces de Hilbert, Math. Studies, 5, North-Holland, Amsterdan, 1973.

HEMIVARIATIONAL INEQUALITIES. EXISTENCE AND APPROXIMATION RESULTS

P.D. Panagiotopoulos
School of Technology
Aristotelian University
Thessaloniki

1. Introduction

The theory of variational inequalities is closely connected to the notion of superpotential introduced by Moreau[1] for convex generally nondifferentiable and nonfinite energy functionals. If Φ is such a functional, a superpotential relation (material law or boundary condition) in the sense of Moreau has the form

$$f \in \partial\Phi(u) \tag{1.1}$$

where f and u are a "force" and a "flux" respectively in the terminology of Onsager's theory and ∂ denotes the subdifferential[2], which is a monotone multivalued operator. Such a law allows the derivation of variational inequalities[3,4] which are expressions of the principle of virtual or complementary virtual work (or power) in static problems and of d' Alembert's principle in dynamic problems.

However, there are large classes of problems in mechanics and engineering involving nonmonotone material laws and boundary conditions which result from nonconvex and in many cases nondifferentiable and nonfinite energy functionals. These laws cannot be studied by means of the methods of convex analysis. To study such problems the superpotential has been generalized[5,6,7,8] by using the notion of generalized gradient[9,10,11] of Clarke, and mechanical laws of the form

$$f \in \overline{\partial}\Phi(u) \tag{1.2}$$

have been considered where $\overline{\partial}$ denotes the generalized gradient. If $u \in X$ and $f \in X'$ where X is a locally convex Hausdorff topological vector space and X' its dual space (the duality pairing is denoted by $\langle .,. \rangle$) then (1) is equivalent to

$$\Phi(v) - \Phi(u) \geq \langle f, v-u \rangle \qquad \forall v \in X \tag{1.3}$$

and (2) to

$$\Phi^{\uparrow}(u, v-u) \geq \langle f, v-u \rangle \qquad \forall v \in X \tag{1.4}$$

where $\Phi^{\uparrow}(.,.)$ denotes the upper-subderivative[11] of Φ.

It is shown[7] that in the case of (1.2) the variational inequalities are replaced by new variational forms which we have called "hemivariational" inequalities. Analogously to the convex case, in which static variational inequalities are under certain conditions equivalent to minimization problems (minimum of potential or complementary energy), in the nonconvex case the hemivariational inequalities lead[5] to substationarity[11] problems.

Here we will study the existence of the solution first of a coercive and then of a semicoercive hemivariational inequality. The solution will be approximated by the solutions of a sequence of regularized problems. We use a weak compactness argument applied by Rauch[12], and McKenna and Rauch[13] in the theory of semilinear differential equations.

2. Formulation of Hemivariational Inequalities.

A Model Problem: The Nonconvex Semipermeability Problem.

Let Ω be an open bounded connected subset of $\mathbb{R}^3$ referred to a fixed Cartesian coordinate system $Ox_1x_2x_3$ and let us consider the equation

$$-\Delta u = f \quad \text{in} \quad \Omega . \tag{2.1}$$

Here u is in the case of a heat conduction problem the temperature and in problems of hydraulics and electrostatics the pressure and the electric potential respectively. The boundary Γ of Ω is assumed to be appropriately regular. By $n = \{n_i\}$ we denote the outward unit normal to Γ. Then $\partial u/\partial n$ represents the heat-, fluid- or electricity flux through Γ. In this context recall in the case of heat conduction problems the Fourier law.+) In the interior semipermeability problem[3] the classical boundary condition

$$u = 0 \quad \text{on} \quad \Gamma \tag{2.2}$$

is assumed to hold, whereas in the boundary-semipermeability problems the boundary conditions are defined as relations between $\partial u/\partial n$ and u. The interior semipermeability conditions are formulated by assuming that $f = \overline{f} + \overline{\overline{f}}$ where $\overline{\overline{f}}$ is given and $\overline{f}$ is a known function of u. If we have pointwise conditions in the form

$$-\overline{f} \in \partial j_1(u) \text{ in } \Omega \quad \text{or} \quad -\frac{\partial u}{\partial n} \in \partial j_2(u) \text{ on } \Gamma \tag{2.3}$$

where j_1 and j_2 are convex, lower semicontinuous and proper functionals on $\mathbb{R}$, then $b_i = \partial j_i$, $i=1,2$, is a maximal monotone operator and the semipermeability variational inequalities[3] result. It is worth noting that besides the semipermeability conditions (let us use in the sequel the language of the heat conduction problems) temperature control problems[3] lead to conditions similar to (2.3).

A natural generalization is to consider nonmonotone multivalued relations. For example the relations

+) $q_i n_i = -k\, \partial u/\partial n$, $k > 0$, where $q = \{q_i\}$ is the heat flux vector and k is the coefficient of thermal conductivity.

$$-\bar{f} \in \hat{b}(u) \quad \text{in} \quad \Omega \tag{2.4}$$

and

$$-\frac{\partial u}{\partial n} \in \hat{b}(u) \text{ on } \quad \Gamma \tag{2.5}$$

where $\hat{b}: \mathbb{R} \to P(\mathbb{R})$ is depicted in Fig. 1, correspond to the behaviour of a semipermeable membrane of finite thickness (e.g. a wall) with heat regulator (Fig. 1a) and to a temperature control problem (Fig. 1b). We now state the following B.V.Ps. (classical formulation)

Problem 1' (resp. Problem 1å): Find u which satisfies (2.1), (2.2), and (2.4) (resp. (2.1) and (2.5)).

We assume that to the multivalued function $\hat{b}$ a measurable selection $b: \mathbb{R} \to \mathbb{R}$ can be determined such that $b(\xi) \in \hat{b}(\xi) \quad \forall \xi \in \mathbb{R}$. For instance, the measurable selection theorem (see Castaing and Valadier [14]) guarantees that if $\hat{b}$ is measurable, closed, nonempty on $\mathbb{R}$ such a b exists. Moreover we assume that b is locally bounded, i.e. $b \in L^{\infty}_{loc}(\mathbb{R})$. This is obvious in the case depicted in Fig. 1. For any $\mu > 0$ and $\xi \in \mathbb{R}$ we define

$$\bar{b}_{\mu}(\xi) = \operatorname*{ess\,sup}_{|\xi_1 - \xi| < \mu} b(\xi_1) \text{ and } \underline{b}_{\mu}(\xi) = \operatorname*{ess\,inf}_{|\xi_1 - \xi| < \mu} b(\xi_1) \ . \tag{2.6}$$

Then we may obviously take the limits as $\mu \to 0$ and obtain the Baire function $\bar{b}(\xi)$ and $\underline{b}(\xi)$.

We assume further that+)

$$\hat{b}(\xi) = [\bar{b}(\xi), \underline{b}(\xi)]. \tag{2.7}$$

Obviously the foregoing assumption is verified in the case of the graphs of Fig. 1a,b. It can be proved[15] that if $b(\xi_{\pm 0})$ exists for every $\xi \in \mathbb{R}$ then a locally Lipschitz function $\varphi : \mathbb{R} \to \mathbb{R}$ can be determined such that

$$\hat{b}(\xi) = \bar{\partial}\varphi(\xi) . \tag{2.8}$$

Here φ is such that

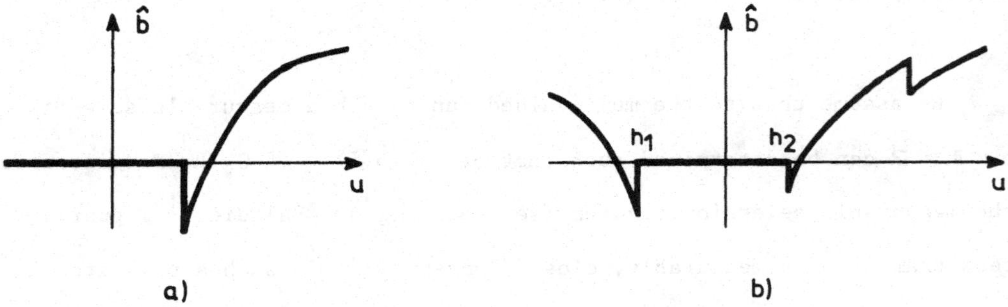

Fig. 1 Nonconvex semipermeability relations

+) It would be sufficient if $[\bar{b}(\xi), \underline{b}(\xi)] \subset \hat{b}(\xi)$

$$\varphi^{\uparrow}(\xi,z) = \limsup_{\substack{h\to 0\\ \lambda\to 0_+}} \frac{1}{\lambda}\int_{\xi+h}^{\xi+h+\lambda z} b(\xi)d\xi \tag{2.9}$$

and $\bar{\partial}$ denotes the generalized gradient of Clarke. Note that since φ is locally Lipschitz $\varphi^{\uparrow}(.,.)$ coincides with the directional derivative $\varphi^{o}(.,.)$ in the sense of Clarke.[9,10]. Thus (2.4) and (2.5) take the forms

$$-\bar{f} \in \partial\varphi(u) \quad \text{in} \quad \Omega \tag{2.10}$$

and

$$-\frac{\partial u}{\partial n} \in \partial\varphi(u) \quad \text{on} \quad \Gamma \tag{2.11}$$

which are equivalent to the hemivariational inequalities

$$\varphi^{\uparrow}(u,v-u) \geq -\bar{f}(v-u) \qquad \forall v \in \mathbb{R} \tag{2.12}$$

and

$$\varphi^{\uparrow}(u,v-u) \geq -\frac{\partial u}{\partial n}(v-u) \qquad \forall v \in \mathbb{R}\ . \tag{2.13}$$

From (2.1) by multiplying by v-u, integrating on Ω and applying formally the Green-Gauss theorem (we set $\int_{\Gamma} uv d\Omega = (u,v)$ and $a(u,v) = \int_{\Omega} \text{grad}\, u \cdot \text{grad}\, v d\Omega$) we obtain

$$a(u,v-u) = \int_\Gamma \frac{\partial u}{\partial n}(v-u)d\Gamma + (f,v-u). \tag{2.14}$$

(2.14) with (2.12) (resp. (2.13)) leads for $\bar{\bar{f}} \in L^2(\Omega)$ to the following variational problems.

Problem 1: Find $u \in \overset{\circ}{H}{}^1(\Omega)$ which satisfies (2.2) and the hemivariational inequality

$$a(u,v-u) + \int_\Omega \varphi^\uparrow(u,v-u)d\Omega \geq (\bar{\bar{f}},v-u) \quad \forall v \in \overset{\circ}{H}{}^1(\Omega) . \tag{2.15}$$

Problem 1a: Find $u \in H^1(\Omega)$ which satisfies the hemivariational inequality

$$a(u,v-u) + \int_\Gamma \varphi^\uparrow(u,v-u)d\Gamma \geq (f,v-u) \quad \forall v \in H^1(\Omega) . \tag{2.16}$$

Note that for φ convex $\overline{\partial\varphi} = \partial\varphi$ and therefore (2.15) and (2.16) reduce to the well-known variational inequalities of convex semipermeability[3] problems. In the next section we will study the existence of the solution of the coercive problem 1.

3. An Existence Result for Problem 1. The Coercive Case.

First a regularized problem (problem 1ε) will be formulated and the solution of problem 1 will be obtained as the limit of the solution of the regularized problem. We consider a function $\rho \in C_o^\infty(-1,+1)$ such that

$$\rho \geq 0 \text{ and } \int_{-\infty}^{+\infty} \rho(\xi)d\xi = 1 \tag{3.1}$$

and we consider the function

$$\rho_\varepsilon(\xi) = \frac{1}{\varepsilon}\rho(\frac{\xi}{\varepsilon}) \quad \varepsilon > 0 \tag{3.2}$$

and the convolution

$$b_\varepsilon = \rho_\varepsilon * b \; , \quad \varepsilon > 0 \; . \tag{3.3}$$

Obviously $b_\varepsilon \in C^\infty(\mathbb{R})$, and is the regularized form of b. We assume also that

$$b(-\infty) < b(+\infty) \; , \tag{3.4}$$

where $b(\infty) = \liminf_{\xi \to \infty} b(\xi)$ and $b(-\infty) = \limsup_{\xi \to -\infty} b(\xi)$, i.e. that b ultimately increases.

We consider now the following problem:

Problem 1_ε: For any $\bar{\bar{f}} \in L^2(\Omega)$ find $u_\varepsilon \in \overset{\circ}{H}{}^1(\Omega)$ such that

$$a(u_\varepsilon,v) + (\frac{d\varphi_\varepsilon}{du}(u_\varepsilon),v) = (\bar{\bar{f}},v) \; \forall \, v \in \overset{\circ}{H}{}^1(\Omega) \; . \tag{3.5}$$

Note that $\frac{d\varphi_\varepsilon}{du}(u(x)) = b_\varepsilon(u(x))$. Problem 1_ε is the regularized form of problem 1. The following proposition holds.

Proposition 1: Suppose that (3.4) holds. Then there exists a solution to problem 1.

Proof: First we prove the existence of a solution u_ε of problem 1_ε. Let us consider a Galerkin basis $\{w_i\}$ of $W = \overset{o}{H}{}^1(\Omega) \cap L^\infty(\Omega)$ and let W_n be a corresponding n-dimensional subspace of W. The finite dimensional problem: Find $u_{\varepsilon n} \in W_n$ such that

$$a(u_{\varepsilon n},v) + (b_\varepsilon(u_{\varepsilon n}),v) - (\bar{\bar{f}},v) = 0 \quad \forall\, v \in W_n \tag{3.6}$$

is considered. This problem has a solution $u_{\varepsilon n}$ as may be shown by means of Brouwer's fixed point theorem[16]: If $\tilde{\Lambda} : \mathbb{R}^n \to \mathbb{R}^n$ is continous such that for $r > 0$ $(\tilde{\Lambda}a)_i a_i \geq 0$ $\forall\, a \in \mathbb{R}^n$ with $|a| = r$, then there exists a_o with $|a_o| \leq r$ such that $\tilde{\Lambda}a_o = 0$.

We write now (3.6) in the form

$$(\Lambda u_{\varepsilon n},v) = 0 \quad \forall\, v \in W_n \tag{3.7}$$

where $\Lambda : W_n \to W'_n$. If $u_{\varepsilon n} = \sum_{i=1}^{n} a_i w_i$ corresponds to $a = \{a_i\}$ then

$$(\tilde{\Lambda}a)_i = a(u_{\varepsilon n},w_i) + (b_\varepsilon(u_{\varepsilon n}),w_i) - (\bar{\bar{f}},w_i) . \tag{3.8}$$

We may determine $\rho_1 > 0$ and $\rho_2 > 0$ such that $b_\varepsilon(\xi) \geq 0$ if $\xi > \rho_1$,

$b_\varepsilon(\xi) \le 0$ if $\xi < -\rho_1$ and $|b_\varepsilon(\xi)| \le \rho_2$, if $|\xi| \le \rho_1$. Then (3.8) implies that

$$
\begin{aligned}
(\Lambda u_{\varepsilon n}, u_{\varepsilon n}) &\ge c_1 \| u_{\varepsilon n} \|^2 + \int_{|u_{\varepsilon n}|>\rho_1} u_{\varepsilon n} b_\varepsilon(u_{\varepsilon n}) d\Omega + \int_{|u_{\varepsilon n}|\le\rho_1} u_{\varepsilon n} b_\varepsilon(u_{\varepsilon n}) d\Omega - \\
&- \| \bar{\bar{f}} \|_{H^{-1}} \| u_{\varepsilon n} \| \ge c_1 \| u_{\varepsilon n} \|^2 - c_2 \| u_{\varepsilon n} \| + c_3
\end{aligned}
\tag{3.9}
$$

where $\| \cdot \|$ denotes the $\overset{o}{H}{}^1$-norm. Thus we can find $r > 0$ such that

$$(\Lambda u_{\varepsilon n}, u_{\varepsilon n}) \ge 0 \quad \text{for} \quad \| u_{\varepsilon n} \| = r \tag{3.10}$$

and therefore (3.6) has a solution with $\| u_{\varepsilon n} \| \le r$. Since $\{u_{\varepsilon n}\}$ is bounded in $\overset{o}{H}{}^1(\Omega)$ we may extract a subsequence again denoted $\{u_{\varepsilon n}\}$ such that for $\varepsilon \to 0$ and $n \to \infty$

$$u_{\varepsilon n} \to u \quad \text{weakly in } \overset{o}{H}{}^1(\Omega) \tag{3.11}$$

and

$$u_{\varepsilon n} \to u \quad \text{strongly in } L^2(\Omega) \tag{3.12}$$

since the imbedding $\overset{o}{H}{}^1(\Omega) \subset L^2(\Omega)$ is compact. Thus

$$u_{\varepsilon n} \to u \quad \text{a.e. in } \Omega \,. \tag{3.13}$$

Subsequently we shall show that $\{b_\varepsilon(u_{\varepsilon n})\}$ is weakly precompact in $L^1(\Omega)$. We have thus to show[17] that for each $\alpha > 0$ there exists a δ such that if mes $\omega < \delta$ then $\int_\omega |b_\varepsilon(u_{\varepsilon n})| d\Omega < \alpha$. We apply to b_ε the inequality[12]

$$\xi_o |p(\xi)| \leq |\xi p(\xi)| + \xi_o \sup_{|\xi| \leq \xi_o} |p(\xi)| \tag{3.14}$$

and we find

$$\int_\omega |b_\varepsilon(u_{\varepsilon n})| d\Omega \leq \frac{1}{\xi_o} \int_\omega |b_\varepsilon(u_{\varepsilon n}) u_{\varepsilon n}| d\Omega + \int_\omega \sup_{|u_{\varepsilon n}| \leq \xi_o} |b_\varepsilon(u_{\varepsilon n})| d\Omega \ . \tag{3.15}$$

From (3.6) we obtain that $\int_\omega |b_\varepsilon(u_{\varepsilon n}) u_{\varepsilon n}| d\Omega$ is bounded. We choose in (3.13) ξ_o large enough, so that the first term in the right-hand side of (3.15) becomes $< \frac{\alpha}{2}$ for all n and ε, and δ such that

$$\operatorname*{ess\,sup}_{|\xi| \leq \xi_o + 1} |b(\xi)| \leq \frac{\alpha}{2\delta} \tag{3.16}$$

and thus we obtain the weak precompactness of $\{b_\varepsilon(u_{\varepsilon n})\}$ in $L^1(\Omega)$. Accordingly a subsequence can be considered such that as $\varepsilon \to 0$ and $n \to \infty$

$$b_\varepsilon(u_{\varepsilon n}) \to \chi \quad \text{weakly in } L^1(\Omega) \ . \tag{3.17}$$

By taking the limit in (3.6) we find (note that $v \in L(\Omega)$ and $\cup W_n$ is dense in $\overset{o}{H}{}^1(\Omega)$) the equality

$$a(u,v) + (\chi, v) - (\bar{\bar{f}}, v) = 0 \quad \forall v \in \overset{o}{H}{}^1(\Omega) \tag{3.18}$$

with $\chi \subset H^{-1}(\Omega) \cap L^1(\Omega)$. Further it will be shown that

$$\chi \in \hat{b}(u) = \overline{\partial}\varphi(u) \quad \text{a.e. in } \Omega . \tag{3.19}$$

Because of (2.7) it suffices to show that

$$\chi \in [\underline{b}(u(x)), \overline{b}(u(x))] \quad \text{a.e. in } \Omega . \tag{3.20}$$

For $\alpha > 0$ we may choose $\omega \subset \Omega$ with $\text{mes}\,\omega < \alpha$ such that

$$u_{\varepsilon n} \to u \qquad \text{uniformly in } \Omega - \omega \tag{3.21}$$

as is obvious from (3.14) where $u \subset L^{\infty}(\Omega - \omega)$. Accordingly for any $\delta > 0$ $n_o > \frac{2}{\delta}$ and ε_o can be determined such that for $n > n_o$ and $\varepsilon < \varepsilon_o$

$$|u_{\varepsilon n}(x) - u(x)| < \frac{\delta}{2} \qquad \forall\, x \in \Omega - \omega . \tag{3.22}$$

For $n > n_o$ and $\varepsilon < \varepsilon_o$ $b_\varepsilon(u_{\varepsilon n}(x))$ is an average value of b for $-\frac{\delta}{2} + u_{\varepsilon n}(x) < \xi < \frac{\delta}{2} + u_{\varepsilon n}(x)$, $x \in \Omega - \omega$, and because the range of ξ is a subinterval of $[-\delta + u(x), \delta + u(x)]$ we conclude that

$$b_\varepsilon(u_{\varepsilon n}(x)) \in \hat{b}_\delta(u(x)) = [\underline{b}_\delta(u(x)), \overline{b}_\delta(u(x))] . \tag{3.23}$$

Similarly we obtain for $e \geq 0$, $e \in L^{\infty}(\Omega - \omega)$, that

$$\int_{\Omega-\omega} \underline{b}_{\delta}(u(x))e\,d\Omega \le \int_{\Omega-\omega} b_{\varepsilon}(u_{\varepsilon n}(x))e\,d\Omega \le \int_{\Omega-\omega} \overline{b}_{\delta}(u(x))e\,d\Omega \tag{3.24}$$

and by taking the limit ($\varepsilon \to 0, n \to \infty$) that

$$\int_{\Omega-\omega} \underline{b}_{\delta}(u(x))e\,d\Omega \le \int_{\Omega-\omega} \chi e\,d\Omega \le \int_{\Omega-\omega} \overline{b}_{\delta}(u(x))e\,d\Omega \ . \tag{3.25}$$

u is bounded on $\Omega-\omega$ and therefore in (3.25) we may take the limit as $\delta \to 0$ und thus (3.25) holds with $\underline{b}_{\delta}$ and $\overline{b}_{\delta}$ replaced by $\underline{b}$ and $\overline{b}$ respectively (Lebesgue theorem). It results that

$$\chi \in \hat{b}(u(x)) = \overline{\partial}\varphi(u(x)) \quad \text{a.e. in } \Omega-\omega \tag{3.26}$$

and for α as small as possible we obtain (3.19), q.e.d.
A proposition analogous to prop. 1 holds also for nonconvex semipermeability problems resulting from the boundary condition (2.11).

Until now we have considered a coercive problem. Further we shall study the semicoercive case for a semicoercive bilinear form a(.,) having a kernel with dimker a = 1 and we will establish a sufficient condition of the Landesman-Lazer[13,18] type.

4. The Semicoercive Case

Suppose that V is a Hilbert space and V' its dual space such that

$$V \subset L^2(\Omega) \subset V' \tag{4.1}$$

where the injections are continuous and dense. Further let us consider a positive, symmetric, bounded, bilinear form $a(u,v) : V\times V \to \mathbb{R}$ and let us denote by $(.,.)$ the duality pairing between V and V'. We assume further that $\dim\ker a = 1$. We choose $r \in \ker a$, $r \neq 0$ and we suppose that $V \cap L^\infty(\Omega)$ is dense in V and that the imbedding $V \subset L^2(\Omega)$ is compact. We pose the following problem.

Problem 2: For any $f \in V'$ find $u \in V$ such that

$$a(u,v-u) + \int_\Omega \varphi^0(u,v-u)d\Omega \geq (f,v-u) \quad \forall v \in V . \tag{4.2}$$

Here φ has the same properties as in problem 1. With respect to (4.2) we consider the regularized problem.

Problem 2_ε: For any $f \in V'$ find $u_\varepsilon \in V$ such that

$$a(u_\varepsilon,v) + (\frac{d\varphi_\varepsilon}{d u}(u_\varepsilon),v) = (f,v) \quad \forall v \in V . \tag{4.3}$$

The following proposition will be proved.

Proposition 2: Suppose that (3.4) holds and that

$$\int_{r>0} b(+\infty)rd\Omega + \int_{r<0} b(-\infty)rd\Omega > (f,r) > \int_{r>0} b(-\infty)rd\Omega + \int_{r<0} b(+\infty)rd\Omega . \tag{4.4}$$

Then problem 2 has a solution.

Proof: Problem 2_ε has a solution $u_\varepsilon \in V$ (with $\| u_\varepsilon \|_V < c$). This results by following the same steps as in the proof of a similar theorem of M^c^Kenna and Rauch[13]. The proof is based on the following estimate resulting by using (4.4). Let

$$b(u_\varepsilon,u_\varepsilon) = a(u_\varepsilon,u_\varepsilon) + (\frac{d\varphi_\varepsilon}{du}(u_\varepsilon),u_\varepsilon)-(f,u_\varepsilon) \ . \tag{4.5}$$

$\| u \|_V$ is equivalent to $\| q \|_V^2 + |p|^2$ where $u = q + pr$ and pr is the orthogonal projection of u onto $\ker a$. Then $b(u_\varepsilon,u_\varepsilon) \leq 0$ implies that

$$\| q \|_V \leq c(|p|^{1/2}+1), \ c \text{ const} > 0 \text{ and } |p| < N \ . \tag{4.6}$$

For the proof of (4.6) the reader is referred to M^c^Kenna and Rauch.[13] Since $\{u_\varepsilon\}$ is bounded in V we may pass to a subsequence such that

$$u_\varepsilon \to u \qquad \text{strongly in } L^2(\Omega) \tag{4.7}$$

due to the compact imbedding $V \subset L^2(\Omega)$. The estimates (3.15) and (3.16) imply that for $\varepsilon \to 0$

$$b_\varepsilon(u_\varepsilon) \to \chi \quad \text{in } L^1(\Omega) \ . \tag{4.8}$$

Further as in prop. 1 we show that (3.19) holds q.e.d.

In prop. 1 and 2 the regularization need not be so smooth. This leads to the approximation of the solution by a finite element scheme. If $r \geq 0$ then (4.4) takes the form

$$\int_\Omega b(-\infty) r d\Omega < (f,r) < \int_\Omega b(+\infty) r d\Omega \tag{4.9}$$

and for $r = c^{st}$ the condition becomes

$$b(-\infty) < \text{average } f < b(+\infty). \tag{4.10}$$

The proofs of props. 1 and 2 are mainly based on the assumption (3.4) which is verified in many one-dimensional boundary conditions. Note that in (4.2) we can consider as V the space $[H^1(\Omega)]^3$ and to study boundary conditions of the pointwise form

$$-S \in \bar{\partial}\varphi(u), \quad S \in \mathbb{R}^3, \quad u \in \mathbb{R}^3 \tag{4.11}$$

arising in elasticity (e.g. in the interface of a crack, or on the boundary of a foundation structure, etc.). Here u (resp. S) denotes the displacement vector (resp. the stress vector) and the proof of prop. 2 needs a slight modification. Then in (4.4) $b(+\infty)$ and $b(-\infty)$ are replaced by appropriate quantities resulting from $\bar{\partial}\varphi$ i.e. $\liminf_{\xi\to+\infty} \bar{\partial}\varphi(\xi)$ and $\limsup_{\xi\to-\infty} \bar{\partial}\varphi(\xi)$. (cf e.g. Rockafellar[19] for the definition of these quantities). The following proposition supplies a necessary condition for the solution of problem 1.

Proposition 3. Problem 2 cannot have a solution unless

$$\int_{r>0} b(+\infty)r d\Omega + \int_{r<0} b(-\infty)r d\Omega \geq (f,r) \geq \int_{r>0} b(-\infty)r d\Omega + \int_{r<0} b(+\infty) d\Omega \,. \quad (4.12)$$

Proof: Let us put $v - u = r$ in (4.2). It results

$$(f,r) \leq \int_{\Omega} \varphi^{o}(u,r) d\Omega \quad (4.13)$$

which implies by means of (2.9) and (3.4) that

$$(f,r) \leq \int_{r>0} b(+\infty) r d\Omega + \int_{r<\infty} b(-\infty) r d\Omega \quad (4.14)$$

and analogously

$$(f,r) \geq \int_{r>0} b(-\infty) r d\Omega + \int_{r<0} b(+\infty) r d\Omega \quad (4.15)$$

q.e.d.

The prop. 3. is based on the assumption (3.4) - actually on the less restrictive assumption

$$b(-\infty) \leq b(\xi) \leq b(+\infty) \quad \forall \xi \in R \,. \quad (4.16)$$

Propositions analogous to the previous ones hold also in the case of multivalued boundary conditions. Note also that due to the lack of convexity no general uniqueness results can be derived.

5. Application: A Nonconvex Semicoercive Problem in Plate Theory.

Proposition 1 has been already derived in the framework of the theory of plates.[20] Props. 2 and 3 can also be applied to the same family of problems. Let us consider a Kirchhoff plate of constant thickness h occupying in its undeformed state an open, bounded, connected subset Ω of $\mathbb{R}^2$. The plate is referred to a fixed Cartesian coordinate system $Ox_1x_2x_3$ which is right-handed. The middle plane of the plate coincides with the Ox_2x_3-plane. Let Γ be the boundary of Ω which is appropriately regular. By ζ is denoted the deflection of the points $x \in \Omega$. The plate is subjected to a distributed load $f = (0,0,f_3)$, $f_3 = f_3(x)$, per unit area of the middle surface. It is assumed further that the plate is subjected to the boundary conditions

$$\frac{\partial \zeta}{\partial n} = 0 \quad \text{on } \Gamma \tag{5.1}$$

and

$$\overline{Q} = 0 \quad \text{on } \Gamma \tag{5.2}$$

where $\overline{Q}$ denotes the total shearing force on Γ in the framework of Kirchhoff's theory. We assume that

$$f_3 = \overline{f}_3 + \overline{\overline{f}}_3 \tag{5.3}$$

where $\overline{\overline{f}}_3 \in L^2(\Omega)$ is a given load distribution and that on $\Omega' \subset \Omega$ with $\overline{\Omega}' \cap \Gamma = \emptyset$ the condition

$$-\overline{f}_3 \in \hat{b}(\zeta) = \overline{\partial}\varphi(\zeta) \tag{5.4}$$

holds. Here $\hat{b}$ satisfies (2.7) and (3.4) and φ is given by (2.9). On $\Omega-\Omega'$ we assume that $\overline{f}_3 = 0$. Relation (5.4) may describe the unilateral contact of the plate with a granular support which causes the nonmonotone law of Fig. 2a, a delamination effect with adhesive contact (Fig. 2b) etc.[7] The dotted line in Figs. 2a,b idealizes the local crushing effect due to the compressive forces at a contact point.

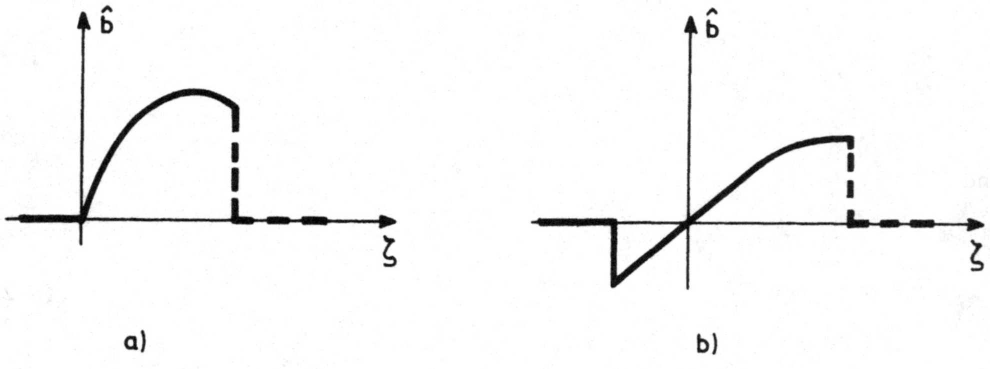

Fig. 2 Nonmonotone unilateral relations in plate theory with contact-crushing effect (dotted line)

All these laws give rise to hemivariational inequalities expressing the principle of virtual work in its inequality form and therefore they are called unilateral laws. Assuming that $\zeta \in H^2(\Omega)$, $\frac{\partial\zeta}{\partial n} = 0$ on Γ, we obtain by applying the Green-Gauss theorem the expression

$$a(\zeta,z) = (f_3,z) + \langle\overline{Q},z\rangle \quad \forall z \in Z \tag{5.5}$$

where $Z = \{z \mid z \in H^2(\Omega),\ \frac{\partial z}{\partial n} = 0 \text{ on } \Gamma\}$, $\frac{1}{2}\,a(\zeta,\zeta)$ is the elastic energy of the plate, $(f_3,z) = \int_\Omega f_3 z\,d\Omega$ and $\langle\overline{Q},z\rangle$ denotes the duality pairing between $H^{-3/2}(\Gamma)$ and $H^{3/2}(\Gamma)$. We have

$$a(\zeta,z) = \frac{E\,h^3}{12(1-\nu^2)} \int_\Omega \left((1-\nu)\frac{\partial^2\zeta}{\partial x_a \partial x_b}\frac{\partial^2 z}{\partial x_a \partial x_b} + \nu\Delta\zeta\Delta z\right)d\Omega \qquad 1\le a,b\le 2,\ 0<\nu<\frac{1}{2} \tag{5.6}$$

where E is the modulus of elasticity and ν is the Poisson ratio. Eq.(5.5) with (5.1-4) allows the formulation of the following problem.

Problem 3: For any $\bar{\bar{f}}_3 \in L^2(\Omega)$ find $\zeta \in Z$ to satisfy the hemivariational inequality

$$a(\zeta,z-\zeta) + \int_{\Omega'} \varphi^o(\zeta,z-\zeta)\,d\Omega' \ge (\bar{\bar{f}}_3,z-\zeta) \qquad \forall z \in Z\,. \tag{5.7}$$

The regularized problem reads:

Problem 3_ε: For any $\bar{\bar{f}}_3 \in L^2(\Omega)$ find $\zeta_\varepsilon \in Z$ to satisfy the variational equality

$$a(\zeta,z) + \int_{\Omega'} \frac{d\varphi_\varepsilon(\zeta_\varepsilon)}{d\zeta} z d\Omega' = (\bar{\bar{f}}_3, z) \quad \forall z \in Z . \tag{5.8}$$

If φ is convex, (5.7) becomes simply a variational equality.[3] The following proposition holds.

Proposition 4: Suppose that $b \in L^\infty_{loc}(\mathbb{R})$ exists such that (2.7), (2.8) and (3.4) hold. Problem 3 admits a solution $\zeta \in Z$ if

$$\int_{r>0} b(+\infty) r d\Omega + \int_{r<0} b(-\infty) r d\Omega > \int_\Omega \bar{\bar{f}}_3 r d\Omega > \int_{r>0} b(-\infty) r d\Omega + \int_{r<0} b(+\infty) r d\Omega . \tag{5.9}$$

Proof: Taking into account the semicoercivity of $a(\zeta,\zeta)$ and since $\dim \ker a = 1$, we can apply prop. 2 q.e.d.

Similar B.V.Ps. can be formulated in the theories of bars and the theory of von Kármán plates.

References

1. Moreau, J.J., La notion de sur-potentiel et les liaisons unilatérales en élastostatique, *C.R.Acad.Sc.Paris*, 267, 954, 1968.

2. Rockafellar, R.T., *Convex Analysis*, Princeton Univers. Press, Princeton, 1970.

3. Duvaut, G. and Lions, J.L., *Les inéquations en Mécanique et en Physique*, Dunod, Paris, 1972, chap.1,2.

4. Fichera, G., Boundary Value Problems in Elasticity with Unilateral Constraints, in *Encyclopedia of Physics VIa/2*, S.Flügge Ed., Springer-Verlag, Berlin, Heidelberg, New York, 1972.

5. Panagiotopoulos, P.D., Nonconvex Superpotentials in the sense of F.H. Clarke and Applications, *Mech.Res.Comm.*,8, 335, 1981.

6. Panagiotopoulos, P.D., Nonconvex Energy Functionals. Application to Nonconvex Elastoplasticity, *Mech.Res.Comm.*, 9, 23, 1982.

7. Panagiotopoulos, P.D., Non-convex Energy Functions. Hemivariational Inequalities and Substationarity Principles, *Acta Mechanica*, 48, 111, 1983.

8. Panagiotopoulos, P.D., *Inequality Problems in Mechanics and Applications. Convex and Nonconvex Energy Functions*, Birkhäuser Verlag, Basel, Boston, 1984 (in press).

9. Clarke, F.H., Generalized gradients and applications, *Trans.A.M.S.*, 205, 247, 1975.

10. Clarke, F.H., *Optimization and Nonsmooth Analysis*, Wiley-Interscience, New York, 1983.

11. Rockafellar, R.T., *La théorie des sous-gradients et ses applications à l'optimisation. Fonctions convexes et non-convexes*, Les presses de l'Université de Montréal, Montréal, 1979.

12. Rauch, J., Discontinuous Semilinear Differential Equations and Multiple Valued Maps, *Proc.A.M.S.*, 64, 277, 1977.

13. M[c]Kenna, P.J. and Rauch, J., Strongly Nonlinear Perturbations of Nonnegative Boundary Value Problems with Kernel, *J.Diff.Eq.*, 28, 253, 1978.

14. Castaing, C. and Valadier, M., *Convex Analysis and Measurable Multifunctions*, Springer-Verlag, Berlin, 1977, 65.

15. Chang, K.C., Variational Methods for Nondifferentiable Functionals and their Applications to Partial Differential Equations, *J.Math. Anal.Appl.*, 80, 102, 1981.

16. Lions, J.L., *Quelques méthodes de résolution des problèmes aux limites nonlinéaires*, Dunod-Gauthier Villars, Paris, 1969, 53.

17. Ekeland, I. and Temam, R., *Convex Analysis and Variational Problems*, North-Holland, Amsterdam, 1976, 239.

18. Landesman, E.M. and Lazer, A.C., Nonlinear Perturbations of Linear Elliptic Boundary Value problems at Resonance, *J.of Math. and Mech.*, 19, 609, 1970.

19. Rockafellar, R.T., Generalized Directional Derivatives and Subgradients of Nonconvex Functions, *Can.J.Math.*, XXXII, 257, 1980.

20. Panagiotopoulos, P.D., Nonconvex Problems of Semipermeable Media and Related Topics. To appear in *ZAMM*, 64, 1984.

UNE THEORIE DES CATASTROPHES POUR CERTAINS PROBLEMES UNILATERAUX?

M. Potier-Ferry
Université Pierre et Marie Curie
Paris

L'objectif de cet exposé est de s'opposer à une idée reçue, à savoir que la théorie de la bifurcation ne s'applique pas pour les problèmes unilatéraux. Il est vrai que pour la mise en oeuvre de la méthode de LYAPUNOV et SCHMIDT ou de développements formels, il faut des propriétés de régularité qui ne sont jamais satisfaites dans un cadre unilatéral. De nombreux auteurs ont tourné la difficulté en utilisant des principes variationnels pour démontrer des résultats de bifurcation (CIMETIERE[1], DO[4,5], MIGNOT et PUEL[10]). Les résultats obtenus avec ces méthodes sont très généraux, mais pas assez précis : par exemple il n'est pas possible de prévoir le nombre de solutions bifurquées, ni leur stabilité. Quant au point de vue de la théorie des catastrophes, qui consiste en particulier à classer des singularités (THOM[18]) il semble encore plus éloigné des applications aux problèmes unilatéraux.

Ce pessimisme ne peut suffire à un mécanicien, car de nombreuses théories mécaniques conduisent à des problèmes unilatéraux, comme le montrent les études publiées dans ce volume. La stabilité des structures plastiques est la plus importante de ces théories associant bifurcation et caractère unilatéral, ce qui a donné lieu à de nombreux travaux, par exemple ENGESSER (1889), VON KARMAN (1910), SHANLEY (1947), HUTCHINSON (1974), KACHANOV (1976). La mécanique de la rupture fragile conduit à des problèmes unilatéraux particulièrement simples, puisqu'ils se posent dans des espaces dont la dimension est le nombre de pointes de fissures. NEMAT-NASSER, KEER et leurs associés[11,12,17] ont élaboré une théorie des bifurcations pour des solides fissurés et fragiles. Ils ont montré notamment un exemple intéressant de rupture de symétrie : au cours de l'extension de deux fissures symétriques, il peut arriver que l'une s'arrête tandis que l'autre progresse plus vite.

Q.S. NGUYEN[13,14] a montré que tous ces problèmes de stabilité peuvent être formulés dans un cadre unique. Dans les systèmes considérés, il y a des seuils au-delà desquels le comportement est dissipatif, ce caractère dissipatif excluant les problèmes de contact de type SIGNORINI. Il a établi un critère de stabilité et un critère de non-bifurcation qui ne sont pas identiques, ce qui est bien connu en plasticité (SHANLEY[16]). Nous nous proposons ici de prolonger son travail en étudiant les problèmes de bifurcation et en essayant de classer toutes les singularités ou catastrophes qui apparaissent sur les courbes de réponse. Comme pour les problèmes réguliers, cette classification dépend des symétries du système. Mais elle dépend aussi de la dimension d'un cone $N_C(A)$, qui, en mécanique de la rupture, est identique au nombre de fissures où le facteur d'intensité des contraintes a atteint sa valeur critique. Nous ne présentons ici que les catastrophes "de codimension un" dans le cas d'un cone de dimension un, puis d'un cone symétrique de dimension deux.

Pour une présentation plus complète, nous renvoyons à POTIER-FERRY[15].

II. - LES SYSTEMES STANDARD

En théorie de la stabilité élastique, les équilibres sont les points stationnaires $u(\lambda)$ d'une énergie potentielle $E(u,\lambda)$ qui est fonction du déplacement u et d'un seul paramètre réel λ représentant le chargement. Lorsque pour une valeur critique λ_c la seconde variation $\delta^2E = D_u^2E(\delta u,\delta u)$ cesse d'être définie positive, on obtient à la fois une limite de stabilité et une singularité (ou catastrophe) sur la courbe de réponse $u(\lambda)$. La théorie classique de la bifurcation permet de montrer que ces singularités sont en général des points limites, c'est-à-dire des points où la charge λ est extrémale. Pour un système symétrique (plus précisément invariant par une réflexion), on trouve également des points de bifurcation symétriques (le fameux "pitchfork"), auquel cas l'instabilité conduit à une rupture de symétrie.

Le problème est de construire une théorie analogue pour les problèmes de stabilité plastique et de mécanique de la rupture. Nous allons introduire maintenant le cadre proposé par Q.S. NGUYEN. L'état du système est caractérisé par un couple (u,α) où u est le déplacement et α un ensemble de paramètres permettant de décrire le comportement dissipatif du système. Lorsque α est constant, le comportement est réversible au sens thermodynamique. En mécanique de la rupture, la position des pointes de fissures est repérée par ce paramètre α, qui parcourt donc un ensemble de dimension finie. En plasticité, α désigne des déformations plastiques, c'est-à-dire une ou plusieurs fonctions selon le modèle. On peut définir une "énergie potentielle" $E(u,\alpha,\lambda)$ qui dépend également de la variable dissipative α. A l'équilibre, l'énergie est stationnaire par rapport aux seules variations de u :

$$\delta E = D_u E(u,\alpha,\lambda)\,(\delta u) = 0 \qquad \forall \delta u \in \mathcal{U}. \tag{1}$$

On fait ensuite l'hypothèse que la seconde variation de l'énergie E par rapport aux variations de u reste définie positive, ce qui n'est pas une restriction pour les applications envisagées. Alors, d'après le thèorème des fonctions implicites, l'équation (1) a, localement, une solution unique en déplacement $u(\alpha,\lambda)$, ce qui permet de définir une énergie potentielle à l'équilibre

$$F(\alpha,\lambda) = E(u(\alpha,\lambda),\alpha,\lambda). \tag{2}$$

Toute la discussion va maintenant être conduite à partir de cette nouvelle énergie $F(\alpha,\lambda)$. On appelle force généralisée associée à α la quantité

$$A(\alpha,\lambda) = -D_\alpha F(\alpha,\lambda). \tag{3}$$

Si on veut prévoir la (ou les) réponse(s) $\alpha(t)$ à une sollicitation $\lambda(t)$, il faut imposer d'autres relations, qui sont ici des inéquations. On suppose que la force A doit appartenir à un certain ensemble convexe C, puis que la vitesse $d\alpha/dt$ obéit à une règle d'écoulement compatible avec le principe de dissipation maximale, c'est-à-dire :

$$A \in C \tag{4}$$

$$\frac{d\alpha}{dt} \in N_C(A), \tag{5}$$

où $N_C(A)$ est le cone des normales extérieures à C en A. Le comportement du matériau, la géométrie du problème, les conditions aux limites sont prises en compte au moyen de l'énergie à l'équilibre $F(\alpha,\lambda)$ et de l'ensemble convexe C des forces admissibles.

Nous appellerons système standard tout système dont l'évolution est gouvernée par un système de relations de la forme (3) (4) (5).

La mécanique de la rupture fragile est un des exemples les plus simples de système standard, si l'on met de côté l'élimination du déplacement, qui demande la solution d'un problème d'élasticité linéaire. Supposons qu'il n'y ait qu'une seule fissure de longueur α. L'énergie à l'équilibre est en général de la forme :

$$F(\alpha,\lambda) = \lambda^2 f(\alpha) \tag{6}$$

La force A définie en (3) s'identifie ici avec le taux de restitution de l'énergie

$$G = - \partial_\alpha F = \lambda^2 g(\alpha) \tag{7}$$

L'ensemble convexe C ne peut être qu'un intervalle $[0,G_c]$ et (4) (5) s'identifie avec la loi de GRIFFITH

$$\begin{aligned} G < G_c &\Longrightarrow d\alpha/dt = 0 \\ G = G_c &\Longrightarrow d\alpha/dt \geqslant 0 \end{aligned} \tag{8}$$

Bien entendu, si le corps élastique possède n pointes de fissures dont on repère les positions à l'aide de n paramètres réels α_j, α sera le vecteur colonne $(\alpha_1, \alpha_2, \dots \alpha_n)$. Cette présentation est valable en élasticité plane avec des fissures qui ne se propagent que de manière rectiligne.

III.- METHODE DE RESOLUTION D'UN SYSTEME STANDARD

Nous étudions les systèmes standard, c'est-à-dire des systèmes d'inéquations différentielles de la forme (3) (4) (5). La résolution de l'équation variationnelle (1) ayant permis d'éliminer le déplacement u, il reste à trouver l'évolution de l'inconnue $\alpha(t)$ (α représente des déformations plastiques ou des longueurs de fissures ...) lorsque le chargement $\lambda(t)$ est donné en fonction du temps. Plus précisément, nous discuterons la stabilité des solutions, puis les bifurcations éventuelles, c'est-à-dire l'existence de plusieurs réponses $\alpha(t)$ à une même sollicitation.

Un système standard (3) (4) (5) peut se mettre sous une forme équivalente et plus facile à résoudre. Q.S. NGUYEN[14] a montré que la vitesse $\dot{\alpha} = d\alpha/dt$ est une solution de l'inéquation variationnelle suivante :

$$D_\alpha^2 F(\dot{\alpha}, \beta - \dot{\alpha}) + \dot{\lambda}\partial_\lambda D_\alpha F\,(\beta - \dot{\alpha}) \geqslant 0 \qquad \forall \beta \in N_C(A)$$

$$\dot{\alpha} \in N_C(A). \tag{9}$$

Cette inéquation variationnelle classique permet de déterminer à chaque instant une ou plusieurs valeurs de la vitesse $\dot{\alpha}$ lorsque α, λ et $\dot{\lambda}$ sont connus. Il est clair que la vitesse $\dot{\alpha}$ est unique lorsque la forme quadratique $D_\alpha^2 F(.,.)$ est définie positive sur l'espace vectoriel engendré par le cone $N_C(A)$, qui sera noté Vect $N_C(A)$:

$$D_\alpha^2 F(\beta,\beta) > 0 \quad \forall \beta \in \text{ Vect } N_C(A),\ \beta \neq 0. \tag{10}$$

On peut remarquer - cf.(5) - que, lorsque le chargement $\lambda(t)$ est croissant, la (ou les) réponse(s) $\alpha(t)$ ne dépendra que de λ, car $N_C(A)$ est un cone. Dans la suite nous ferons toujours cette hypothèse d'un chargement croissant. Alors $d\alpha/d\lambda$ est une solution de l'inéquation variationnelle

$$D_\alpha^2 F\left(\frac{d\alpha}{d\lambda}, \beta - \frac{d\alpha}{d\lambda}\right) + \partial_\lambda D_\alpha F\left(\beta - \frac{d\alpha}{d\lambda}\right) \geqslant 0 \qquad \forall \beta \in N_C(A)$$

$$\frac{d\alpha}{d\lambda} \in N_C(A). \tag{11}$$

Supposons que l'on sache trouver au moins une solution de (11)

$$\frac{d\alpha}{d\lambda} = a(\alpha,\lambda) \tag{12}$$

qui dépend forcément de α et de λ puisque la forme bilinéaire $D_\alpha^2 F(.,.)$, la forme linéaire $\partial_\lambda D_\alpha F(.)$ et le cone $N_C(A)$ varient avec α et λ. On est alors ramené à une équation différentielle (12) qui aura une solution unique $\alpha(\lambda)$ si $a(\alpha,\lambda)$ est une fonction régulière de α et λ, ce qui sera vérifié dans les problèmes considérés ici. Dans ce cas, à chaque solution du problème en vitesses (11), il correspond une et une seule courbe de réponse $\alpha(\lambda)$.

La coercivité de la forme quadratique $D_\alpha^2 F(.,.)$ est une condition suffisante pour qu'il existe au moins une solution de l'inéquation variationnelle (11) :

$$D_\alpha^2 F(\beta,\beta) > 0 , \quad \forall \beta \in N_C(A) , \quad \beta \neq 0 . \tag{13}$$

Cette condition suffisante d'existence (13) est importante, car un argument de Q.S. NGUYEN[14] permet de l'interpréter aussi comme une condition de stabilité. Sauf, dans le cas d'un cone de dimension un, la condition de stabilité (13) est donc plus restrictive que la condition (11) d'unicité locale de la réponse $\alpha(t)$ (ou condition de non-bifurcation). Contrairement au cas de l'élasticité, il peut donc y avoir des bifurcations sans que la solution fondamentale perde sa stabilité (voir [16,9 ...)

L'étude des bifurcations ne se présente pas de la même manière que dans le cas élastique. Si l'énergie $F(\alpha,\lambda)$ joue toujours un rôle important, la structure géométrique du cone $N_C(A)$ et son évolution avec le paramètre λ seront fondamentales dans la discussion. Reprenons l'exemple du corps élastique possèdant n pointes de fissures repérées par $\alpha = (\alpha_1,\ldots,\alpha_n)$. La force sera le vecteur $A = (G_1,\ldots,G_n)$ où

$$G_i = - \partial F/\partial\alpha_i = - \lambda^2 \partial f/\partial\alpha_i \tag{14}$$

est le taux de restitution de l'énergie associé à l'extension de la i-ème pointe de fissure. L'ensemble des forces admissibles est

$$C = \{ A = \{(G_1,\ldots,G_n) : 0 \leqslant G_i \leqslant G_c\}$$

Si pour une valeur donnée du chargement, on a

$$G_i = G_c \qquad \text{pour } 1 \leqslant i \leqslant p$$
$$G_i < G_c \qquad \text{pour } p+1 \leqslant i \leqslant n ,$$

seules les p premières pointes de fissures peuvent avancer, d'après la loi (8). Le cone sera de dimension p :

$$N_C(A) = \{\beta = (\beta_1,\ldots;\beta_n) : \quad \beta_i \geqslant 0 \quad i \leqslant p \text{ et } \quad \beta_j = 0 \quad j \geqslant p+1\}$$

La dimension de ce cone est donc le nombre de pointes de fussures qui peuvent progresser. Pour les problèmes analogues de plasticité discrète (modèles de type Shanley[16,8,14,15]), la dimension du cone est le nombre de barres en charge plastique.

IV.- CATASTROPHES AVEC UN CONE DE DIMENSION UN

Nous discuterons ici le cas d'un cone $N_C(A)$ de dimension un. En mécanique de la rupture, cela signifie que, pour les chargements considérés, il n'y a qu'une pointe de fissure où le taux de restitution de l'énergie puisse atteindre sa valeur maximale. Nous n'étudierons que les singularités (ou catastrophes) les plus générales, que nous appelerons catastrophes de codimension un, selon la terminologie classique[18].

Dans le cadre des systèmes standard (3) (4) (5), nous faisons les hypothèses suivantes :

H1 C est un ensemble convexe fermé de $\mathbb{R}^n$, d'intérieur non vide.

H2 Pour $\lambda < \lambda_1$, la force $A(\alpha_1,\lambda)$ est à l'intérieur de C. Pour $\lambda = \lambda_1$, elle atteint la frontière de C en un point régulier.

Un point A_1 de la frontière ∂C sera appelé point régulier pour $\lambda = \lambda_1$ s'il existe une fonction régulière à valeurs réelles $\mathcal{F}(A,\lambda)$ telle que, au voisinage de (A_1,λ_1), on ait

$$C = \{A \in \mathbb{R}^n : \mathcal{F}(A,\lambda) \leqslant 0\}$$

$$\partial C = \{A \in \mathbb{R}^n : \mathcal{F}(A,\lambda) = 0\} .$$

En un point régulier de la frontière, le cone est de dimension un et il est engendré par la normale

$$n = D_A\mathcal{F} \tag{15}$$

supposée non nulle. La dérivée de $\mathcal{F}$ le long des chemins constants $\alpha = \alpha_1$ est désignée par

$$\left(\frac{d\mathcal{F}}{d\lambda}\right)_\alpha = \frac{\partial \mathcal{F}}{\partial \lambda} + n \cdot \frac{\partial A}{\partial \lambda} . \tag{16}$$

D'après l'hypothèse H2, cette dérivée est positive ou nulle. Nous faisons une hypothèse de transversalité :

$$\text{H3} \quad \left(\frac{d\mathcal{F}}{d\lambda}\right)(\alpha_1,\lambda_1) > 0 .$$

Seule l'hypothèse de point régulier H2 est restrictive puisqu'elle conduit à un cone de dimension un. Le théorème suivant montre que dans ce cas, il n'y a en général que deux sortes de catastrophes possibles.

THEOREME

Soit un système standard (3) (4) (5) avec $\lambda(t)$ croissant. La force $A(\alpha_1,\lambda)$ est à l'intérieur de C pour λ inférieur à λ_1 et atteint la frontière ∂C en $\lambda = \lambda_1$. Les hypothèses H1, H2, H3 sont supposées réalisées.

(i) Si l'état (α_1,λ_1) est instable, c'est-à-dire si

$$D^2_\alpha F(n,n) < 0 \qquad \text{pour} \quad (\alpha,\lambda) = (\alpha_1,\lambda_1) \tag{17}$$

la solution constante $\alpha = \alpha_1$ ne peut être prolongée pour λ plus grand que λ_1 (voir Figure 1 ; nous appelons cette singularité avancée instable à cause de son interprétation en mécanique de la rupture).

(ii) Si l'état (α_1,λ_1) est stable, c'est-à-dire si

$$D^2_\alpha F(n,n) > 0 \qquad \text{pour} \quad (\alpha,\lambda) = (\alpha_1,\lambda_1) \tag{18}$$

la solution constante $\alpha = \alpha_1$ a un prolongement unique et stable $\alpha(\lambda)$ pour λ plus grand que λ_1. Il n'y a donc pas de singularité en (α_1, λ_1). (En mécanique de la rupture, on parlera d'avancée stable ou d'extension stable de fissure).

(iii) Considérons cette courbe de solutions $\alpha(\lambda)$ et supposons qu'elle atteint un point (α_2,λ_2) tel que pour $(\alpha,\lambda) = (\alpha_2,\lambda_2)$

H4 $\quad D^2_\alpha F(n,n) = 0$

H5 $\quad D^3_\alpha F(n,n,n) - 2\, D^2_\alpha \mathcal{F}(D_\alpha F\, n,\, D_\alpha F\, n) \neq 0.$

Alors λ_2 est une charge maximum et le point (α_2,λ_2) est un point limite (Voir Figure 2).

(iv) Le point limite et l'avancée instable sont les seules singularités de codimension un, tant que le cone normal $N_C(A)$ reste de dimension un.

Figure 1 : avancée instable

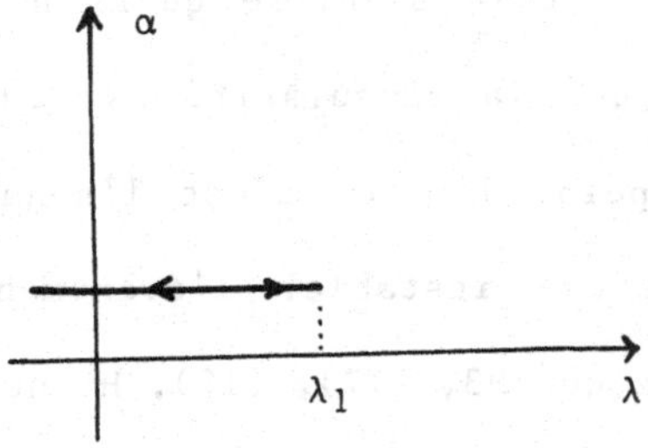

Figure 2 : avancée stable, puis point limite.

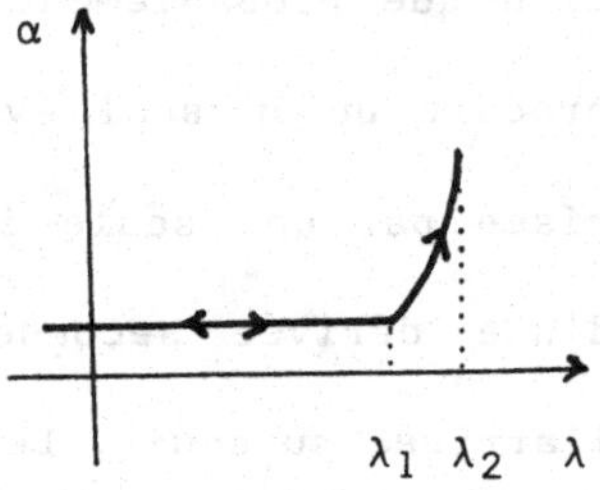

Le point limite est une singularité connue dans le cas des systèmes réguliers, en particulier en élasticité où c'est d'ailleurs la seule singularité de codimension un. L'avancée instable est une singularité nouvelle, qui est caractéristique des systèmes unilatéraux dissipatifs étudiés ici.

Ce théorème est applicable en mécanique de la rupture lorsqu'une seule pointe de fissure peut avancer. On obtient une avancée instable de la fissure si

$$\frac{\partial G}{\partial \alpha}(\alpha_1, \lambda_1) > 0.$$

Si le signe est opposé, il y a avancée stable. Les hypothèses H4 et H5 caractérisant le point limite s'écrivent ici

$$\frac{\partial G}{\partial \alpha}(\alpha_2,\lambda_2) = 0 \quad , \quad \frac{\partial^2 G}{\partial \alpha^2}(\alpha_2, \lambda_2) > 0.$$

La notion de singularité de codimension un est facile à comprendre bien que probablement délicate à définir. Elle signifie qu'il ne s'est produit qu'un seul évènement ou encore que la singularité est caractérisée par une seule équation. Pour le point limite, c'est l'annulation d'une dérivée seconde (H4). Pour l'avancée instable, l'évènement est l'arrivée au seuil. Les conditions telles que H3, (17), (18), H5 ne comptent pas dans le calcul de la codimension car elles persistent lorsqu'on perturbe l'énergie potentielle à l'équilibre F.

V.- CATASTROPHES AVEC UN CONE SYMETRIQUE DE DIMENSION DEUX

Dans ce paragraphe, nous cherchons à classer des singularités qui sont connues depuis longtemps pour le flambement plastique, plus récemment pour la rupture fragile. Nous en donnons une présentation unifiée. Pour le flambement plastique, il s'agit des modèles de type SHANLEY[16,8,14] à deux ou quatre degrés de liberté. En mécanique de la rupture, SUMI *et al*[17] ont étudié l'extension de deux fissures symétriques sous l'effet de chargements thermiques. Ils ont montré qu'après une extension stable et symétrique de deux fissures, l'une pouvait s'arrêter tandis que l'autre avançait deux fois plus vite. Les notions de symétrie et de rupture de symétrie seront donc fondamentales dans cette présentation.

Comme au paragraphe précédent, l'ensemble C est convexe fermé d'intérieur non vide (H1). Pour prendre en compte la symétrie du problème, nous supposons que l'énergie à l'équilibre $F(\alpha,\lambda)$, l'ensemble convexe C et la règle d'écoulement sont invariants par une réflexion R, c'est-à-dire un opérateur linéaire involutif (R^2 = Id, R $\neq$ Id) :

H6 $\quad F(\alpha,\lambda) = F(R\alpha,\lambda)$

H7 $\quad A \in C \Longrightarrow RA \in C$

H8 $\quad RA_1.RA_2 = A_1.A_2$ (R est donc selfadjoint).

Nous dirons qu'un vecteur A est symétrique si RA = A. Nous supposons que :

H9 L'état initial (α_1,λ_0) est symétrique. Pour $\lambda_0 \leqslant \lambda < \lambda_1$, la force $A(\alpha_1,\lambda)$ est à l'intérieur de C. Pour $\lambda = \lambda_1$, elle atteint le bord en un "coin régulier de dimension deux".

D'après H6, la force $A(\alpha_1,\lambda)$ est symétrique. Un point A_1 de ∂C est appelé <u>coin régulier de dimension deux</u> s'il existe deux fonctions régulières à valeurs réelles $\mathcal{F}(A,\lambda)$, $\mathcal{F}'(A,\lambda)$ telles que, localement, on ait :

$$C = \{ A \in R^n : \mathcal{F}(A,\lambda) \leqslant 0 \ , \ \mathcal{F}'(A,\lambda) \leqslant 0\}$$

$$\partial C = \{ A \in R^n : \mathcal{F}(A,\lambda) = 0 \text{ ou } \mathcal{F}'(A,\lambda) = 0\}$$

$$\mathcal{F}(A_1,\lambda_1) = \ '(A_1,\lambda_1) = 0$$

Lorsque, en mécanique de la rupture, deux fissures peuvent progresser, la force $A = (G_1,\ldots,G_n)$ se trouve en un coin de dimension deux (cf §3). Pour rester cohérent avec H7, les deux fonctions seuils doivent être symétriques l'une de l'autre :

$$\text{H'7} \qquad \mathcal{F}(RA,\lambda) = \mathcal{F}'(A,\lambda).$$

La dernière hypothèse est l'analogue de H3 :

$$\text{H10} \qquad \left(\frac{d\mathcal{F}}{d\lambda}\right)_\alpha (\alpha_1,\lambda_1) > 0 .$$

Ces hypothèses abstraites reviennent à supposer que le problème est symétrique (H6,H'7,H8) et que le cone $N_C(A)$ est de dimension deux. Ce cone est engendré par les deux normales extérieures

$$n = D_A\mathcal{F} \quad , \quad n' = D_A\mathcal{F}' = Rn .$$

$$N_C(A) = \{kn + k'n' \ , \ k \geqslant 0 \ , \ k' \geqslant 0\} \tag{19}$$

Nous avons montré au paragraphe 3 que chaque branche $\alpha(\lambda)$ correspond à une solution de l'inéquation variationnelle (11), que nous réexprimons en fonction de k et k' :

$$(k,k') \in R_+^2 \tag{20}$$

$$p\{k(\ell-k) + k'(\ell'-k')\} + q\{k(\ell'-k') + k'(\ell-k)\} - r(\ell-k + \ell'-k') \geqslant 0$$

$$(\ell,\ell') \in R_+^2$$

où les constantes p, q, r dérivent de l'énergie à l'équilibre

$$p = D_\alpha^2 F(n,n) \quad , \quad q = D_\alpha^2 F(n,n') \tag{21}$$

$$r = - \partial_\lambda D_\alpha F(n) > 0 \quad \text{(d'après H10).} \tag{22}$$

L'origine $k = k' = 0$ ne peut être solution de (20) à cause de l'inégalité (22). Toute solution (k,k') située à l'intérieur du cone R_+^2 est symétrique

$$k = k' = r/(p + q) \tag{23}$$

sauf si $p = q$. Les solutions situées au bord du cone sont

$$k = r/p \quad , \quad k' = 0 \quad \text{et} \quad k = 0 \quad , \quad k' = r/p \tag{24}$$

Dans le cas exceptionnel $p = q > 0$, il y a un intervalle de solutions

$$k + k' = r/p \quad , \quad k \geqslant 0 \ , \ k' \geqslant 0 \ .$$

dont nous ne retenons que les valeurs extrêmes (24) ou médiane (23) qui seules peuvent évoluer continument lorsque λ varie.

Après une discussion détaillée des solutions (23) (24) (voir[15]), on montre qu'il n'y a que cinq catastrophes de codimension un, ce qui est précisé par le théorème suivant.

THEOREME

Soit un système standard (3) (4) (5) avec $\lambda(t)$ croissant. La force $A(\alpha_1, \lambda)$ se trouve à l'intérieur de C pour $\lambda_0 \leqslant \lambda < \lambda_1$ et atteint le bord ∂C pour $\lambda = \lambda_1$ dans les conditions décrites par les hypothèses H1, H6, H'7, H8, H9, H10. Nous discutons les prolongements de la courbe de solutions triviale $\alpha = \alpha_1$ en fonction des paramètres $p(\lambda)$, $q(\lambda)$ définis en (21).

(i) Si $p(\lambda_1) + q(\lambda_1)$ est négatif, la courbe triviale n'a aucun prolongement. La singularité est l'avancée instable représentée à la Figure 1.

(ii) Si $q(\lambda_1) > -p(\lambda_1) > 0$,
il existe un prolongement unique et symétrique, mais ces solutions sont instables. (Voir Figure 3 ; la singularité correspond à un autre type d'avancée instable).

(iii) Si $q(\lambda_1) > p(\lambda_1) > 0$,
il existe un prolongement symétrique et stable $\alpha_s(\lambda)$ pour $\lambda > \lambda_1$. Deux branches de solutions non symétriques et stables bifurquent à partir de n'importe quel état $\alpha_s(\lambda)$ (Figure 4).

(iv) Si $p(\lambda_1) > q(\lambda_1) > 0$,
il y a un seul prolongement $\alpha_s(\lambda)$ qui est stable et symétrique. Le point (α_1, λ_1) n'est pas singulier.

Il reste à discuter les singularités de cette branche de solutions symétriques $\alpha_s(\lambda)$.

(v) Si $p(\lambda) + q(\lambda)$ tend vers zéro pour la première fois en $\lambda = \lambda_2$ et si H5 est réalisé, le point (α_2, λ_2) est un point limite (Figure 2)

(vi) Si $p(\lambda) - q(\lambda)$ tend vers zéro et change de signe pour la première fois en $\lambda = \lambda_2$, la branche symétrique $\alpha_s(\lambda)$ se prolonge au delà de λ_2 et reste stable. Deux branches de solutions non symétriques et stables bifurquent à partir de n'importe quel état $\alpha_s(\lambda)$ (Figure 5)

(vii) Il n'y a pas d'autres catastrophes de codimension un tant que A reste en un coin régulier de dimension deux.

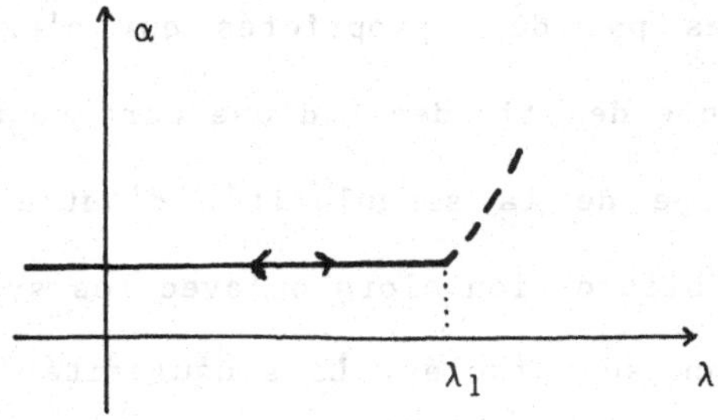

Figure 3 : avancée instable (deuxième type)

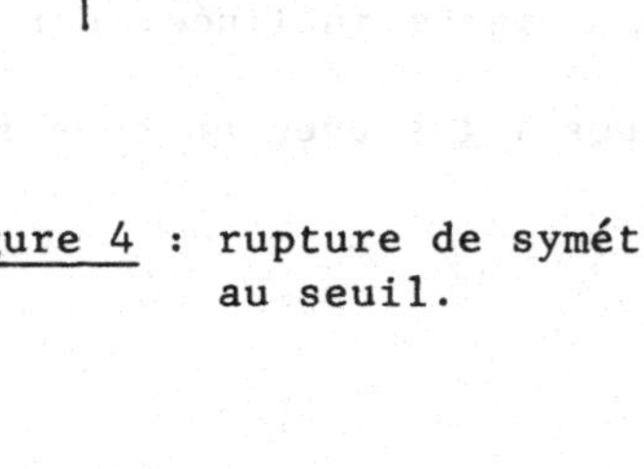

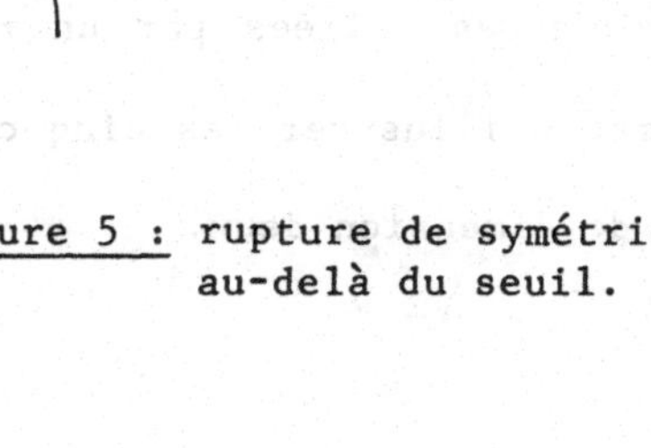

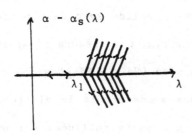

Figure 4 : rupture de symétrie au seuil.

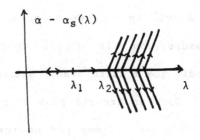

Figure 5 : rupture de symétrie au-delà du seuil.

Deux de ces cinq catastrophes élémentaires existent avec un cone de dimension un : l'avancée instable de type (i) et le point limite (v). D'un point de vue pratique, l'avancée instable de type (ii) n'est pas très différente de celle de type (i), puisque dans les deux cas, il n'y a aucune solution stable au delà du seuil. Cette singularité peut exister dans le modèle de SHANLEY si la courbe charge-déformation a une pente $E_t = d\sigma/d\varepsilon$ discontinue au seuil et si, à cause de cette discontinuité, la charge dépasse dès le seuil la charge critique du "module réduit".

Des bifurcations avec rupture de symétrie sont possibles dès le seuil (iii) ou après une évolution irréversible (vi). Ces deux singularités sont caractérisées par deux propriétés qui n'existent jamais en élasticité ou en mécanique des fluides : d'une part toutes les solutions sont stables au voisinage de la singularité, d'autre part il y a un intervalle de points de bifurcation alors qu'avec les systèmes réguliers, les points de bifurcation sont isolés. La singularité (vi) a été découverte par SHANLEY[16].

Si on compare les deux vitesses (23) (24) dans le cas de la singularité (vi) (p = q) et qu'on l'applique à un solide avec deux pointes de fissures, on voit qu'après rupture de symétrie, une fissure s'est arrêtée tandis que l'autre avance deux fois plus vite (Voir SUMI et al[17]).

Dans un travail plus complet[15], nous avons étudié le glissement de deux masses reliées par un ressort sur une pente inclinée. Cet exemple permet d'illustrer les cinq catastrophes possibles avec un cone symétrique de dimension deux.

REFERENCES

1. CIMETIERE, A., Un problème de flambement unilatéral en théorie des plaques, J. Mécanique 19, 183, 1980.

2. CIMETIERE,A., Flambement naissant dans les plaques élastoplastiques minces. Preprint.

3. CONSIDERE, A., Résistance des pièces comprimées, in Congrès International des procédés de construction 3, 371, librairie polytechnique, Paris, 1981.

4. DO, C., Bifurcation theory for elastic plates subjected to unilateral conditions, J. Math. Anal. Appl., 60, 435, 1977.

5. DO, C., Flambement élastoplastique d'une plaque mince, J. Math. pures et Appl., 62, 11, 1983.

6. ENGESSER, F., Ueber die knickfestigkeit gerader stäbe, Z. Architektur und Ingenieurwesen, 35, 445, 1889.

7. HILL, R., A general theory of uniqueness and stability in elastic plastic solids, J. Mech. Phys. Solids, 6, 336, 1958.

8. HUTCHINSON, J.W., Plastic buckling, Advances in Appl. Mech., 14, 67, 1974.

9. KACHANOV, M., Foundations of Plasticity, North-Holland, Amsterdam, 1976.

10. MIGNOT F. et PUEL J.P. Flambage de plaques élastoplastiques, Annali Mat. Pura Appl., 128, 51, 1980.

11. NEMAT-NASSER, S., KEER, L.M., and PARIHAR K.S., Unstable growth of thermally induced interacting cracks in brittle solids, Int. J. Solids Structures, 14, 409, 1978.

12. NEMAT-NASSER, S., SUMI, Y. and KEER, L.M., Unstable growth of tension cracks in brittle solids : stable and unstable bifurcations, snap through and imperfection sensitivity, Int. J. Solids Structures, 16, 1017, 1980.

13. NGUYEN Q.S., Stabilité et bifurcation en rupture et en plasticité, C. Rend. Acad. Sci. Paris, 2-292, 817, 1981.

14. NGUYEN, Q.S., Bifurcation et stabilité des systèmes irréversibles obéissant au principe de dissipation maximale, J. Mécanique Théo. Appl., 2, 1983.

15. POTIER-FERRY, M., Towards a catastrophe theory for the mechanics of plasticity and fracture, submitted to Int. J. Engng. Sci.

16. SHANLEY, F.R., Inelastic column theory, J. Aeronautical Sci. 14, 261, 1947.

17. SUMI, Y., NEMAT-NASSER, S. and KEER, L.M., A new combined analytical and finite-element solutions method for stability of the growth of interacting tension cracks in brittle solids, Int. J. Engng. Sci., 18, 211, 1980.

18. THOM, R., Stabilité structurelle et morphogénèse, Benjamin, Reading, 1972.

19. VON KARMAN, T., Untersuchungen über Knicksfestigkeit. Mitteilungen über Forschungarbeiten, Z.Verein Deutscher Ingenieure, 81, 1910.

CONTACTS UNILATERAUX AVEC FROTTEMENT EN VISCOELASTICITE

M. Raous
Laboratoire de Mécanique et d'Acoustique
Marseille

Résumé- On généralise les résultats concernant les phénomènes de contact avec ou sans frottement établis en Elasticité à des lois de comportement plus complexes comme la Viscoélasticité à paramètres variables. Une extension au cas de la Viscoplasticité sera donnée. On pose le problème en terme d'inéquation variationnelle couplée à une équation différentielle non linéaire. Dans le cadre de discrétisation par la méthode des éléments finis, on propose un algorithme de résolution utilisant une méthode de surrelaxation avec projection pour traiter l'inéquation variationnelle, une méthode de point fixe pour traiter le modèle de frottement à seuil de glissement fonction de la contrainte normale, et des méthodes d'intégration numérique explicites ou implicites pour traiter l'équation différentielle.

Abstract- We generalize the results about the contact phenomena with or without friction in Elasticity to more complex behavior laws such as Viscoelasticity with variable coefficients. It will be given an extension to the viscoplastic case. The problem is set under the form of a variational inequation coupled with a nonlinear differential equation. We use a finite element discretization. We use an overrelaxation method with projection to solve the variational inequation, a fixed point method to solve the problem with the normal stress depending on the friction born, and explicit or implicit numerical integration methods to treat the differential equation.

1. INTRODUCTION

Nous nous intéressons aux problèmes unilatéraux posés par les phénomènes de contact avec ou sans frottement dans le cadre de lois de comportement viscoélastiques. Parmi les phénomènes mécaniques envisagés, on s'intéresse tout d'abord au comportement d'un solide fissuré sous charges alternatives ou sous chargements thermiques non homogènes : les conditions de non interpénétration des lèvres de la fissure s'écrivent sous forme de conditions unilatérales. Il s'agit également des phénomènes de contact entre le matériau et l'outil dans les problèmes de mise en forme au sens large : emboutissage, formage, filage ... Cette étude s'inscrit dans le cadre des activités du GRECO "Grandes Déformations et Endommagement". Toutefois il ne sera ici question, pour l'instant, que de modèles en petites déformations.

Parmi les traitements numériques envisagés en Elasticité dans le contexte des méthodes d'éléments finis, on peut distinguer essentiellement trois approches.

On trouve tout d'abord des méthodes utilisant des techniques de chargement incrémental avec un contrôle de l'état de contact a posteriori à chaque pas et correction itérative (voir Francavilla-Zienkiewicz[1], Sachdeva-Kamakishnan[2]). Ces méthodes consistent essentiellement à introduire des forces extérieures de contact quand il y a interpénétration ou recouvrement du solide déformé et de l'obstacle, et à relacher des parties de frontière supposées liées à l'obstacle dans le cas d'apparition de forces de traction.

Il existe ensuite des méthodes qui consistent à introduire des éléments d'interface appartenant à un milieu fictif évanescent lors du contact. Des lois de comportement diverses de ce milieu sont envisagées afin de tenir compte du frottement (Nsangou-Batoz-Touzot[3], Cristescu-Loubignac[4]).

Enfin, une troisième classe de méthodes relève de l'approximation du problème posé en terme d'inéquation variationnelle. C'est sous cette forme que nous poserons le problème. On peut en effet noter que les conditions unilatérales de contact (problème de Signorini) et les conditions de frottement conduisent à un ensemble d'inégalités constituant un problème dit de complémentarité. On montre que la forme variationnelle de ce problème peut s'écrire de manière équivalente, soit en terme d'inéquation variationnelle, soit en terme de point selle de Lagrangien (méthode mixte) (Duvaut-Lions[5], Glowinski-Lions-Trémolières[6], Campos-Oden-Kikuchi[7], B.Frekidsson-G.Rydholm-P.Sjöblom[8], D.Talaslidis-

P.D.Panagiotopoulos[9], Degueuil-Lancelle[10]).

Dans le cas de loi de comportement non élastique, on a déjà étudié le problème de la fermeture d'une fissure sans frottement des lèvres, en Viscoélasticité (Raous[11], Bouc-Geymonat-Jean-Nayroles[14]) et en Viscoplasticité (Raous[11], Notin[15]). Nous montrons ici comment on peut poser le problème de contact avec frottement en Viscoélasticité et en Viscoplasticité ; nous proposons une méthode de résolution dont l'algorithme sera précisé. Il s'agit d'adopter une présentation de ces lois de comportement permettant de généraliser et d'adapter les résultats obtenus en Elasticité. Nous sommes alors conduits à une inéquation variationnelle couplée à une équation différentielle linéaire en Viscoélasticité, non linéaire en Viscoplasticité. La méthode s'applique à une loi de frottement à seuil de glissement fixé, et le traitement du cas général où le seuil de glissement est fonction de la contrainte normale est réalisé à l'aide d'une méthode de point fixe.

2. LE PROBLEME MECANIQUE ET SA FORMULATION

On considère un solide occupant un domaine Ω de $\mathbb{R}^3$ de frontière suffisamment régulière $\delta\Omega=\delta_1\Omega \cup \delta_2\Omega \cup \delta_3\Omega$. Des conditions aux limites homogènes de déplacement sont imposées sur la partie $\delta_1\Omega$ de la frontière ($u = 0$ sur $\delta_1\Omega$).

Le solide est soumis à une densité volumique de force ϕ_1 définie sur Ω et à une densité surfacique de force ϕ_2 définie sur $\delta_2\Omega$.

Il sera éventuellement soumis à un champ de déformation imposée e^{o} défini sur Ω : il s'agira de dilatation thermique

$$e^{o} = -(\tilde{T} - T_{ref})\chi \qquad (1)$$

où $\tilde{T}$ désigne un champ de température imposé (fonction du temps et de la variable d'espace), T_{ref} le champ de température de référence, et χ le tenseur des coefficients de dilatation.

Le solide est soumis sur la partie $\delta_3\Omega$ de sa frontière à un contact unilatéral avec frottement avec un obstacle fixe. Ce modèle concerne également un solide fissuré si le domaine et les chargements sont symétriques par rapport à l'axe de la fissure. On trouvera dans Raous[12] une généralisation au cas de fissure quelconque.

2.1. Modèle du contact unilatéral.

Précisons tout d'abord les notations (voir Duvaut-Lions[16]). Soit **u** le champ de déplacement, on notera $\gamma\mathbf{u}$ la trace de **u** sur $\delta_3\Omega$ et en désignant par **n** le vecteur de composante n_i, normal à la frontière en un point de celle-ci, on écrira :

$$\gamma\mathbf{u} = u_N.\mathbf{n} + \mathbf{u}_T \qquad \text{avec} \quad u_N = \{\gamma\mathbf{u}\}_i\, n_i \qquad (2)$$
$$\{\mathbf{u}_T\}_i = \{\gamma\mathbf{u}\}_i - u_N n_i$$

Les vecteurs notés en lettres grasses sont des vecteurs de $\mathbb{R}^3$ dans le cas général et des vecteurs de $\mathbb{R}^2$ pour les problèmes plans ou de révolution.

Soit **F** la densité de force exercée par l'obstacle sur le solide. On écrit :

$$\mathbf{F} = F_N\mathbf{n} + \mathbf{F}_T \qquad (3)$$

avec :

$$\mathbf{F} = \sigma\mathbf{n} \qquad \text{soit } F_i = \sigma_{ij}n_j$$

où σ désigne le tenseur des contraintes. On a :

$$F_N = \mathbf{F}.\mathbf{n} = \sigma_{ij}n_in_j$$

$$\mathbf{F}_T = \mathbf{F} - F_N\mathbf{n} \iff \{\mathbf{F}_T\}_i = \sigma_{ij}n_j - F_Nn_i$$

Les conditions de contact unilatéral s'écrivent alors en tenant compte de l'orientation du vecteur **n** vers l'extérieur de la surface.

$$u_N \leq 0 \tag{4}$$

$$F_N \leq 0 \tag{5}$$

$$u_N.F_N = 0 \tag{6}$$

L'interprétation mécanique est la suivante : en l'absence de contact $u_N < 0$ et (6) implique que $F_N = 0$ (pas de force de contact) et lors du contact $u_N = 0$ et $F_N \leq 0$ (force de compression).

2.2. Modèle de frottement.

Les lois de frottement les plus simples sont des lois à seuil. On écrit que lorsque la contrainte tangentielle atteint un certain seuil, fonction de la contrainte normale, il y a glissement en ce point et que la vitesse de déplacement est alors colinéaire à la force tangentielle. Il s'agit de la loi de Coulomb classique : (μ désigne le coefficient de frottement).

<u>Loi I : loi de Coulomb.</u>

$$|\mathbf{F}_T| \leq -\mu F_N \tag{7}$$

$$\text{avec} \quad |\mathbf{F}_T| < -\mu F_N \implies \frac{\partial \mathbf{u}_T}{\partial t} = 0 \tag{8}$$

$$|\mathbf{F}_T| = -\mu F_N \implies \frac{\partial \mathbf{u}_T}{\partial t} = -\lambda \mathbf{F}_T \text{ avec } \lambda > 0 \tag{9}$$

On pourra se reporter aux travaux de J.J. Moreau[17,18] pour une formulation du problème de Coulomb dans le cadre de la théorie de la convexité.

En statique, Duvaut[16] utilise la formulation suivante posée en termes de déplacement et non de vitesse de déplacement.

<u>Loi II : pseudo-loi de Coulomb.</u>

$$|\mathbf{F}_T| \leq -\mu F_N \tag{7}$$

$$\text{avec } |\mathbf{F}_T| < -\mu F_N \Longrightarrow \mathbf{u}_T = 0 \tag{10}$$

$$|\mathbf{F}_T| = -\mu F_N \Longrightarrow \mathbf{u}_T = -\lambda' \mathbf{F}_T \text{ avec } \lambda' > 0 \tag{11}$$

Dans le cas général, l'équivalence des deux modèles (lois I et II) est fausse. En particulier, pour des problèmes d'évolution, il est clair qu'en cas de changement de signe de la composante tangentielle F_T de la force de frottement, c'est bien la vitesse du déplacement tangentiel qui changera de signe et non le déplacement tangentiel mesuré dans la configuration de référence. Cette difficulté sera réglée de la manière suivante. La solution complète va être construite à l'aide d'un procédé incrémental et nous allons utiliser une configuration de référence actualisée bien qu'il s'agisse ici de petites déformations. Cette procédure n'est pas très pénalisante dans l'optique de la généralisation de la méthode proposée au cas des grandes déformations où il est usuel de travailler sur une configuration de référence actualisée.

Nous étudions ici des phénomènes quasi statiques pour des problèmes plans (contraintes planes, déformations planes, symétrie de révolution) en petites déformations. Pour ces problèmes, moyennant la remarque précédente concernant le référentiel, la loi II constitue effectivement une loi de frottement de Coulomb.

La modélisation simultanée des conditions unilatérales et du frottement dans le contexte de lois de comportement non élastiques implique un choix entre une formulation en vitesse ou en déplacement. La formulation naturelle en vitesse bien adaptée aux lois de type viscoélastique ou viscoplastique et au frottement s'adapte mal au traitement du contact unilatéral. C'est pourquoi nous utilisons la formulation ci-dessus qui constitue une première approche de traitement numérique des contacts unilatéraux avec frottement en Viscoélasticité et Viscoplasticité qui ne paraissent pas avoir été traités jusqu'à présent.

2.3. Les relations cinématiques

Soit **u** le champ de déplacement, e le champ des tenseurs de déformation et e^o le champ de tenseur de déformation imposée (relation (1)), on écrira sous l'hypothèse des petites déformations :

$$e = Du + e^o \iff e_{ij} = \frac{1}{2}(u_{i,j} + u_{j,i}) + e^o_{ij} \qquad (12)$$

où D est l'opérateur gradient symétrique.

2.4. Les équations d'équilibre.

Soit ϕ_1 des charges volumiques données définies sur Ω, soit ϕ_2 des charges surfaciques données définies sur la partie $\partial_2\Omega$ de la frontière, soit **F** la force de contact inconnue définie sur la partie $\partial_3\Omega$ de la frontière, les équations d'équilibre s'écrivent :

$$
{}^t D\sigma = \phi + {}^t L\, F \iff \begin{cases} \operatorname{div}\sigma = -\phi_1 \text{ dans } \Omega & \sigma_{ij,j} = -\phi_{1i} \\ \sigma n = \phi_2 \text{ sur } \partial_2\Omega & \sigma_{ij}n_j = \phi_{2i} \\ \sigma n = F \text{ sur } \partial_3\Omega & \sigma_{ij}n_j = F_i \end{cases} \qquad (13)
$$

où σ désigne le tenseur des contraintes, $\phi = {}^t\{\phi_1, \phi_2\}$ et ${}^tL$ est un opérateur de relèvement (voir Raous[11]). Des précisions concernant le cadre fonctionnel du problème seront trouvées dans Bouc-Geymonat-Jean-Nayroles[14] et Raous[11]. En particulier, on montre que l'opérateur d'équilibre est le transposé de l'opérateur gradient symétrique pour des produits de dualité convenablement choisis entre l'espace des déplacements et celui des charges d'une part, entre l'espace des déformations et celui des contraintes d'autre part, et pour des choix de structure appropriée pour ces différents espaces fonctionnels.

2.5. Loi de comportement.

Le traitement numérique proposé est adapté à toute loi de comportement présentée sous forme d'équation différentielle linéaire ou non linéaire. Nous traiterons ici essentiellement le cas du modèle viscoélastique de Maxwell (loi étant éventuellement à coefficients variables pour tenir compte des effets de température ou de vieillissement).

<u>Loi III : Viscoélasticité de Maxwell.</u>

$$\sigma = K(e - \xi) \qquad (14)$$

$$\sigma = \eta\dot{\xi} \qquad (15)$$

où ξ est le tenseur des déformations visqueuses. Il s'agit d'un paramètre caché que nous conserverons dans la formulation. Nous nous placerons dans la suite sous l'une des hypothèses simplificatrices "contraintes planes", "déformations planes" ou "problème à symétrie de révolution" : σ, e, ξ seront alors identifiés à des vecteurs de $\mathbb{R}^3$ ou $\mathbb{R}^4$, et K et η seront respectivement la matrice de raideur (caractérisée par le module d'Young E et le module de Poisson ν) et la matrice de viscosité (caractérisée par E, ν et le temps de relaxation τ). Lorsque les effets de température seront significatifs pour le matériau, les différentes caractéristiques E, τ et χ seront fonctions du champ de température, lui-même fonction du temps. Les matrices de raideur et de viscosité seront donc fonctions du temps.

3. FORMULATION VARIATIONNELLE DU PROBLEME

3.1. Les équations du problème.

On montre comme en Elasticité (voir Duvaut-Lions[16]) que le problème (7)(10)(11) conduit à une inéquation quasi variationnelle. Par contre le problème à seuil de glissement fixe g conduit à un problème d'inéquation variationnelle. Il s'agit du modèle de frottement de Tresca:

Loi IV : frottement de Tresca.

$$|\mathbf{F}_T| \leq g \tag{16}$$

$$\text{avec } |\mathbf{F}_T| < g \implies \mathbf{u}_T = 0 \tag{17}$$

$$|\mathbf{F}_T| = g \implies \mathbf{u}_T = -\lambda \mathbf{F}_T \quad \text{avec } \lambda > 0 \tag{18}$$

où g est une fonction définie sur $\partial_3\Omega\times[0,T]$.

Le problème initial posé avec la loi II de frottement va être résolu à l'aide de la loi IV par une suite d'approximations successives sur le seuil de glissement g de la loi de Tresca. Pour une fonction g fixée, nous calculons la solution $(\mathbf{u},\sigma)$ du problème suivant.

Les fonctions ϕ_1, e^o, $\mathbf{u}$, σ, e, sont des fonctions de type suivant:

$$f : (x,t) \longmapsto f(x,t) \qquad x \in \Omega \, , \, t \in [o,T]$$

la fonction ϕ_2 est du type : $(x,t) \longmapsto \phi_2\,(x,t) \quad x \in \partial_2\Omega \, , \, t \in [o,T]$

et la fonction g du type : $(x,t) \longmapsto g(x,t) \quad x \in \partial_3\Omega \, , \, t \in [o,T]$

Problème P1 :

Soit les charges $\phi = {}^t\{\phi_1, \phi_2\}$ et la déformation imposée e^o ; soit une fonction positive donnée g (seuil de glissement) ; trouver le déplacement $\mathbf{u}$, la contrainte σ et la déformation e tels que :

$$e = D\mathbf{u} + e^o \tag{12}$$

$${}^tD\sigma = \phi + {}^tL\,\mathbf{F} \tag{13}$$

$$\sigma = K(e - \xi) \tag{14}$$

$$\sigma = \eta\dot{\xi} \qquad \text{et} \quad \xi(0) = \xi_o \tag{15}$$

$$u_N \leq 0 \tag{4}$$

$$F_N \leq 0 \tag{5}$$

$$u_N . F_N = 0 \tag{6}$$

$$|\mathbf{F}_T| \leq g \tag{16}$$

$$\text{avec si } |\mathbf{F}_T| < g \implies \mathbf{u}_T = 0 \tag{17}$$

$$\text{si } |\mathbf{F}_T| = g \implies \mathbf{u}_T = -\lambda\mathbf{F}_T \quad \text{avec } \lambda > 0 \tag{18}$$

On va calculer F_N sur $\partial_3\Omega$ à partir de cette solution $(\mathbf{u},\sigma)$. Notons S l'application qui associe F_N à g : $S(g) = F_N$. La solution $(\bar{\mathbf{u}},\bar{\sigma})$ du problème réel où les relations (16)(17)(18) du problème P1 sont remplacées par les relations (7)(10)(11) de la loi II, sera le point fixe de l'application

$$g = -\mu S(g) \tag{19}$$

Les conditions théoriques de convergence de ce procédé de point fixe ont été établies dans le cas d'une loi de frottement non locale (voir Duvaut[22]).

3.2. Forme variationnelle.

L'écriture variationnelle du problème P1 peut s'exprimer sous l'une des deux formulations équivalentes suivantes. La première est la recherche de la solution d'une inéquation variationnelle (point de vue des travaux virtuels). La deuxième est un problème aux variations qui consiste à rechercher le minimum d'une fonctionnelle (point de vue énergétique).

Nous choisissons une formulation en déplacement, nous conserverons le paramètre caché ξ (déformation visqueuse ou viscoplastique), et nous éliminerons les variables σ et e qui pourront être déterminées à partir de la solution $\mathbf{u}$ calculée.

On introduit le cône $\mathbf{K}$ des contraintes sur la variable $\mathbf{v}$.

$$\mathbf{K} = \{\mathbf{v} \in H_0^1(\Omega)^2 \mid v_N = L\mathbf{v} \leq 0 \quad \text{sur } \partial_3\Omega\} \tag{20}$$

L'opérateur L associe à $\mathbf{v}$ la composante normale de la trace de $\mathbf{v}$ sur $\partial_3\Omega$.

Problème P2 : Trouver $\mathbf{u} : (x,t) \longmapsto \mathbf{u}(x,t) \in \mathbb{K}$

tel que : $\forall\, \mathbf{v} \in \mathbb{K} \quad (\mathbf{v} : x \longmapsto \mathbf{v}(x) \quad \mathbf{x} \in \Omega),\ \forall t \in [o,T]$

$$a(\mathbf{u}(.,t),\mathbf{v}(.)-\mathbf{u}(.,t))-(\phi_1(.,t),\mathbf{v}(.)-\mathbf{u}(.,t))-(Ke^\circ(.,t),D(\mathbf{v}(.)-\mathbf{u}(.,t)))+$$
$$(K\xi(.,t),D(\mathbf{v}(.)-\mathbf{u}(.,t)))-\int_{\partial_2\Omega}{}^t[\phi_2(x,t)].[\gamma\mathbf{v}(x)-\gamma(\mathbf{u}(x,t))]dx+$$
$$\int_{\partial_3\Omega} g(x,t).|\,\mathbf{v}_T(x)|\,dx-\int_{\partial_3\Omega} g(x,t).|\,\mathbf{u}_T(x,t)|\,dx \geq 0 \tag{21}$$

et $\forall x \in \Omega \quad \forall t \in [o,T]$

$$\dot{\xi}(x,t)=\eta^{-1}K(D\mathbf{u}(x,t)+e^\circ(x,t)-\xi(x,t)) \tag{22}$$

avec $\xi(x,o)=\xi_o(x)$ donné.

où $a(\mathbf{u},\mathbf{v}) = (D\mathbf{u},\ KD\mathbf{v})$ est une forme bilinéaire symétrique (${}^tDKD$ est l'opérateur d'élasticité). $(.,.)$ désigne le produit scalaire ordinaire dans $L^2(\Omega)^3$, c'est-à-dire $(\phi,\mathbf{u}) = \int_\Omega {}^t[\phi].[u]d\omega$, le point désigne le produit scalaire dans $\mathbb{R}^2$, $\gamma\mathbf{v}$ désigne la trace de $\mathbf{v}$ sur la partie $\partial_2\Omega$ de la frontière.

Problème P3 : trouver $\mathbf{u} : (x,t) \longmapsto u(x,t) \in \mathbb{K}\ ,\ x \in \Omega\ ,\ t \in [o,T]$

tel que : $\forall \mathbf{v} : x \longmapsto \mathbf{v}(x) \in \mathbb{K} \quad x \in \Omega$

$$J(\mathbf{u}(.,t),t) \leq J(\mathbf{v}(.),t) \qquad t \in [o,T] \tag{23}$$

et $\quad \forall\, x \in \Omega \quad \forall t \in [o,T]$

$$\dot{\xi}(x,t)=\eta^{-1}\ K(D\mathbf{u}(x,t) + e^\circ(x,t) - \xi(x,t))$$

avec $\xi(x,o) = \xi(x)$ donné,

avec :

$$J(\mathbf{v}(.),t) = \tfrac{1}{2}\, a(\mathbf{v}(.),\mathbf{v}(.))-(\phi_1(.,t),\mathbf{v}(.))-(Ke^\circ(.,t),D\mathbf{v}(.)) \tag{24}$$
$$+(K\xi(.,t),D\mathbf{v}(.))+\int_{\partial_2\Omega}{}^t[\phi_2(x,t)].[\gamma\mathbf{v}(x)]dx+\int_{\partial_3\Omega} g(x,t).|\,\mathbf{v}_T(x)|\,dx$$

3.3. La question de l'existence et unicité des solutions

La question de l'existence et unicité des solutions pour le problème de contact unilatéral avec frottement en Viscoélasticité reste un problème ouvert. Des résultats de Duvaut[22] d'une part et de Demkowiz-Oden[21] d'autre part ont établi ces propriétés pour le problème élastique avec une loi de frottement non locale. Cette loi de frottement non locale est écrite sous la forme d'une convolution avec une fonction f régulière positive dont l'intégrale est égale à l'unité.

$$F_N^* = f(s,.) * F_N(u) \qquad (25)$$ où

$*$ désigne le produit de convolution sur $\partial_3\Omega$ et F_N^* est une régularisée de F_N. La démonstration de Duvaut s'appuie sur la méthode de point fixe évoquée précédemment qui sera utilisée pour la résolution numérique du problème tandis que la démonstration de Demkowicz-Oden s'appuie sur une théorie de perturbation de l'inéquation variationnelle.

3.4. Extension à la viscoplasticité.

Le problème viscoélastique a été posé sous la forme d'une inéquation variationnelle couplée à une équation différentielle. L'ayant posé sous cette forme, nous allons pouvoir utiliser, pour sa résolution, des méthodes standards d'analyse numérique. L'intérêt de cette présentation est la possibilité d'étendre les schémas numériques proposés à toute loi de comportement formulée sous forme d'équation différentielle. C'est ainsi que nous avons traité dans le passé de problème de contact unilatéral sans frottement en Viscoplasticité (Raous[12,13]). Ces schémas s'étendent au cas avec frottement étudié ici.

Il s'agit d'une forme simplifée des lois viscoplastiques proposées par Lemaître[19] et Chaboche[20]. C'est une loi de Norton-Hoff à laquelle est ajoutée un modèle d'écrouissage. Elle s'écrit :

$$\dot{\mathbf{X}}' = c(a\dot{\xi} - \mathbf{X}'\dot{p}) \tag{26}$$

$$\dot{\xi} = \left(\frac{J}{\eta}\right)^n \frac{1}{J} \left(K' \, (e-\xi) - \mathbf{X}'\right) \tag{27}$$

où $\mathbf{X}$ est un champ de pseudo-contrainte caractérisant l'écrouissage, ξ la déformation plastique, $\dot{p}$ est le deuxième invariant du tenseur des vitesses de déformation, $\tilde{J}$ le deuxième invariant du déviateur $(\sigma' - \mathbf{X}')$ où σ' et $\mathbf{X}'$ désignent les déviateurs des tenseurs σ et $\mathbf{X}$. K' est la matrice d'élasticité concernant les déviateurs de contrainte et de déformation, η est le paramètre caractérisant la viscosité, a, c et n sont des coefficients.

On montre que le problème se présente encore sous la forme d'une inéquation variationnelle couplée à une équation différentielle qui est maintenant fortement non linéaire. L'écriture du problème de minimum associé conduit à la formulation suivante.

<u>Problème P4</u> : trouver $\mathbf{u}(x,t) \in \mathbb{K}$ tel que :

$$J(\mathbf{u}(.,t),t) \leq J(\mathbf{v}(.),t) \qquad t \in [0,T] \tag{28}$$

$$\dot{\xi}(x,t) = \left(\frac{J(x,t)}{\eta}\right)^n \frac{1}{J(x,t)} \left(K'(D\mathbf{u}(x,t) + e^{o}(x,t) - \xi(x,t)) - \mathbf{X}'(x,t)\right) \tag{26}$$

$$\dot{\mathbf{X}}'(x,t = c(a\,\dot{\xi}(x,t) - \mathbf{X}'(x,t).\dot{p}(x,t)) \tag{27}$$

et $\xi(x,0) = \xi_o(x)$ et $\mathbf{X}'(x,0) = \mathbf{X}'_0(x)$ donnés

où la fonctionnelle $J(\mathbf{v}(.),t)$ est la même que dans le cas viscoélastique.

4. SCHEMA DE DISCRETISATION ET ALGORITHME

Nous utilisons une discrétisation de type éléments finis sur l'espace. Dans ce contexte, la loi de comportement est traitée à l'aide de méthodes d'intégration numérique explicites, semi implicites ou implicites suivant les cas.

4.1. Semi discrétisation temporelle.

Nous avons utilisé dans le passé des méthodes de type explicite : Euler, Heunn (méthode de Runge Kutta d'ordre 2)...(voir Raous[11,12,13] et Notin[15]). Ces méthodes présentent l'avantage de la simplicité de mise en oeuvre numérique et l'inconvénient d'une condition de stabilité concernant le choix du pas d'intégration. Dans un travail d'analyse numérique (Geymonat-Raous[23]), nous avons exhibé cette condition de stabilité dans le cas de la Viscoélasticité sans contact unilatéral. Cette condition est directement liée au temps de relaxation τ caractérisant la viscosité du matériau. Elle s'écrit :

$$\Delta t < 2\tau \tag{28}$$

Si cette condition de stabilité est souvent acceptable, elle est en Viscoplasticité beaucoup plus contraignante et le calcul correct de la solution est très coûteux car une discrétisation temporelle fine est nécessaire (Raous[13], Notin[15]).

Nous avons récemment développé avec P. Chabrand et C. Licht (Raous,Chabrand,Licht[24]) des méthodes implicites ou semi implicites de type θ-méthodes. Elles s'écrivent sous la forme suivante :

$$\dot{y} = f(t,y) \tag{29}$$

$$y_{k+1} = y_K + \Delta t[(1 - \theta)\, f(t_k,y_k) + \theta f(t_{k+1},y_{k+1})] \tag{30}$$

On retrouve la méthode d'Euler explicite pour $\theta = 0$, la méthode d'Euler implicite pour $\theta = 1$, la méthode de Crank Nickolson pour $\theta = \frac{1}{2}$. On établit dans le cadre de la Viscoélasticité sans contact unilatéral que la méthode est inconditionnellement stable si $\theta \geq \frac{1}{2}$ et que la condition de stabilité lorsque $\theta < \frac{1}{2}$ s'écrit :

$$\Delta t < \frac{2\tau}{1 - 2\theta} \tag{31}$$

La méthode de Cranck Nickolson s'est avérée très efficace lors de la mise en oeuvre. Afin de simplifier l'écriture, nous écrirons ici le problème approché par la méthode d'Euler implicite appliquée à l'équation (24).

$$\xi_{k+1} - \xi_k = \Delta t\ \eta^{-1}K(Du_{k+1} + e^{\circ}_{k+1} - \dot{\xi}_{k+1}) \tag{32}$$

La discrétisation temporelle est supposée régulière ; Δt est le pas de discrétisation et l'indice k concerne le temps $k\Delta t$ $(k = 1, M)$.

On pose :

$$K^* = K(\mathbb{I} + \Delta t\ \eta^{-1}K)^{-1} \tag{33}$$

où $\mathbb{I}$ est la matrice unité.

En remarquant que $(\mathbb{I}+\Delta t\ \eta^{-1}K)^{-1}\Delta t\ \eta^{-1}K = \mathbb{I} - (\mathbb{I}+ \Delta t\ \eta^{-1}K)^{-1}$ et en utilisant les propriétés de symétrie de η^{-1} et K, on montre que la relation (32) peut s'écrire :

$$\xi_{k+1} = (\mathbb{I} - \Delta t\ \eta^{-1}K^*)\ \xi_k + \Delta t\eta^{-1}K^*(Du_{k+1} + e^{\circ}_{k+1}) \tag{34}$$

$$\xi_{k+1} = \xi_k + \Delta t\ \eta^{-1}K^*(Du_{k+1} + e^{\circ}_{k+1} - \xi_k) \tag{35}$$

Grâce au changement de variable (33), nous retrouvons ici une formulation analogue à celle résultant de l'utilisation de la méthode d'Euler explicite : la matrice K est remplacée par la matrice K^* et il y a un glissement d'indice pour certains termes. Cette formulation permet de passer très facilement de la méthode explicite à la méthode implicite lors de la mise en oeuvre. On trouvera dans la thèse de P. Chabrand[25] les changements de variable qui conduisent à la formulation (35) dans le cas général où $\theta \in [0,1]$. En viscoélasticité grâce à la linéarité de la forme différentielle, le coût de calcul à chaque instant t_{k+1} à l'aide d'une θ-méthode est le même qu'avec une méthode explicite. C'est donc une θ-méthode que nous utiliserons : la valeur $\theta = \frac{1}{2}$ qui correspond à la méthode de Crank Nickolson est la plus souvent utilisée.

Par ailleurs, nous écrivons le problème de minimum (26) à l'instant t_{k+1}, dans lequel nous introduisons la relation (35) exprimant ξ_{k+1}. Nous obtenons alors le problème de minimum suivant :

<u>Problème P5</u> : trouver $\mathbf{u}_{k+1}(x) \in \mathbb{K}$ tel que

$$J_{k+1}(\mathbf{u}_{k+1}(.)) \leq J_{k+1}(\mathbf{v}(.)) \qquad \forall \mathbf{v}(x) \in \mathbb{K}$$

avec :

$$J_{k+1}(\mathbf{v}(.)) = \frac{1}{2}\, a^*(\mathbf{v}(.),\mathbf{v}(.)) - (\boldsymbol{\phi}_{1_{k+1}}(.),\mathbf{v}(.)) - (K^* e^{\circ}_{k+1}(.),D\mathbf{v}(.))$$
$$+(K^*\xi_k(.),D\mathbf{v}(.))-\int_{\partial_2\Omega} {}^t[\boldsymbol{\phi}_{2_{k+1}}(x)]\,[\gamma\mathbf{v}(x)]dx+\int_{\partial_3\Omega} g_{k+1}(x).|\mathbf{v}_T(x)|\,dx \qquad (36)$$

avec :

$$a^*(\mathbf{v}(.),\mathbf{v}(.)) = (K^*\, D\mathbf{v}(.),D\mathbf{v}(.)).$$

Le schéma numérique se précise donc : nous avons à résoudre à chaque instant t_{k+1} le problème de minimum P5 où ξ_k est donné par la relation récurente (35) écrite à l'instant t_k. Ce schéma est complété par la condition initiale $\xi(0) = \xi_0$.

Remarque 1- la méthode d'Euler implicite est évidemment inconditionnellement stable. Toutefois, il faut remarquer que la forme bilinéaire $a^*(\mathbf{v}(.),\mathbf{v}(.))$ intervenant dans (36) est directement liée à K^* défini en (33). On montre que la constante de coercivité de cette forme bilinéaire est directement proportionnelle au rapport $\tau/\Delta t$. Aussi si on choisit un pas de temps très grand devant le temps de relaxation τ, la coercivité devient faible et le conditionnement du problème devient plus mauvais.

Remarque 2- suivant la régularité temporelle des fonctions données ϕ_1,ϕ_2, e^o, nous sommes amenés à utiliser un pas de temps Δt variable ; dans ce cas, il faut reconstruire à chaque pas la matrice qui est associée à la forme discrète de la forme $a^*(\mathbf{v}(.),\mathbf{v}(.))$ (voir paragraphe suivant). Dans ce cas, une méthode explicite est souvent préférable à condition que le temps de relaxation τ ne soit pas trop petit devant T la période d'observation du phénomène. Ceci est d'ailleurs raisonnable d'un point de vue physique : si on considère un matériau de temps de relaxation τ soumis à un échelon de déformation, le phénomène de relaxation qui s'ensuit est pratiquement achevé au bout d'un temps de l'ordre de 10τ et il n'est pas naturel de calculer la solution sur un intervalle [0,T] où T serait égal à $100\times\tau$ ou $1000\times\tau$.

En conclusion, les méthodes implicites seront essentiellement utilisées en Viscoélasticité dans le cas de pas de temps constant ou de modèle à caractéristiques mécaniques variables (fonctions de la température par exemple) où il faut alors reconstruire la matrice associée à $a^*(\mathbf{v}(.),\mathbf{v}(.))$ à chaque pas.

Remarque 3- cas de la Viscoplasticité. La forte non linéarité des équations différentielles (16) (17) ne permet pas une formulation simple analogue au schéma discret (35). Un processus itératif est alors nécessaire pour traiter l'équation différentielle : des tests ont été effectués avec la méthode de prédiction correction (Notin[15]) ; des itérations de point fixe sont également envisagées. Dans l'attente d'améliorations futures, les calculs en Viscoplasticité ont été effectués à l'aide de méthodes explicites du 1er ordre ou de 2e ordre. M. Djaoua[26] propose un schéma implicite pour la loi de Norton Hoff s'appuyant sur une discrétisation en contrainte. Ceci s'adapte mal au traitement unilatéral envisagé ici.

4.2- Discrétisation spatiale.

Soit une discrétisation par éléments finis P1 (triangles à 3 noeuds) : q désigne le nombre total de noeuds. On notera U_{2q} le sous-ensemble des vecteurs colonnes de $\mathbb{R}^{2q}$ et K_{2q} le sous-ensemble convexe fermé de U_{2q} défini par :

$$K_{2q} = \{\bar{v} \in U_{2q} \ / \ \bar{\bar{v}}_N \leq 0\} \tag{37}$$

Si I désigne le nombre de noeuds de la frontière concernée par le contact unilatéral, on désigne avec une barre dessus les vecteurs de $\mathbb{R}^{2q}$ et avec deux barres dessus les vecteurs de $\mathbb{R}^I$.

Le problème de minimum (36) peut alors s'écrire :

Problème P6- trouver $\bar{u}_{k+1} \in K_{2q}$ tel que $\forall k = 1,...M$

$$J^h_{k+1}(\bar{u}_{k+1}) \leq J^h_{k+1}(\bar{v}) \quad \forall\, \bar{v} \in K_{2q}$$

où $J^h_{k+1}(\bar{v})$ est la forme discrète associée à $J_{k+1}(\mathbf{v})$

$$J^h_{k+1}(\bar{v}) = \frac{1}{2}\,{}^t\bar{v}A^*\bar{v} - {}^t\bar{F}_{k+1}\bar{v} + {}^t\bar{\bar{g}}_{k+1}\,B|\bar{\bar{v}}_T| - {}^t[C(\bar{\varepsilon}^o_{k+1} - \tilde{\xi}_k)].\bar{v} \qquad (38)$$

Notations : A^* est la matrice de rang 2q associée à l'opérateur ${}^tDK^*D$. $\bar{F}$ est le vecteur associé aux charges volumiques ϕ_1 dans Ω et aux charges surfaciques ϕ_2 sur $\partial_2\Omega$. B est une matrice diagonale à coefficients positifs : elle résulte de l'intégration numérique du terme $\int_{\partial_3\Omega} g_{k+1}(x).|\mathbf{v}_T(x)|\,dx$ par la formule des trapèzes. Cette approximation est grossière non pas en raison de l'ordre de la méthode d'intégration choisie mais par le fait de ne conserver que les valeurs aux noeuds de la fonction $|v_T|$, ce qui est important pour la cohérence de la formulation par éléments finis. Une approximation plus fine de ce terme est très pénalisante d'un point de vue numérique, mais est toutefois envisagée. $\tilde{\xi}_k$, $\bar{\varepsilon}^o_{k+1}$ sont des vecteurs de $\mathbb{R}^{3p}$ où p est le nombre d'éléments de la discrétisation. Nous nous plaçons ici sous l'hypothèse des contraintes planes ou des déformations planes. C est la forme discrète de l'opérateur ${}^tDK^*$. C'est une matrice creuse de dimension (2q,3p) composée en fait de blocs indépendants. Elle sera manipulée seulement sous la forme de ces blocs élémentaires.

Ainsi, le problème P6 est analogue à un problème d'élasticité associé à l'opérateur ${}^tDK^*D$, auquel s'ajoute un terme de charge supplémentaire faisant intervenir ξ_k déterminé à l'instant précédent.

4.3- Résolution du problème à seuil de glissement fixé.

Il s'agit de la résolution à chaque instant t_{k+1}, du problème P6 où ξ_k est donné par la relation (35). Nous utilisons un algorithme de surrelaxation avec projection de la solution à chaque itération sur le cône $\mathbb{K}$ (méthode de Cryer-Christopherson). La fonctionnelle J est ici non différentiable à cause du terme en valeur absolue qui ne concerne que les composantes du déplacement des nœuds de contact. Lors du traitement de ces composantes, on considére les deux éventualités possibles concernant le signe de la solution et on se ramène ainsi à la minimisation d'une fonctionnelle quadratique. On trouvera dans Glowinski-Lions-Trémolières[6] les résultats de convergence de cette méthode adaptée à cette forme particulière de la fonctionnelle J. Dans les exemples considérés le coefficient de relaxation était de l'ordre de 1,9 et le nombre d'itérations nécessaires a toujours été inférieur à 2q. Nous pouvons déduire du champ de déplacement calculé, le champ de déformation et celui des contraintes. Nous calculerons la composante F_N sur $\partial_3\Omega$ en utilisant directement le résidu vectoriel issu de la méthode de Cryer où on retrouve la projection des composantes de la force de réaction dans la base des fonctions de base associées aux éléments finis retenus.

4.4- Résolution du problème avec la loi de frottement (7)(10)(11).

Il s'agit du problème général où le seuil de glissement est fonction de la contrainte normale. Nous utilisons le procédé de point fixe décrit au paragraphe 3.1. Pour un seuil de glissement fixé $g_{k+1}^{(\alpha)}$, nous calculons la solution du problème P6. Nous calculons la force normale $F_{Nk+1}^{(\alpha)}$ et nous posons $g_{k+1}^{(\alpha+1)} = -\mu F_{Nk+1}^{(\alpha)}$ dans une nouvelle résolution du problème P6. La solution cherchée est obtenue lorsque $g_{k+1}^{(\alpha)}$ et $g_{k+1}^{(\alpha+1)}$ sont suffisamment voisins.

On remarquera que nous avons une bonne condition initiale pour $g_{k+1}^{(\alpha)}$ en utilisant la contrainte normale calculée à l'instant t_k précédent.

Cet algorithme est en fait amélioré par l'utilisation d'un procédé diagonal entre les itérations de relaxation et les itérations de point fixe. Au cours des itérations de surrelaxation on actualise la valeur au seuil de glissement $g_{k+1}^{(\alpha)}$ en utilisant les valeurs intermédiaires de F_N. Les résultats de convergence théorique de ce procédé n'ont pas été établis à notre connaissance et ce problème mathématique reste ouvert. On notera que les difficultés numériques dues à l'introduction du frottement ne se cumulent pas arithmétiquement avec celles provenant de la non linéarité de contact et de la loi de comportement grâce au choix de ce procédé particulièrement efficace.

4.5- Algorithme.

La résolution du problème complet peut se résumer de la façon suivante.

a- soit la condition initiale $\tilde{\xi}_k = \tilde{\xi}_0$, k = 0,

b- soit une condition initiale sur le seuil de glissement $\bar{\bar{g}}_{k+1}^{(\alpha)} = \bar{\bar{g}}_0$, avec $\alpha = 0$, c- minimisation de la fonctionnelle $J^h_{k+1}(\overline{v})$ sur le cône K_{2q} pour $\tilde{\xi}_k$ connu ; $\overline{\overline{g}}^{(\alpha)}_{k+1}$ est actualisé plusieurs fois au cours des itérations en calculant $\overline{\overline{F}}^{(\alpha)}_{Nk+1}$ et en posant $\overline{\overline{g}}^{(\alpha+1)}_{k+1} = -\mu\, \overline{\overline{F}}^{(\alpha)}_{Nk+1}$.

On en déduit une solution $\overline{u}_{k+1}$, $\overline{\overline{F}}_{Nk+1}$;

d- test sur la convergence dans la recherche du point fixe. On pose $\overline{\overline{g}}^{(\alpha+1)}_{k+1} = -\mu\overline{\overline{F}}_{N\ k+1}$. Si $|\overline{\overline{g}}^{(\alpha+1)}_{k+1} - \overline{\overline{g}}^{(\alpha)}_{k+1}| > \varepsilon$, on pose $\overline{\overline{g}}_o = \overline{\overline{g}}^{(\alpha+1)}_{k+1}$, $\alpha = 0$, GO TO (b) ;

e- connaissant $\overline{u}_{k+1}$ et $\tilde{\xi}_k$, on calcule $\tilde{\xi}_{k+1}$ à l'aide de la relation (35) ;

si $k < M$ (M est la borne temporelle supérieure), on pose $k = k+1$, $\overline{\overline{g}}_o = -\mu\, \overline{\overline{F}}_{Nk}$, $\alpha = 0$, GO TO (b),

si $k = M$ - Fin.

Remarque : le schéma implicite décrit ne nécessite pas le calcul de $\overline{u}_o$, aussi le schéma est complété par une résolution pour k=0 d'un problème analogue au problème P6 correspondant en fait à un problème élastique avec une contrainte imposée $C\xi_o$. On montre que cela revient à remplacer dans (38) les indices k et k+1 par zéro et A^* par A qui est la forme discrète de l'opérateur ${}^tDKD$.

5.APPLICATIONS NUMERIQUES

Nous donnons ici deux exemples illustrant l'efficacité de la méthode en ce qui concerne un matériau fissuré sous charges thermiques complexes, et un problème de contact avec frottement où on notera la bonne détermination des forces de réaction suivant la nature du contact : décollement, glissement, blocage. Afin de limiter le volume de cette présentation, la discussion détaillée de la mise en oeuvre numérique ainsi que des exemples d'application en Mécanique plus nombreux seront présentés dans un article plus technique en cours de rédaction.

5.1. Solide viscoélastique fissuré

Il s'agit d'un modèle simplifié du comportement d'une aube de turbine de réacteur. La plaque fissurée considérée est soumise à un gradient thermique évolutif périodique au cours du temps. L'élasticité et la viscosité du matériau dépendent fortement de la température. Au cours du temps, la fissure est tantôt ouverte tantôt fermée. Le contact est supposé sans frottement. On trouvera sur la figure 1 les courbes isobares du deuxième invariant du déviateur des contraintes à un instant où la fissure est fermée : on notera la continuité des lignes isobares sur la ligne de fissuration qui met en évidence la bonne détermination des forces d'action-réaction. Il s'agit dans cet exemple d'un problème non symétrique par rapport à l'axe de fissuration, c'est à dire un cas où le convexe $\mathbb{K}$ n'est pas le demi plan positif, et où l'opération de projection est plus compliquée.

5.2. Mise en forme

Il s'agit de l'écoulement forcé d'un matériau <u>viscoélastique</u> sur une matrice. Le solide est un barreau allongé d'acier à haute température de section rectangulaire (étude de la demi-section) posé sur un obstacle en créneau et soumis à une pression sur sa face supérieure. Cet exemple académique simple montre les qualités de la méthode en mettant en évidence l'évolution au cours du temps des zones de contact et de leur nature.

On trouvera sur les figures 2 et 3, l'état déformé (amplifié 5 fois) du solide, ainsi que le tracé des forces de contact au temps t=0 (déformation élastique initiale instantanée) et au temps $t = 3\tau$ (où τ est le temps de relaxation) : il y a eu fluage et le matériau s'est écoulé sur l'obstacle. La zone de contact s'est étendue et la nature du contact (blocage ou glissement) a évolué. Le coefficient de frottement μ est égal à 0.2. On précise dans le tableau 1 les différentes zones du contact et leur nature, pour les résultats présentés sur les figures 2 et 3, c'est-à-dire pour t = 0 et $t = 3\tau$ (après fluage).

zone / temps	non contact $F_N = \lvert F_T \rvert = 0$	contact avec glissement $\lvert F_T \rvert = -\mu \ F_N$	contact avec avec blocage $\lvert F_T \rvert < -\mu \ F_N$
t = 0	zone [c d]	zone [b c] et zone [d e]	zone [a b] et zone [e f]
$t = 3\tau$	zone [c'e']	zone [b'c']	zone [e'f']

Tableau 1 : évolution du contact durant le phénomène de fluage.

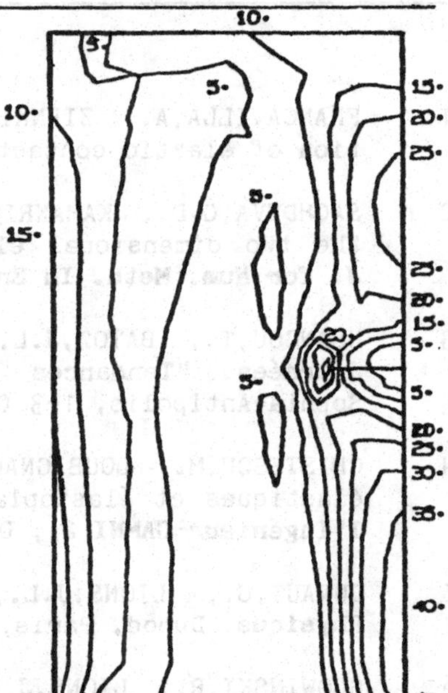

FIGURE 1

Deuxième invariant de déviateur des contraintes dans un solide viscoélastique fissuré soumis à un chargement thermique.

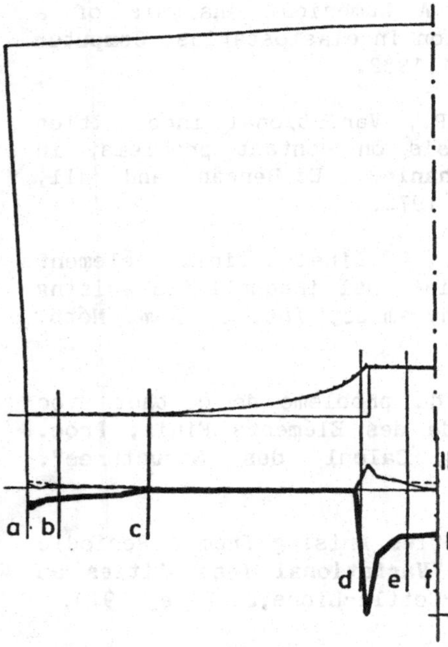

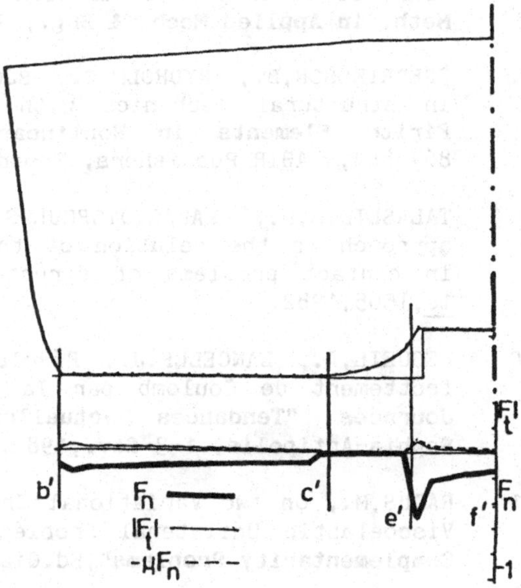

FIGURE 2

Déformée grossie 5 fois et forces de contact à t = 0.

FIGURE 3

Déformée grossie 5 fois et forces de contact à t = 3τ.

REFERENCES

1 FRANCAVILLA,A., ZIENKIEWICZ,O.C., A note on numerical computation of elastic contact problem, **Num. Meth. in Eng.**, 9, 913,1975.

2 SACHDEVA,G.D., KAMAKRISHNAN,C.V., A finite element solution for the two dimensional elastic contact problems with friction, **Int. J. for Num. Meth. In Eng.**, 17, 1257,1981.

3 NSANGOU,T., BATOZ,J.L., TOUZOT,G., Eléments de contact, Proc. Journées **"Tendances actuelles en calcul des structures"**, Sophia-Antipolis, 1-3 fév. 1982.

4 CRISTESCU,M., LOUBIGNAC,G., Résolution de contact entre solides élastiques et élastoplastiques, in **Méth. Num. dans les Sciences de l'Ingénieur-GAMNI 2.**, Dunod, Paris, 1980.

5 DUVAUT,G., LIONS,J.L., **Les inéquations en Mécanique et en Physique**, Dunod, Paris, 1972.

6 GLOWINSKI,R., LIONS,J.L., TREMOLIERES,R., **Analyse Numérique des Inéquations Variationnelles**, Dunod, Paris, 1976.

7 CAMPOS,L.T., ODEN,J.T., KIKUCHI,N., A Numerical Analysis of a class of contact problems with friction in elastostatics, **Computer Meth. in Applied Mech. & Eng.**, 34, 821,1982.

8 FREDRIKSSON,B., RYDHOLM,G., SJÖBLOM,P., Variational inequalities in structural mechanics with emphasis on contact problems, in **Finite Elements in Nonlinear Mechanics**, Ed.Bergan and all, 863-884, TABIR Publishers, Trondheim, 1978.

9 TALASLIDIS,D., PANAGIOTOPOULOS,P.D., A linear finite element approach to the solution of the variational inequalities arising in contact problems of structural dynamics, **Int. J. Num. Méth.** 18,1505,1982.

10 DEGUEIL,A., LANCELLE,J., Résolution du problème de contact avec frottement de Coulomb par la méthode des Eléments Finis, **Proc. Journées "Tendances actuelles en Calcul des Structures"**, Sophia-Antipolis, 1-3 fév.,1982.

11 RAOUS,M., On two Variational Inequalities Arising from a Periodic Viscoelastic Unilateral Problem, In **"Variational Inequalities an Complementarity Problems"**,Ed.Gianessi-Cottle-Lions,J. Wiley,1979.

12 RAOUS,M., Comportement d'un solide fissuré sous charges alternatives en viscoélasticité non linéaire avec écrouissage, **Journal de Mécanique Théorique et Appliquée**, N° spécial, 125,1982.

13 RAOUS,M., Fissuration sous contraintes alternées en viscoélasticité et viscoplasticité,**Thèse**,Univ.de Provence,Marseille,1980.

14 BOUC,R., GEYMONAT,G., JEAN,M., NAYROLES,B., Cauchy and Periodic Unilateral Problems for Aging Linear Viscoelastic Materials, **Journal of Math. Analysis & Applications**, 61, n°1,7,1977.

15 NOTIN,C., Résolution numérique d'un problème périodique de viscoplasticité non linéaire,**Thèse Doct.Ing.**,Univ.de Provence,1979.

16 DUVAUT,G., LIONS,J.L., **Les inéquations en Mécanique et en Physique**, Dunod, Paris, 1972.

17 MOREAU,J.J., Convex analysis and friction, in **Trands in Applications of Pure Mathematics to Mechanics**, Vol. II, H. Zorski Ed., Pitman Publ.,London, 263,1979.

18 MOREAU,J.J., Convexité et frottement. Univ.de Montréal, Dép. d'Informatique, Publication n°32,1970.

19 LEMAITRE,J. Sur la détermination des lois de comportement des matériaux élastoviscoplastiques, **Thèse**, Orsay (Publ.ONERA N° 135).

20 CHABOCHE,J.L., Description thermodynamique phénoménologique de la viscoplasticité cyclique avec endommagement, **Thèse**, Paris VI,1978.

21 DEMKOWICZ,L., ODEN,J.T., On some existence an uniqueness results in contact problems with non local friction, **Nonlinear Analysis, Theory, Methods and Applications**, 6,10, 1075,1982.

22 DUVAUT,G., Equilibre d'un solide élastique avec contact unilatéral et frottement de Coulomb,**C.R.A.S.**,Paris,Série A,t.290,263,1980.

23 GEYMONAT,G., RAOUS,M., Méthodes d'éléments finis en viscoélasticité périodique, **Lect. Notes in Mathematics**, 606, Springer,1977.

24 RAOUS,M., CHABRAND,P., LICHT,C., Méthodes implicites et semi implicites en viscoélasticité (à paraître).

25 CHABRAND,P., Grandes variations de température pour un corps viscoélastique, **Thèse 3e Cycle**, Univ. de Provence, Marseille,1984.

26 DJAOUA,M., Analyse mathématique et numérique de quelques problèmes en mécanique de la rupture, **Thèse**, Univ. Paris VI, 1983.

ON DELAMINATION IN PLATES: A UNILATERAL CONTACT APPROACH

J.N. Reddy
Department of Engineering, Science and Mathematics
Virginia Polytechnic Institute and State University
Blacksburg

A. Grimaldi
Dipartimento di Ingegneria Civile Edile
II University of Rome

ABSTRACT: *The unilateral contact of plates on elastic foundation is considered. Both vanishing and finite tensile contact (or adhesion) strength between the plate and the foundation are considered. The approach is used to study delamination due to transverse loads in two layer plates. The relationship with the brittle fracture mechanics approach is also discussed. Numerical results are presented to validate the present approach.*

SOMMARIO: *Si esamina il problema di contatto di piastre su una fondazione elastica, costituita da molle caratterizzate da resistenza a trazione nulla o finita. Il secondo caso è usato per modellare l'adesione tra gli strati di piastre composite, ed è applicato allo studio della delaminazione di un pannello a due strati in presenza di forze trasversali. La formulazione proposta viene confrontata con il modello di frattura fragile, e vengono infine presentati alcuni risultati numerici.*

1. INTRODUCTION

Plates laminated of orthotropic layers are increasingly used in aerospace, civil, and mechanical engineering structures.

If the edges of laminates are not secured properly, or if defects in bonding are present, delamination of layers can take place under applied loads. The present study deals with the modelling of delamination growth using the unilateral contact approach [1, 3].

The shear deformation theory of layered composites plates [4] is used to model the bending response, and the contact between layers is modeled by springs with both vanishing and finite tensile strength. The case of finite strength can be used to model [5] delamination between two elastic plates. A finite element formulation developed by the authors [6] is applied to investigate the propagation of debonding of two-layer plates with specified defect in the lamination. The approach is valitated by comparing the present finite-element results with analytical solutions for isotropic plates. Further the relation between the present approach and the fracture mechanics approach [7-9] for study of delamination in plates is discussed.

2. THEORY AND FORMULATION

Consider an elastic plate resting on an elastic foundation (see Fig. 1), and subjected to transversely distributed load, q. Let w denote the transverse deflection and p the reaction of the foundation. The foundation is modeled by a set of elastic springs. The displacement field in the plate is assumed to be of the form

$$u_1(x, y, z) = z\,\psi_x(x, y)$$
$$u_2(x, y, z) = z\psi_y(x, y) \qquad (1)$$
$$u_3(x, y, z) = w(x, y)$$

where u_i denotes the displacement in the x_i-direction ($x_1 = x$, $x_2 = y$, $x_3 = z$), and ψ_x and ψ_y are the bending slopes. The theory accounts for the transverse shear strains (i.e., it is assumed that plane sections remain plane after deformation – non necessarily normal to the midplane). The effect of shear deformation on the plate response is quite significant, especially in composite-material plates. The assumed displacement field not only accounts for the transverse shear effects, it also leads to lower order equations that facilitate the development of C°-elements (see [4]).

Neglecting the body moments and surface shearing forces, we write the equations of equilibrium in the presence of applied transverse forces, q, as

$$Q_{x,x} + Q_{y,y} = q(x, y) - P(w)$$

$$M_{x,x} + M_{xy,y} - Q_x = 0 \tag{2}$$

$$M_{xy,x} + M_{y,y} - Q_y = 0$$

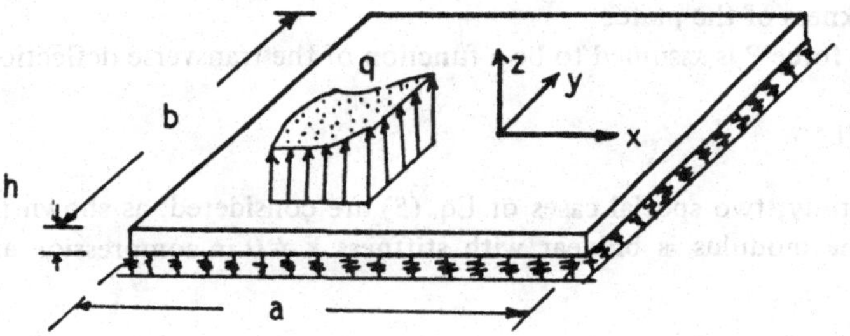

Figure 1 - The geometry of a rectangular plate on an elastic foundation (modulus k) and subjected to distributed load (q).

where Q_i and M_i are the shear stress and moment resultants given by

$$Q_x = D_{44}\left(\frac{\partial w}{\partial x} + \psi_x\right) \quad , \quad Q_y = D_{55}\left(\frac{\partial w}{\partial y} + \psi_y\right)$$

$$M_x = D_{11}\frac{\partial \psi_x}{\partial x} + D_{12}\frac{\partial \psi_y}{\partial y} \ ,$$

$$M_y = D_{12}\frac{\partial \psi_x}{\partial x} + D_{22}\frac{\partial \psi_y}{\partial y}$$

$$M_{xy} = D_{33}\left(\frac{\partial \psi_x}{\partial y} + \frac{\partial \psi_y}{\partial x}\right) \tag{3}$$

where D_{ij} are material stiffnesses of an orthotropic plate

$$D_{11} = \frac{E_1 h^3}{12(1-\nu_{12}\nu_{21})}, \quad D_{22} = \frac{E_2 D_{11}}{E_1}$$

$$D_{12} = \nu_{12} D_{22}, \quad \nu_{21} = \frac{\nu_{12} E_2}{E_1}, \quad D_{33} = \frac{G_{12} h^3}{12}$$

$$D_{44} = D_{55} = G_{13} h \tag{4}$$

and h is the thickness of the plate.

The springs force P is assumed to be a function of the transverse deflection:

$$P(w) = k(w) \cdot w . \tag{5}$$

In the present study, two special cases of Eq. (5) are considered, as shown in Fig. 2. In the first case, the modulus is bilinear with stiffness $k \neq 0$ in compression and $k = 0$ in

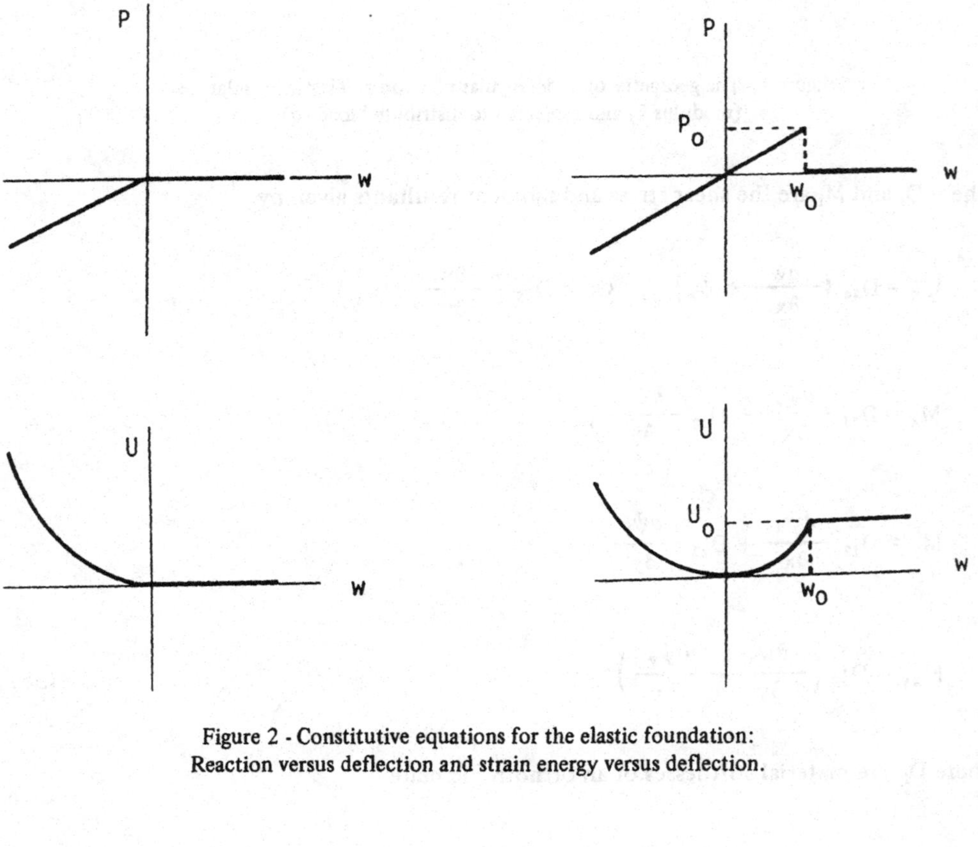

Figure 2 - Constitutive equations for the elastic foundation:
Reaction versus deflection and strain energy versus deflection.

tension. In the second case, the modulus is equal to k in compression and tension upto some specified displacement, w_0, and zero for $w > w_0$. Figure 2 also contains qualitative plots of the associated strain energy versus deflection. In the first case, the strain energy function is convex, and therefore one can prove the existence and uniqueness of solutions. In the second case, the convexity of the strain energy is not present, and therefore, uniqueness cannot be proved in a general case.

Finite-element model

The finite-element model of Eq. (2) can be derived by assuming interpolation of the form

$$w = \sum_{i=1}^{n} w_i \psi_i , \quad \psi_x = \sum_{i=1}^{n} X_i \psi_i , \quad \psi_y = \sum_{i=1}^{n} Y_i \psi_i \tag{6}$$

over an element Ω^e. Substituting Eq. (6) into the variational form associated with Eq.(2) we obtain:

$$\begin{bmatrix} [K^{11}] & [K^{12}] & [K^{13}] \\ & [K^{22}] & [K^{23}] \\ \text{symm.} & & [K^{33}] \end{bmatrix} \begin{Bmatrix} \{W\} \\ \{X\} \\ \{Y\} \end{Bmatrix} = \begin{Bmatrix} \{F^1\} \\ \{F^2\} \\ \{F^3\} \end{Bmatrix} \tag{7}$$

where

$$[K^{11}] = [\bar{K}^{11}] + [\hat{K}^{11}] ,$$

$$\bar{K}^{11}_{ij} = \int_{\Omega^e} (D_{44} \frac{\partial \psi_i}{\partial x} \frac{\partial \psi_j}{\partial x} + D_{55} \frac{\partial \psi_i}{\partial y} \frac{\partial \psi_j}{\partial y}) \, dx\, dy$$

$$\hat{K}^{11}_{ij} = \int_{\Omega^e} k(w) \psi_i \psi_j \, dx\, dy$$

$$K^{12}_{ij} = \int_{\Omega^e} D_{44} \frac{\partial \psi_i}{\partial x} \psi_j dx\,dy , \quad K^{13}_{ij} = \int_{\Omega^e} D_{55} \frac{\partial \psi_i}{\partial y} \psi_j dx\,dy$$

$$K^{22}_{ij} = \int_{\Omega^e} (D_{11} \frac{\partial \psi_i}{\partial x} \frac{\partial \psi_j}{\partial x} + D_{33} \frac{\partial \psi_i}{\partial y} \frac{\partial \psi_j}{\partial y} + D_{44} \psi_i \psi_j) \, dx\,dy$$

$$K_{ij}^{23} = \int_{\Omega^e} \left(D_{12} \frac{\partial \psi_i}{\partial x} \frac{\partial \psi_j}{\partial y} + D_{33} \frac{\partial \psi_i}{\partial y} \frac{\partial \psi_j}{\partial x}\right) dx\,dy\,,$$

$$K_{ij}^{33} = \int_{\Omega^e} \left(D_{33} \frac{\partial \psi_i}{\partial x} \frac{\partial \psi_j}{\partial x} + D_{22} \frac{\partial \psi_i}{\partial y} \frac{\partial \psi_j}{\partial y} + D_{55}\, \psi_i \psi_j\right) dx\,dy$$

$$F_i^1 = \int_{\Gamma^e} q\,\psi_i\, dx dy + \int_{\Gamma^e} Q_n\, \psi_i ds\,, \qquad Q_n = Q_x n_x + Q_y n_y$$

$$F_i^2 = \int_{\Gamma^e} M_n\, \psi_i ds\,, \qquad M_n = M_x n_x + M_{xy} n_y$$

$$F_i^3 = \int_{\Gamma^e} M_{ns}\, \psi_i ds\,, \qquad M_{ns} = M_{xy} n_x + M_y n_y\,. \tag{8}$$

The element stiffness matrix in (8) is of the order 3n by 3n, where n is the number of nodes per element. For example, when the four-node rectangular element is used, then the element stiffness matrix is of the order 12 by 12.

The element equations in (7) can be assembled, boundary conditions can be imposed, and the equations can be solved in the usual manner [11]. The main difference between the present model and ordinary plate model is in $[\hat{K}^{11}]$, which depends, in general, on the sign of $(w - w_0)$. The computational scheme involves an iterative procedure to update $[\hat{K}^{11}]$ according to the sign of $(w - w_0)$ at Gauss points.

Discussion of numerical results

Numerical results of four example problems are discussed here:

1. Bilateral contact problem of a long plate subjected to line load along the short edge.
2. Unilateral contact problem of a long plate subjected to line load parallel to the short edge at the center.
3. A square plate subjected to a point load at the center.
4. Delamination of a two-layer plate subjected to line load along the short edge.

Example 1. - The classical problem of a beam on an elastic foundation (see Fig. 3) is modeled with the plate bending element discussed in this paper. A uniform of 1 X 8 four and nine 'node rectangular elements is used, and the results for transverse deflection, $\bar{w} = WD/qa^3$, are presented in Table 1. The parameters used are

$$\frac{a}{b} = 8\,, \qquad \frac{a}{h} = 16\,, \qquad a/\lambda = 3.14\,, \qquad \nu = 0.1\,, \qquad \lambda = \left(\frac{4D}{k}\right)^{1/4}\,,$$

$$D = \frac{Eh^3}{12(1-\nu^2)}$$

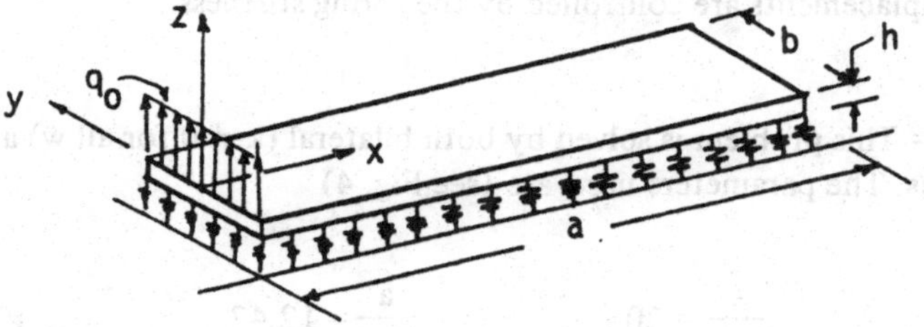

Figure 3 - A rectangular plate on an elastic foundation and subjected to an edge line load at x = 0.

Table 1 - Dimensionless deflections $\frac{wD}{q_o a^3} \times 10^3$ along beam axis

(F = full integration, R = reduced integration).

x/a	Analytical solution	FEM 4 nodes-F	FEM 4 nodes-R	FEM 9 nodes-F	FEM 9 nodes-R
0.00	16.15	13.13	15.97	16.27	0.15
0.125	10.05	9.25	10.05	10.10	−5.95
0.250	5.20	5.81	5.20	5.22	−10.83
0.375	1.90	3.16	1.87	1.90	−14.14
0.500	0.001	1.16	−0.08	0.002	−16.06
0.625	−0.86	−0.29	−1.02	−0.96	−16.98
0.750	−1.08	−1.37	−1.36	−1.31	−17.32
0.875	−0.95	−2.25	−1.40	−1.38	−17.40
1.00	−0.70	−3.08	−1.36	−1.38	−17.40

The effect of integration on the accuracy is also investigated. In all cases, the bending and spring stiffness terms are integrated using the standard Gauss rule (2 X 2 for linear and 3 X 3 for quadratic elements) and full or reduced integration (i.e., one point less than the usual in each direction) is used for shear energy terms (i.e., terms involving

D_{33} and D_{44}). The finite element results obtained by using the linear element with reduced integration and the quadratic element with full integration agree well with theoretical solution of a beam on elastic foundation [12]. However, the quadratic element with reduced integration is found to be unstable (i.e., gives completely erroneous results). This latter observation contradicts that found in the analysis of ordinary plates (i.e., plates without elastic foundation [4]). It should be pointed out that in the present example no essential (or geometric) boundary conditions are specified and therefore the rigid body displacements are controlled by the spring stiffness.

Example 2. - This problem is solved by both bilateral ($k \neq 0$ for all w) and unilateral contact conditions. The parameters used are (see Fig. 4)

$$\frac{a}{b} = 10\,, \qquad \frac{a}{h} = 20\,, \qquad \frac{a}{\lambda} = 12.42\,, \qquad \nu = 0.1\,.$$

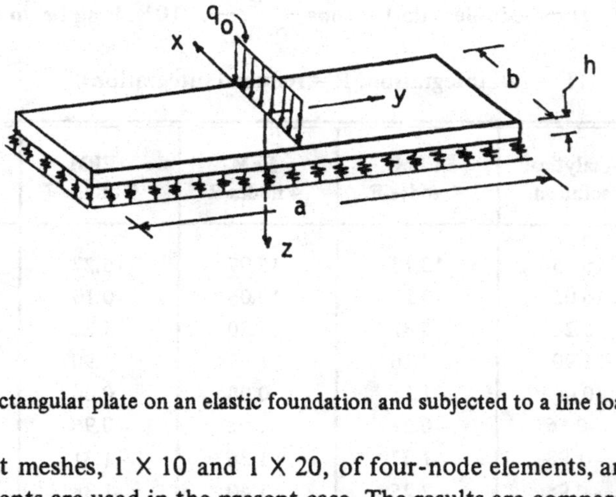

Figure 4 - A rectangular plate on an elastic foundation and subjected to a line load at y = 0.

Two different meshes, 1 X 10 and 1 X 20, of four-node elements, and 1 X 10 mesh of nine-node elements are used in the present case. The results are compared (for bilateral case) with the analytical solution in Table 2. Once again it is found that the nine-node element with reduced integration is unstable (hence the results are not presented here). The finite element results for the nondimensionalized deflection are in good agreement with the analytical solution. Also, note that the effect of the transverse shear deformation is to increase the nondimensionalized deflection with decreasing values of the side-to-thickness ratio.

Table 2. - Dimensionless deflections ($\frac{wD}{q_0 a^3} \times 10^6$) along the beam axis.

y/a	Analytical Solution	Bilateral FEM [1] (1 X 10) 4 nodes-R	Bilateral FEM [1] (1 X 10) 9 nodes-F	Bilateral FEM [1] (1 X 20) 4 nodes-R	Bilateral FEM [2] (1 X 20) 4 nodes-R	Unilateral FEM [1] (1 X 20) 4 nodes-R	Unilateral FEM [1] (1 X 10) 4 nodes-F
0.0	65.21	68.18	70.39	71.65	64.71	78.60	76.07
0.1	23.90	26.96	22.69	22.66	24.95	16.10	16.54
0.2	–0.98	–4.73	–1.08	–1.99	–1.96	–61.24	–54.66
0.3	–2.17	–4.35	–1.98	–2.32	–2.72	–138.90	–123.73
0.4	–0.32	–0.105	–0.23	–0.25	–0.26	–216.22	–.193.81
0.5	0.12	1.27	0.44	0.55	0.67	–293.56	–263.89

(1) a/h = 20
(2) a/h = 100

Example 3. - Due to the biaxial symmetry, only one quarter of the plate is modeled (see Fig. 5), using a 8 X 8 uniform mesh of linear elements or 4 X 4 mesh of quadratic elements. In both cases the full integration is found to give better results (see Table 3). Due to lack of other results in the literature, no assessment of accuracy can be made.

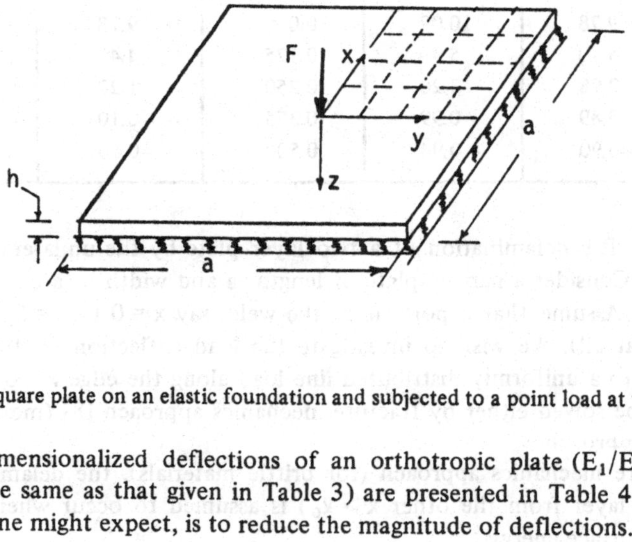

Figure 5 - Square plate on an elastic foundation and subjected to a point load at the center.

The nondimensionalized deflections of an orthotropic plate ($E_1/E_2 = 10$, and all other data is the same as that given in Table 3) are presented in Table 4. The effect of orthotropy, as one might expect, is to reduce the magnitude of deflections.

Table 3 - Dimensionless deflections $\frac{wD}{Fa^2} \times 10^4$

(isotropic plate, a/h = 20, a/λ = 7, ν = 0,1).

x/a (y = 0)	FEM (8 X 8) - 4 nodes-F bilateral	FEM (4 X 4) - 9 nodes-F bilateral	FEM (8 X 8) - 4 nodes-F unilateral	FEM (4 X 4) - 9 nodes-F unilateral
0.0	12.62	14.36	13.26	15.23
0.125	6.23	6.75	6.66	7.27
0.250	1.97	1.85	1,53	1.09
0.375	0.034	0.030	–2.11	–3.22
0.500	–0.95	–0.62	–5.21	–6.86

Table 4 - Dimensionless deflections $\frac{wD}{Fa^2} \times 10^4$

(anisotropic plate, $E_1 = 10E_2$, a/h = 20, a/λ = 7, ν = 0.1).

x/a (y = 0)	FEM (8 X 8) - 4 nodes-F bilateral	FEM (8 X 8) - 4 nodes-F unilateral	y/a (x = 0)	FEM (8 X 8) - 4 nodes-F bilateral	FEM (8 X 8) - 4 nodes-F unilateral
0.0	9.78	10.09	0.0	9.78	10.09
0.125	5.70	5.99	0.125	4.41	4.50
0.250	2.98	3.20	0.250	1.27	5.60
0.375	0.89	0.99	0.375	–0.10	– 2.02
0.500	–0.90	– 0.94	0.500	–0.80	– 3.61

Example 4. - The delamination of a two-layer plate by the unilateral contact approach is studied. Consider a narrow plate of length a and width b ($a/b \gg 1$) is welded to a similar plate. Assume that a portion of the weld, say $x = 0$ to $x = \ell_0$, is defective (i.e., not welded at all). We wish to investigate the load-deflection relation when the plate is subjected to a uniformly distributed line load along the edge $x = 0$ (see Fig. 6). The problem can be solved either by fracture mechanics approach [8] (mode I fracture) or by the present approach.

In the fracture mechanics approach (for brittle materials), the delamination (i.e., separation of one layer from the other $x \geq \ell_0$) is assumed to occur when the energy release equals the surface energy :

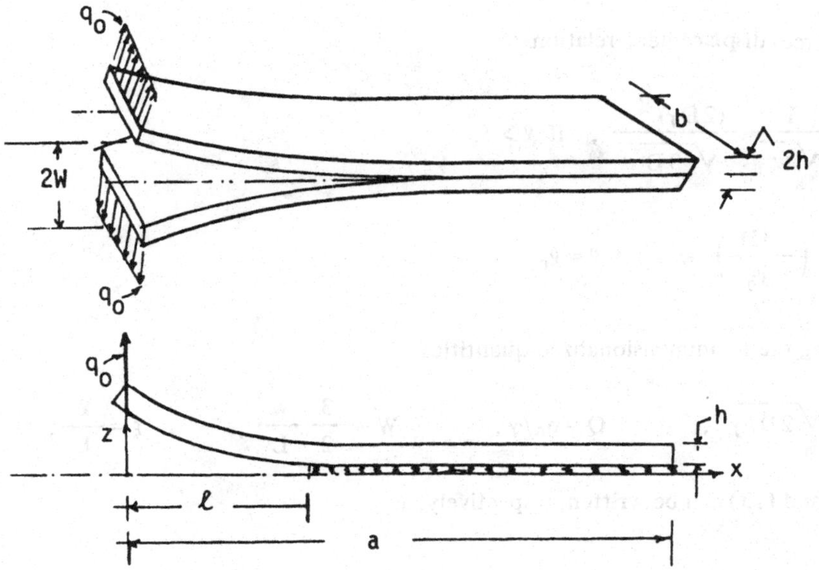

Figure 6 - Delamination of a two-layer rectangular plate under transverse edge load (ℓ denotes initial delamination).

$$G = \gamma b \tag{9}$$

Here G denotes the energy release per unit length of opening and γ is the surface energy per unit area (of opening). The energy release is given by

$$G = -\frac{dE}{d\ell} \tag{10}$$

where E is the total potential energy, and $d\ell$ is the infinitesimal length along the opening. In the present case, we have (see Fig. 6)

$$E = -\frac{1}{2}\,(\text{force})\,(\text{displ.}) = -\frac{1}{2}\,(q_0 b)\left(\frac{q_0 b \ell^3}{3Db}\right) = -\frac{1}{6}\,\frac{q_0^2 \ell^3 b}{D} \tag{11}$$

where D is the flexural rigidity of an isotropic plate (see Eq. (4)).

From Eqs. (9)-(11), we obtain

$$q_0 \ell = \sqrt{2D\gamma}\ , \tag{12}$$

and the force-displacement relation

$$q_0 \frac{1}{\sqrt{w}} \frac{(2D\gamma)^{3/4}}{\sqrt{3D}}\ , \quad \text{if } \ell > \ell_0$$

$$q_0 = \left(\frac{3D}{\ell_0^3}\right) w\ , \quad \text{if } \ell = \ell_0 \tag{13}$$

Introducing the nondimensionalized quantities

$$L = \sqrt{2D/\gamma}\ , \qquad Q = q_0/\gamma\ , \qquad W = \frac{3}{2}\frac{w}{L}\ , \qquad \xi = \frac{\ell}{L}\ , \tag{14}$$

Eqs. (12) and (13) can be written, respectively, as

$$Q = \frac{1}{\xi} \tag{15}$$

$$Q = \begin{cases} \dfrac{1}{\sqrt{W}}\ , & \ell > \ell_0 \\[2ex] \dfrac{W}{\xi_0^3}\ , & \ell = \ell_0 \end{cases} \tag{16}$$

Plots of Q versus W and ξ_0 versus W are shown in Fig. 7.

In the unilateral approach, the interface of the two plates is modeled by an elastic foundation with zero stiffness between $x = 0$ and $x = \ell_0$. The strain energy is assumed not to exceed $U_0 = 1/2\, k w_0^2$ in tension. Here w_0 denotes the maximum elongation of the springs. When the strain energy of the foundation equals U_0, the delamination of the layers occurs. The relation between U_0 in the present approach and γ in the fracture mechanics approach is given by $U_0 = \gamma$. Thus, U_0 is a property of the material. For a given material, if $k \to \infty$, we must have $w_0 \to 0$. The solution of the unilateral problem is given by superposing the solutions of the plate segments between $x = 0, \ell$, and $x = \ell$, a:

$$Q = \frac{1}{\xi + \lambda/L} \tag{17}$$

$$W = \frac{\xi^2}{1 + \frac{\lambda}{L}\frac{1}{\xi}} + 3\left(\frac{\lambda}{L}\right)^2 \left(1 + \frac{L}{\lambda}\xi\right), \quad \ell > \ell_0$$

$$W = Q\xi_0^3 \left[1 + 3\left(\frac{\lambda}{L\xi_0}\right)^2 \left(2 + \frac{\lambda}{L\xi_0} + \frac{L\xi_0}{\lambda}\right)\right], \quad \ell = \ell_0 \tag{18}$$

where λ is defined in Example 1. Note that as $k \to \infty$, we have $\lambda \to 0$, and from (17) and (18) we obtain (15) and (16). Thus, the unilateral solution yields, in the limit as $k \to \infty$, the fracture mechanics solution to the equation. The unilateral contact approach is more convenient to analyze the two-dimensional delamination. In fact the fracture mechanics approach is more complicated when the delamination is two-dimensional.

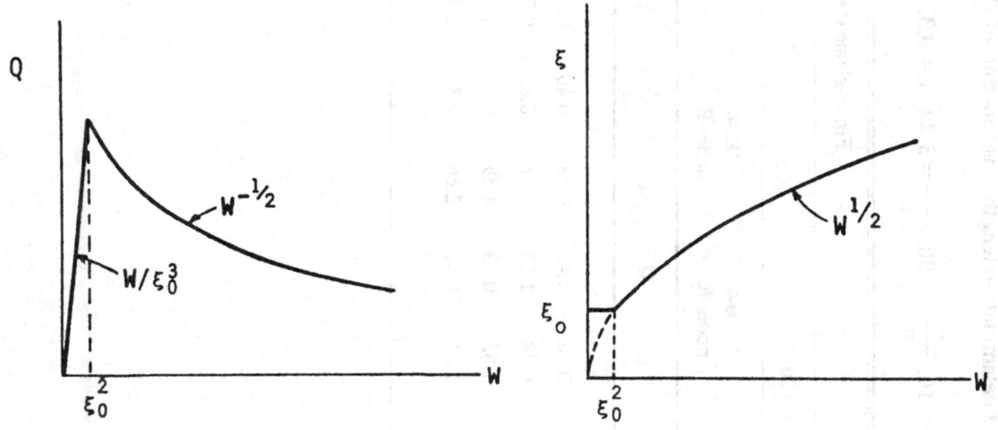

Figure 7 - Nondimensionalized load-deflection and tip deflection versus delamination curves for the problem in Fig. 6.

The finite-element approximation based on the unilateral contact approach involves the use of the finite element model in (7), with w_0 given by (from the limit energy)

$$w_0 = \sqrt{\frac{2U_0}{k}} = \sqrt{\frac{2\gamma}{k}} . \tag{19}$$

Note also that the element matrices corresponding to the no contact (i.e., defective welding) region have zero contribution from K_{ij}^{11}.

Table 5 - Load Q and delamination length ξ as function of the opening displacement W

$$\left(\frac{a}{b} = 10,\ \frac{a}{h} = 100,\ \frac{a}{L} = 5.22,\ \nu = 0.3,\ \xi_0 = 0.78\right).$$

Analytical solution (eqs. (15) - (16))			Finite-element solution											
			a/λ = 20						a/λ = 40					
			1 X 20-4 nodes-R		1X40-4 nodes-R		1X80-4 nodes-R		1X20-4 nodes-R		1X40-4 nodes-R		1X80-4 nodes-R	
W	ξ	Q	ξ	Q	ξ	Q	ξ	Q	ξ	Q	ξ	Q	ξ	Q
0.5	0.78	1.04	0.78	0.44	0.78	0.41	0.78	0.40	0.78	0.70	0.78	0.63	0.78	0.60
1	1	1	0.78	0.88	0.78	1.23	0.78	0.81	0.78	1.40	0.78	1.53	0.78	1.21
5	2.24	0.45	1.57	0.80	1.83	0.55	1.89	0.52	1.57	1.27	1.67	0.84	1.83	0.64
10	3.16	0.32	2.35	0.61	2.61	0.41	2.66	0.37	2.09	0.88	2.35	0.61	2.61	0.46

In the finite-element analysis (20, 40 and 80 linear elements along the x-direction and one element along the y-direction are used), the deflections at the nodes on the line $x = 0$ are specified to be W (equivalent to specifying uniformly distributed load there). For different values of W, the corresponding applied load Q and the delamination length ℓ are computed, and the results are presented in Table 5. The finite element results are compared with the analytical solution in (15) and (16) for two different values of k. The difference between the finite-element solution and the analytical solution is attributed to the slow convergence for a fixed λ. Further note that, the analytical solution is based on the assumption that $\lambda = 0$. For smaller values of λ, the finite element mesh should be refined to capture the small wave lengths in the solution.

Some numerical results obtained with a more refined mesh and an application of the present approach to two-dimensional propagation of delamination is given in [10].

CONCLUSIONS

The unilateral contact approach is employed to model delamination and determine the growth of the delamination. A shear deformation plate theory is used to develop a finite element model for symmetrically laminated plates which have defective bonding at the midplane. In the case of one-dimensional delamination, the fracture mechanics approach and the present approach are related and the results are compared. Application of the approach to laminated composite plates is straight forward.

REFERENCES

1. Signorini, A., Sopra alcune questioni di elastostatica, in *Atti della Soc. Ital. per il Progresso delle Scienze*, 1933.

2. Fichera, G., Boundary value problems of elasticity with unilateral constraints, in *Encyclopedia of Physics*, Vol. VIa/2, Springer-Verlag, Berlin, 1972.

3. Duvaut, G., and Lions, J.L., *Inequalities in Mechanics and Physics*, Springer-Verlag, Berlin, 1976.

4. Reddy, J.N., A Penalty plate-bending element for the analysis of laminated anisotropic composite plates, *Int. J. Numer. Meth. Engng.*, Vol. 15, pp. 1187-1206, 1980.

5. Fremond, A., Adhesion de Solids Elastiques, Lecture delivered at *Euromech Symposium on Unilateral Problems in Mechanics*, CISM-Udine, Italy, May, 1982.

6. Grimaldi, A., and Reddy, J.N., On delamination in plates: a unilateral contact approach, in Report No. VPI-E-82-83, Virginia Polytechnic Institute and State University, Blacksburg, VA 24061 (1982).

7. Kassir, M.K., and Sih, G.C., *Three-Dimensional Crack Problems*, Noordhoff, Leyden, The Netherlands, 1975.

8. Early, J.W., Compression induced delamination in a unidirectional graphite/epoxy composite, Report MM-372A-81-14, Texas A&M University, College Station, Texas, December 1981.

9. Burridge, R., and Keller, J.B., Peeling, slipping, and cracking - some one-dimensional free-boundary problems in mechanics, in *SIAM Review*, Vol. 20, No. 1, 1978.

10. Ascione, L., and Bruno, D., On the delamination problem of two-layer plates, This meeting (1983).

11. Reddy, J.N., *An Introduction to the Finite Element Method*, McGraw-Hill, New York, 1983

12. Selvadurai, A.P.S., *Elastic Analysis of Soil-Foundation Interaction*, Elsvier, Amsterdam, 1979.

ELASTOSTATICS OF STRUCTURES WITH UNILATERAL CONDITIONS ON STRESS AND DISPLACEMENT FIELDS

G. Romano
Istituto di Scienza delle Costruzioni
University of Naples

M. Romano
Istituto di Scienza delle Costruzioni
University of Catania

Summary

A general analysis of the elastostatic problem for structures with unilateral conditions on the stress distributions and on the displacement fields is developed.

The unilateral external constraints are assumed to define a convex conical manifold of admissible displacement fields.

Linear elastic materials with a convex constitutive condition on the stress are considered.

Anelastic strain are assumed to develop according to a convex coniugacy rule which generalizes the standard normality rule of perfect plasticity.

A complete theoretical scheme of the constitutive properties of the material is developed on this basis.

The existence of a convex and differentiable elastic strain energy is proved and the expression of the complementary elastic energy is given.

Two general results yielding the equilibrium and the geometric compatibility conditions under external and internal convex constraints are invoked to formulate the basic variational principles governing the elastostatic problem.

The minimum principles for the potential and the complementary energy functionals and the related error bounding techniques, extending the classical results in linear elasticity, are established.

It is shown that, under suitable regularity assumptions, namely the additivity of the involved subdifferentials, the stress formulation yields the existence of the solution for the elastostatic problem.

1. INTRODUCTION

In structural mechanics the analysis of models in which unilateral conditions are imposed by external and internal constraints is of the greatest interest.

As a matter of fact, in the reality, geometric external constraints on the displacement fields are often of a unilateral type.

Moreover constitutive models with convex conditions on the internal stress distributions may simulate in an effective way the behaviour of a number of interesting structural materials.

Among these we way mention the "no tension" materials and the "no compression" membrane type models.

Rock mechanics problems and the analysis of concrete and masonry structures are important fields of application in structural engineering.

We present here a general theory of structural models in which the unilateral external constraints are assumed to define a convex conical manifold of admissible displacement fields.

The material properties are characterized by a convex yield condition on the stress distributions.

An elastic constitutive model is considered by splitting the total strain response into the sum of a linear elastic and an anelastic part.

For the latter a convex conjugacy rule, which generalizes the standard normality rule of perfect plasticity, is assumed.

In this respect it is worth noting that the violation of such a rule can be shown to lead to generally non consistent physical models (G.Romano, M.Romano.[1]).

The constitutive scheme is analyzed in the general context of convex analysis and the existence of a convex and differentiable elastic strain energy is proved.

The properties of the proximity operators, introduced by J.J. Moreau.[2], are the basic tools in this investigation.

The displacement and the stress formulations of the elastostatic problem are developed on the basis of general results concerning the equilibrium and the geometric compatibility under external and internal convex constraints (G.Romano, M.Romano.[3]).

The minimum principles for the potential and the complementary energy are shown to be necessary and sufficient conditions for the solution of the elastostatic problem and the related error bounding techniques are established.

The existence of the solution of the elastostatic problem is proved under the assumption that the subdifferentials of the indicator functions of the convex sets of the admissible and of the equilibrated stress distributions do have the additivity property.

The proof is founded upon a previous result, due to one of the authors (G.Romano.[8]), establishing the sufficiency of the principle of the minimum complementary energy, for linear elastic materials under unilateral external constraints.

2. GENERALITIES

We shall develop the analysis of the structural model in the geometrical context of the small displacements theory in which non linear geometric effects are neglected and hence velocity fields and displacements from a reference configuration can be identified.

The linear spaces of displacement and strain fields will be respectively denoted by V and W and the dual spaces of external forces and of internal stress distributions will be denoted by V' and W'.

The duality pairings representing the external and the internal virtual works will both be denoted by the symbol $\langle \cdot , \cdot \rangle$.

The deformation operator : $T : V \rightarrow W$

mapping the displacement fields into the corresponding strain fields, and the dual equilibrium operator : $T' : W' \rightarrow V'$

mapping the internal stress distributions into the corresponding external force distributions, are related by the virtual work identity:

$$\langle \sigma, Tv \rangle = \langle T'\sigma, v \rangle \qquad \text{for any} \qquad v \in V \ , \quad \sigma \in W'$$

The following orthogonality relations will be assumed to hold :

$$R(T) = N(T')^{\perp} ,$$

$$R(T') = N(T)^{\perp} ,$$

where R and N denote the range and the null space, respectively.

When endowing the spaces V and W with a Hilbert space topology, the above orthogonality relations can be proved under the assumption that the range R(T) of T is closed.

3. GEOMETRIC CONSTRAINTS

We consider the structure to be subject to external unilateral constraints such that the set of the admissible displacements is a convex conical manifold : $C = w_o + C_o \quad , \quad w_o \in V$,

where C_o is a closed convex cone with vertex at the origin. Denoting by L the largest affine manifold included in C, the displacements $w \in L$ will correspond to maximally constrained configurations (m.c.c.).

The mechanical definition of perfect constraints states that the admissible reactions at a configuration $u \in C$ must perform a non-negative virtual work for any variation of displacement compatible with the unilateral external constraints.

In geometric terms, this means that the admissible reactions r must belong to the inward cone to C at u, that is :

$$- r \in \theta\chi_C(u) \ , \tag{3.1}$$

where the symbol θ denotes the subdifferential operator and χ is the indicator function of C. The condition (3.1) is then equivalent to :

$$\langle r, v - u \rangle \geq o \quad \text{for any} \quad v \in C \ .$$

We remark that at an m.c.c. :

$$- \theta\ \chi_C(w) = C_o^+ ,$$

with C_o^+ the positive polar cone of C_o.

4. CONSTITUTIVE PROPERTIES

We shall analyze a constitutive model for materials in which the admissible stress distributions are required to belong to a closed convex set.

The material response is assumed to be the sum of a linear elastic strain field and of an anelastic one which satisfies the standard normality rule.

The theory will be developed in the general context of convex analysis and the basic properties of the constitutive model will be derived by a suitable extension of some ideas and results due to J.J.Moreau.[2].

Let us consider a pair of conjugate convex functions f and g, defined on the strain and on the stress space, respectively.

By assuming that f and g are the pointwise suprema of their affine minorants, the conjugacy relation can be written as :

$$f\,(\delta) = \sup \{ \langle\tau,\delta\rangle - g(\tau) \;/\; \tau \in W' \} \qquad \delta \in W, \tag{4.1}$$

$$g\,(\sigma) = \sup \{ \langle\sigma,\eta\rangle - f(\eta) \;/\; \eta \in W \} \qquad \sigma \in W' \,. \tag{4.2}$$

We recall that σ and δ are conjugate with respect to f and g if one of the following equivalent relations holds :

i) $f(\delta) + g(\sigma) = \langle\sigma,\delta\rangle$,

ii) $\sigma \in \partial f(\delta)$ i.e. $f(\eta) - f(\delta) \geq \langle\sigma,\eta - \delta\rangle$ for any $\eta \in W$,

iii) $\delta \in \partial g(\sigma)$ i.e. $g(\tau) - g(\sigma) \geq \langle\tau - \sigma,\delta\rangle$ for any $\tau \in W'$.

Denoting by $S : W \to W'$ the elastic stiffness operator assumed to be linear, positive definite and symmetric, and by $A = S^{-1}$ the elastic compliance operator, the material response is assumed to be :

$$\varepsilon = A\sigma + \delta \quad \text{or} \quad \sigma = S(\varepsilon - \delta) \tag{4.3}$$

where the anelastic strain δ satisfies a generalized normality rule :

$$\delta \in \partial g(\sigma) \tag{4.4}$$

A suitable extension of Moreau's definition of the proximity operator (J.J.Moreau.[2]) allows to derive the general properties of the constitutive model.

To this end let us define the following proximity operators[(1)] :

a) $\sigma = \text{prox}_{gA}(S\varepsilon)$ is the solution of the minimum problem :

$$\min \{\tfrac{1}{2} \, \|S\varepsilon - \tau\|_A^2 + g(\tau) \; / \; \tau \in W'\} ,$$

(1) It can be proved that the minimum problems below admit an unique solution (J.J.Moreau.[2]).

i.e., σ satisfies

$$\tfrac{1}{2} \| S\varepsilon - \sigma\|_A^2 + g(\sigma) = \xi(S\varepsilon) \ , \tag{4.5}$$

b) $\delta = \text{prox}_{fS}(\varepsilon)$ is the solution of the minimum problem :

$$\min \{ \tfrac{1}{2} \| \varepsilon - \eta\|_S^2 + f(\eta) \ / \ \eta \in W \}$$

i.e. δ satisfies

$$\tfrac{1}{2} \| \varepsilon - \delta \|_S^2 + f(\delta) = \phi(\varepsilon) \tag{4.6}$$

where $\| \cdot \|_S$ and $\| \cdot \|_A$ denote the norm in the energy of S and A respectively and ξ and ϕ are the functionals that associate to ε the value of the minima (4.5) and (4.6)

We remark that the proximity operator is in fact a generalization of the orthogonal projector and reduces to it when the involved convex function is the indicator of a closed convex set.

By virtue of a classical result of J.L.Lions.[4], (C.Baiocchi, A.Capelo.[5]), the minimum problems (4.5) and (4.6) are equivalent to the variational inequalities :

$$g(\tau) - g(\sigma) \geq <\tau - \sigma, \varepsilon - A\sigma> \quad \text{for any} \quad \tau \in W', \tag{4.7}$$

$$f(\eta) - f(\delta) \geq S(\varepsilon - \delta), \eta - \delta> \quad \text{for any} \quad \eta \in W \ , \tag{4.8}$$

which, by the definition of the subdifferential operator, can be written as :

$$\delta = \varepsilon - A\sigma \in \theta \ g(\sigma) \ ,$$

$$\sigma = S(\varepsilon - \delta) \in \theta\, f(\delta).$$

The additive decomposition of the total strain field in the elastic part $A\sigma$ and the anelastic part δ, which satisfies the generalized normality rule (4.4), is thus uniquely defined by the proximity operators a) and b).

The functionals $\phi(\varepsilon)$ and $\xi(S\varepsilon)$, defined by (4.5) and (4.6), result to be convex and differentiable and their gradients are the proximity operators :

$$\sigma = \text{grad}\ \phi(\varepsilon) = \text{prox}_{gA}(S\varepsilon), \tag{4.9}$$

$$\delta = \text{grad}\ \xi(S\varepsilon) = \text{prox}_{fS}(\varepsilon). \tag{4.10}$$

Since from (4.9) σ is the gradient of ϕ with respect to ε, the potential $\phi(\varepsilon)$ has the meaning of the elastic strain energy of the material.

We shall give hereafter an explicit proof of (4.9) which provides some inequalities useful in the sequel.

To this end we first observe that, for any pair of strain fields ε and $\varepsilon_\circ$, denoting by δ, $\delta_\circ$ the associated anelastic strains, the following identity holds :

$$\phi(\varepsilon) - \phi(\varepsilon_\circ) = \phi(\varepsilon) + \phi(\varepsilon_\circ) - 2\phi(\varepsilon_\circ) =$$

$$= \phi(\varepsilon) + \phi(\varepsilon_\circ) - \|\ \varepsilon_\circ - \delta_\circ \|_S^2 - 2f(\delta_\circ) =$$

$$= \phi(\varepsilon) + \phi(\varepsilon_\circ) - 2f(\delta_\circ) +$$

$$+ \langle S(\varepsilon_\circ - \delta_\circ), (\varepsilon - \varepsilon_\circ) - (\delta - \delta_\circ) - (\varepsilon - \delta)\rangle .$$

Now, from (4.6), we have :

$$\phi(\varepsilon) + \phi(\varepsilon_\circ) - 2f(\delta_\circ) = \tfrac{1}{2} \| (\varepsilon - \delta) - (\varepsilon_\circ - \delta_\circ) \|_S^2 +$$
$$+ \langle S(\varepsilon_\circ - \delta_\circ), \varepsilon - \delta\rangle + f(\delta) - f(\delta_\circ),$$

and then, setting :

$$\sigma = S(\varepsilon - \delta) \quad \text{and} \quad \sigma_\circ = S(\varepsilon_\circ - \delta_\circ),$$

we get :

$$\phi(\varepsilon) - \phi(\varepsilon_\circ) = \tfrac{1}{2} \| \sigma - \sigma_\circ \|_A^2 - \langle \sigma_\circ, \delta - \delta_\circ\rangle +$$
$$+ f(\delta) - f(\delta_\circ) + \langle \sigma_\circ, \varepsilon - \varepsilon_\circ \rangle \tag{4.11}$$

From (4.11) and the inequality (4.8) finally it follows that :

$$\phi(\varepsilon) - \phi(\varepsilon_\circ) \geq \tfrac{1}{2} \| \sigma - \sigma_\circ \|_A^2 + \langle \sigma_\circ, \varepsilon - \varepsilon_\circ \rangle , \tag{4.12}$$

and , a fortiori :

$$\phi(\varepsilon) - \phi(\varepsilon_\circ) \geq \langle \sigma_\circ, \varepsilon - \varepsilon_\circ \rangle , \tag{4.13}$$

and, interchanging the roles of ε and $\varepsilon_\circ$:

$$\phi(\varepsilon) - \phi(\varepsilon_\circ) \leq \langle \sigma, \varepsilon - \varepsilon_\circ\rangle , \tag{4.14}$$

From (4.13) and (4.14) we get :

$$0 \leq \phi(\varepsilon) - \phi(\varepsilon_\circ) - \langle \sigma_\circ, \varepsilon - \varepsilon_\circ\rangle \leq \langle \sigma - \sigma_\circ, \varepsilon - \varepsilon_\circ \rangle \leq$$
$$\leq \| \varepsilon - \varepsilon_\circ \|_S^2 \tag{4.15}$$

where the last inequality follows from the non expansion property of the proximity operators (J.J.Moreau.[2]).

From the inequalities (4.15) we see that : $\sigma_\circ = \text{grad}\, \phi(\varepsilon_\circ)$.

Then by (4.13) we get the convexity of ϕ.

The convex conjugate functional of ϕ is defined by :

$$\psi(\sigma) = \sup \{ \langle\sigma,\eta\rangle - \phi(\eta) \;/\; \eta \in W \}.$$

The concave and differentiable functional : $\langle\sigma,\eta\rangle - \phi(\eta)$

attains its maximum at $\eta = \varepsilon = A\sigma + \delta$ since its gradient at this point is : $\sigma - \text{grad}\, \phi(\varepsilon) = 0$.

The explicit expression of the functional ψ is thus given by :

$$\psi(\sigma) = \langle\sigma,\varepsilon\rangle - \phi(\varepsilon) = \| \sigma \|_A^2 + \langle\sigma,\delta\rangle - \tfrac{1}{2}\| \sigma \|_A^2 - f(\delta) =$$

$$= \tfrac{1}{2}\| \sigma \|_A^2 + g(\sigma) \; . \tag{4.16}$$

By the conjugacy relation and (4.16) we get :

$$\varepsilon \in \theta\psi(\sigma) = A\sigma + \theta\, g(\sigma), \tag{4.17}$$

whence we infer that $\psi(\sigma)$ is the complementary elastic energy of the material. We remark that, while ϕ is differentiable, ψ is differentiable if and only if the function g is.

Let us now consider the special constitutive model in which the stress distributions are assumed to belong to a closed convex set Q. We may infer the basic properties from the general scheme developed above by setting :

$$g(\sigma) = \chi_Q(\sigma) = \begin{cases} 0 & \text{if } \sigma \in Q , \\ +\infty & \text{if } \sigma \notin Q , \end{cases} \tag{4.18}$$

where χ_Q is the indicator function of the convex Q.

The conjugate function of g turns out to be the support function f of the convex Q :

$$f(\delta) = \sup \{ <\tau,\delta> \ / \ \tau \in Q \},$$

and the conjugacy relations give :

$$\sigma \in Q ,$$

$$\delta \in \theta\chi_Q(\sigma) ,$$

where $\theta\chi_Q(\sigma)$ is the outward normal cone to Q at σ.

By substituting (4.18) into (4.5) we get :

$$\inf \{ \tfrac{1}{2} \| \tau - S\varepsilon \|_A^2 \ / \ \tau \in Q \} = \tfrac{1}{2} \| \sigma - S\varepsilon \|_A^2 = \xi(S\varepsilon) ,$$

and then :

$$\sigma = \mathrm{proj}_Q(A;S\varepsilon) ;$$

namely, the stress distribution is the orthogonal projection of $S\varepsilon$, in the energy of A, on Q.

An interesting special case is met when the set Q of admissible stress distributions is a closed convex cone.

Such a model can effectively be adopted to simulate the response of materials without tensile strength and with a very high compressive strength. Rock mechanics problems and the analysis of concrete and masonry structures are fields of application in structural engineering.

In this case the conjugate function f turns out to be the indicator function of the negative polar Q^- of Q, and we have :

$$\delta = \mathrm{proj}_{Q^-}(S;\varepsilon);$$

namely, the anelastic strain is the orthogonal projection of the total strain ε, in the energy of S, on Q^-.

5. THE ELASTOSTATIC PROBLEM

The elastostatic problem for the structural model defined above is formulated as follows :

Given a load distribution $\ell \in V'$,

and a prescribed strain field $\varepsilon \in W$

Find an admissible displacement field $u \in C$, (5.1)

and an admissible stress distribution $\sigma \in Q$, (5.2)

satisfying the constitutive property : $\sigma = \text{grad}\,\phi(Tu - \varepsilon)$ (5.3)

and the equilibrium condition : $<\ell, v> \geq <\sigma, Tv>$

for any $v \in \{\theta\, \chi_C(u)\}^-$, (5.4)

where $\{\theta\, \chi_C(u)\}^-$ is the negative polar cone of the outward normal cone to C at u, that is, the closed cone generated by the admissible variations of displacements from u.

The convex set of all stress distributions satisfying (5.4) will be denoted by $\Sigma_\ell(u)$.

Then (5.2) and (5.4) are equivalent to : $\sigma \in \Sigma_\ell(u) \cap Q$.

The condition to be imposed to the load distribution so that the convex set $\Sigma_{\ell}(u) \cap Q$ be not empty, is given by (G.Romano, M.Romano[3]) :

$$<\ell,v> \leq \operatorname{supp}_Q(Tv) \quad \text{for any} \quad v \in \{\theta\, \chi_C(u)\}^- \tag{5.5}$$

where supp (•) denotes the support function :

$$\operatorname{supp}_Q(Tv) = \sup \{<\sigma,Tv> \;/\; \sigma \in Q\}.$$

The equilibrium condition (5.5) can be stated as follows : for any admissible variation of displacement field, the external virtual work must be not greater than the maximum internal virtual work.

5.1 Displacement formulation

By substituting (5.3) into (5.4) we get the variational condition to be satisfied by the displacement field $u \in C$:

$$<\operatorname{grad} \phi(Tu - \varepsilon), Tv> \geq <\ell,v> \quad \text{for any} \quad v \in \{\theta\, \chi_C(u)\}^- . \tag{5.6}$$

Defining the potential energy functional :

$$\Phi(u) = \phi(Tu - \varepsilon) - <\ell,u> \tag{5.7}$$

the condition (5.6) can be written as :

$$- \operatorname{grad} \Phi(u) \in \theta\, \chi_C(u) \; ; \tag{5.8}$$

namely : the gradient of the potential energy functional at u must belong to the inward cone to C at u.

It is easily seen that (5.8) is equivalent to the minimum problem :

$$\Phi(u) = \min \{ \Phi(v) \ / \ v \in C \}. \tag{5.9}$$

A direct proof of the existence of the solution for the minimum problem above was not available up to now.

A proof for the finite dimensional case, under linear geometric constraints has been given in (G.Romano,M.Romano.[9]).

We shall give an existence proof for the solution of the elastostatic problem, for the infinite dimensional case, in the next paragraph.

5.2 Stress formulation

The elastostatic problem may alternatively be set in terms of the stress distributions which are in equilibrium, under the applied load $\ell \in V'$, in a maximally constrained configuration $w \in L$.

In fact it can be proved that (G.Romano.[8]) :

The condition : $\quad - \varepsilon + Tw \in \partial \chi_{\Sigma_\ell}(\sigma) \ , \quad \sigma \in \Sigma_\ell = \Sigma_\ell(w),$ (5.10)

or, equivalent : $\quad <\tau - \sigma, \varepsilon - Tw> \geq 0 \quad$ for any $\quad \tau \in \Sigma_\ell$,

implies that :

i) the strain field ε is compatible, that is, there exists an admissible displacement field $u \in C$ such that $\varepsilon = Tu$.

ii) the stress distribution $\sigma \in \Sigma_\ell$ is in equilibrium at $u \in C$ under the action of the load ℓ, i.e., $\sigma \in \Sigma_\ell(u)$.

Now, by setting :

$$\varepsilon = \varepsilon^* + A\sigma + \delta \quad \text{with} \quad \sigma \in Q \,, \quad \delta \in \theta\, \chi_Q(\sigma),$$

we get :

$$\varepsilon \in \varepsilon^* + \theta\psi(\sigma) \,. \tag{5.11}$$

Moreover, setting :

$$r(\sigma) = T'\sigma - \ell$$

and, defining the complementary energy functional :

$$\Psi(\sigma) = \psi(\sigma) - \langle\sigma,\varepsilon^*\rangle - \langle r(\sigma),w\rangle \tag{5.12}$$

the conditions (5.10) and (5.11) can be written as :

$$0 \in \theta(\Psi + \chi_{\Sigma_\ell})(\sigma) \,. \tag{5.13}$$

This is equivalent to the minimum problem :

$$\Psi(\sigma) = \min \{ \Psi(\tau) / \tau \in \Sigma_\ell \} \tag{5.14}$$

which can also be written as :

$$\Psi(\sigma) = \min \{ \tfrac{1}{2} \|\tau\|_A^2 - \langle\tau,\varepsilon^*\rangle - \langle r(\tau),w\rangle / \tau \in \Sigma_\ell \cap Q \} \tag{5.15}$$

This minimum problem admits an unique solution if and only if the convex set $\Sigma_\ell \cap Q$ is not empty, i.e. iff the condition (5.5) is satisfied.

To prove that the solution of (5.15) is also a solution of the elastostatic problem, we can follow the above steps backwards.

It has to be remarked that, to deduce (5.12) from (5.13), we need the validity of the additive property of the subdifferentials.

Sufficient conditions can be found in (J.J.Moreau.[6]) and an attitude to overcome this difficulty has been proposed in (B.Nayroles.[7]).

5.3 Error bound

From the inequality (4.12) we get :

$$\Phi(u) - \Phi(u_\circ) = \phi(u) - \phi(u_\circ) + <\ell, u - u_\circ> \geq$$

$$\geq \tfrac{1}{2} \|\sigma(u) - \sigma(u_\circ)\|_A^2 + <\ell, u - u_\circ> + <\sigma_\circ, T(u-u_\circ)>.$$

Analogous developments for the functional Ψ yield :

$$\Psi(\sigma) - \Psi(\sigma_\circ) = \psi(\sigma) - \psi(\sigma_\circ) - <r(\sigma) - r(\sigma_\circ), w> \geq$$

$$\geq \tfrac{1}{2} \|\sigma - \sigma_\circ\|_A^2 + <\sigma - \sigma_\circ, Tu_\circ> - <r(\sigma) - r(\sigma_\circ), w> ,$$

where $\sigma(u) = \text{grad}\, \phi(Tu)$, $u \in C$ and $\sigma \in \Sigma_\ell$.

If $u_\circ$, $\sigma_\circ$ is the solution of the elastostatic problem, then $\sigma(u_\circ) = \sigma_\circ$, and the last terms in the above inequalities, by virtue of (5.4) and (5.10), are non-negative. Moreover :

$$\Phi(u_\circ) + \Psi(\sigma_\circ) = <\sigma_\circ, Tu_\circ> - <\ell, u_\circ> - <r(\sigma_\circ), w> = <r(\sigma_\circ), u_\circ - w> = 0 .$$

Hence, adding the above inequalities, we get the extimate :

$$\Phi(u) + \Psi(\sigma) \geq \tfrac{1}{2} \|\sigma(u) - \sigma_\circ\|_A^2 + \tfrac{1}{2} \|\sigma - \sigma_\circ\|_A^2 ,$$

for any $u \in C$, $\sigma \in \Sigma_\ell$.

REFERENCES

1. Romano, G., Romano, M., On the foundations of limit analysis (to appear).

2. Moreau, J.J., Proximitē et dualitē dans un espace hilbertien Bull. Soc. Math. France, 93, 1965.

3. Romano, G., Romano, M., Equilibrium and compatibility under external and internal convex constraints, Atti Ist. Scienza Costruzioni, Catania, 1983.

4. Lions, J.L., Sur le Controle Optimal de Systèmes Governès par des Equations aux Dèrivèes partielles, Dunod, Gauthier Villars, 1968.

5. Baiocchi, C., Capelo, A., Disequazioni Variazionali e Quasivariazionali, Pitagora Editrice, Bologna, 1978.

6. Moreau, J.J., Fonctionelles Convexes, Sēminaire sur les Equations aux Dérivées Partielles, Collège de France, Paris, 1966-67.

7. Nayroles, B., Point de vue algebrique, convexitē et integrandes convexes en mēcanique de solides, in New Variational Technique in Mathematical Physics, Ed. Cremonese, Roma, 1974.

8. Romano, G., The complementary energy principle in elastostatics with unilateral constraints, Seminar on Problems in Mechanics of Materials and Structures, Rome 4-7 May, 1982.

9. Romano, G., Romano, M., Sulla soluzione di problemi strutturali in presenza di legami costitutivi unilaterali, Rend. Acc. Naz. Lincei, Serie VIII, Vol. LXVII, Ferie 1979.

LOCKING MATERIALS AND HYSTERESIS PHENOMENA

P.M. Suquet
Mécanique des Milieux Continus
Université Montpellier II

Abstract. A modelling of mechanical hysteresis phenomena, accounting for internal locking of materials is proposed. A mathematical discussion of ideal locking materials is given. A special emphasis is set on the locking limit analysis.

Résumé. On propose un modèle d'hystérésis mécanique, tenant compte des effets de blocage interne de la matière. Le cas des matériaux à blocage est discuté sous un angle mathématique. On porte une attention particulière à l'analyse limite de blocage.

1. SYNOPSIS

Locking materials have been introduced by PRAGER in 1957-1958 in order to account for internal unilateral constraints in the mechanics of continua [1, 2, 3]. For this type of materials the stress-strain curve exhibits an hardening part revealing an internal locking of the matter. This hardening effect can be purely elastic (rubber) or accompanied by plastic effects (cristals). In the last case, hysteresis phenomena similar to those observed in electro-magnetism, take place.

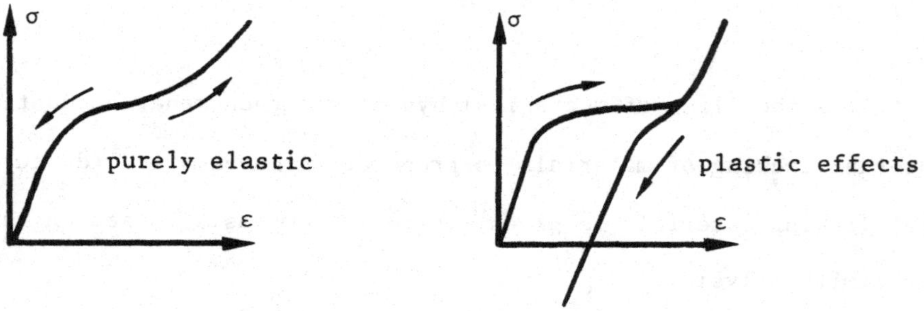

- Figure 1 -

The present work, devoted to a discussion of a few aspects of locking and hysteresis phenomena, is twofold :

- the first part proposes a possible modelling of hysteresis phenomena. Constitutive laws are derived and their structure is discussed. A few open mathematical problems are addressed.

- the second part is devoted to ideal locking materials, as considered by PRAGER. We focus the attention on what is called here the locking limit analysis, the aim of which is to determine the set of admissible imposed displacements before complete locking. The example of torsion of cylindrical bars is discussed : it shows that stress singularities are likely to occur.

ACKNOWLEDGMENTS

The second part of this work is partly taken from a joint study with F. DEMENGEL [4], whose help is gratefully acknowledged.

2. CONSTITUTIVE LAWS AND MECHANICAL HYSTERESIS

2.1. Rheological models.

The classical rheological models are well known : spring, dash-pot, glider. We introduce a locking model which exhibits the following constitutive law

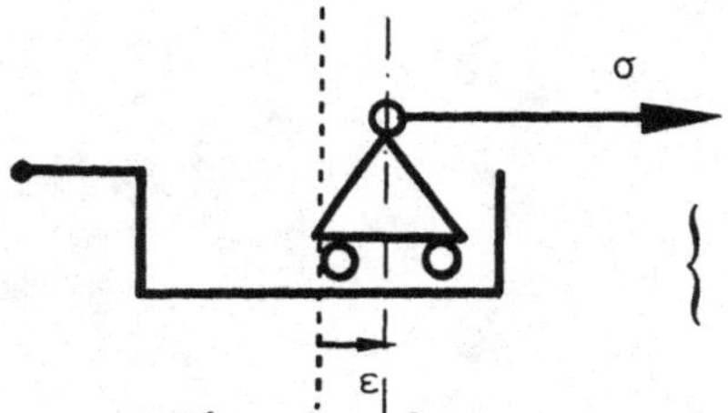

$$|\varepsilon| \leqslant \varepsilon_o \quad \text{and}$$

$$\begin{cases} |\varepsilon| < \varepsilon_o \Rightarrow \sigma = 0 \\ |\varepsilon| = \varepsilon_o \Rightarrow \sigma = \lambda \dfrac{\varepsilon}{|\varepsilon|} \quad \lambda \geqslant 0 \end{cases}$$

- Figure 2 -

This element called a *lock*, is used in more complex models.

Ideal locking model

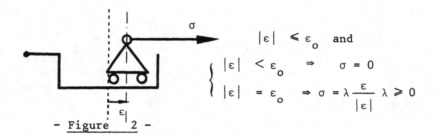

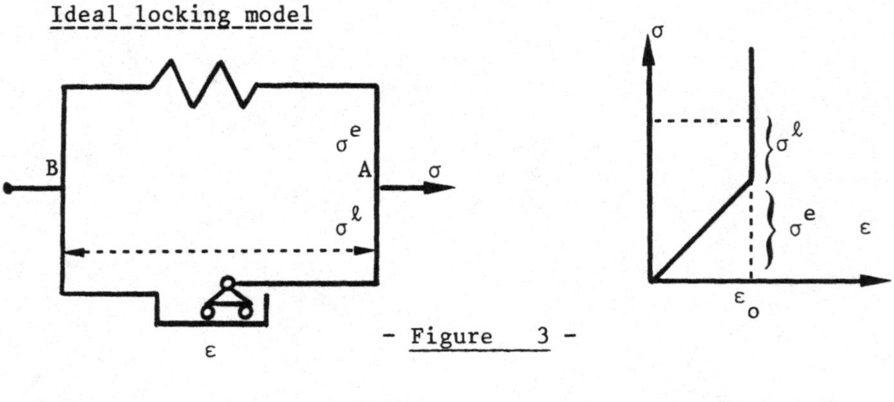

- Figure 3 -

The constitutive law of ideal locking materials is :

$$\begin{cases} \sigma = \sigma^{e} + \sigma^{\ell} \quad , \quad \sigma^{e} = E\,\varepsilon \\ |\varepsilon| \leqslant \varepsilon_{o} \; , \; \sigma^{\ell} = 0 \text{ if } |\varepsilon| < \varepsilon_{o} \; , \; \sigma^{\ell} = \lambda \dfrac{\varepsilon}{|\varepsilon|} \text{ if } |\varepsilon| = \varepsilon_{o} \; , \; \lambda \geqslant 0 \end{cases} \tag{2.1}$$

Instead of imposing a given stress in A , one can impose a given displacement u^{d} . This given displacement must obey

$$|u^{d}| \leqslant \varepsilon_{o}$$

Therefore, a locking material cannot undergo *any* imposed displacement. The determination of the admissible imposed displacements is the object of the *locking limit analysis* (cf. § 3.2) .

Model with hysteresis

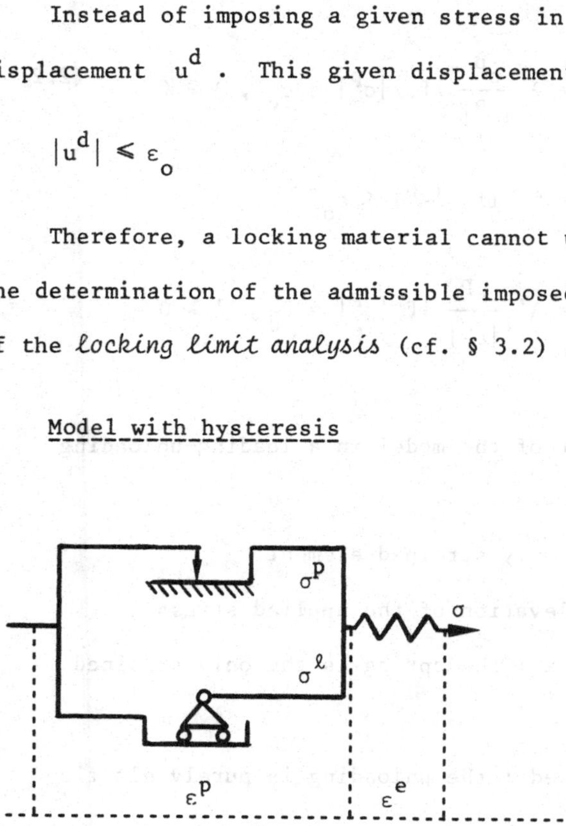

a. Model

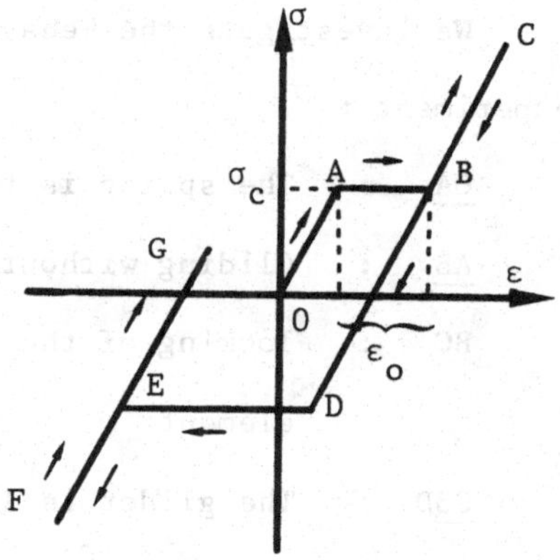

b. Loading-Unloading Test

- Figure 4 -

The constitutive law of the model is the following

$$\varepsilon = \varepsilon^e + \varepsilon^p \qquad \sigma = \sigma^p + \sigma^\ell$$

$$\varepsilon^e = \frac{\sigma}{E} \; ,$$

$$|\sigma^p| \leq \sigma_c^{(\dagger)} \quad \text{and} \quad \begin{cases} \dot{\varepsilon}^p = 0 \quad \text{if} \quad |\sigma^p| < \sigma_c \\ \dot{\varepsilon}^p = \lambda \dfrac{\sigma^p}{|\sigma^p|} \quad \text{if} \quad |\sigma^p| = \sigma_c \; , \; \lambda \geq 0 \end{cases} \tag{2.2}$$

$$|\varepsilon^p| \leq \varepsilon_o \quad \text{and} \quad \begin{cases} \sigma^\ell = 0 \quad \text{if} \quad |\varepsilon^p| < \varepsilon_o \\ \sigma^\ell = \lambda' \dfrac{\varepsilon^p}{|\varepsilon^p|} \quad \text{if} \quad |\varepsilon^p| = \varepsilon_o \; , \; \lambda' \geq 0 \end{cases}$$

We investigate the behavior of the model in a loading-unloading experiment :

OA : The spring is the only strained element

AB : Gliding without elevation of the applied stress

BC : locking of the lock : the spring is the only strained element

CBD : The glider is locked : the unloading is purely elastic until the stress reaches $-\sigma_c$

DE : The glider slips without stress modification

(†) σ_c is the yield limit of the glider.

EF : locking of the lock

FEG : The glider is fixed

The model exhibits an *hysteresis behavior.*

2.2. 3-dimensional case.

Ideal locking material

The natural generalization of (2.1) goes as follows :
There exists a convex set B in the strain space, which the strain tensor is constrained to stay in :

$\varepsilon(u) \in B$.

Moreover

$$\begin{cases} \sigma_{ij} = \sigma^{e}_{ij} + \sigma^{\ell}_{ij} \\ \sigma^{e}_{ij} = a_{ijkh}\, \varepsilon_{kh}(u) \\ \sigma^{\ell} \in \partial I_B(\varepsilon(u)) \end{cases} \tag{2.3}$$

where I_B is the indicator function of the set B in the space E of 3 × 3 symmetric tensors of order 2 .

Remark . A typical example of set B is the following

$B = \{e \in E \mid -k_1 \leqslant e_{ii} \leqslant k_o\}$

Only volumic changes are constrained. The class of locking materials described by this choice of B is that of materials with limited compressibility. The case of incompressible materials is recovered with a special choice $k_o = k_1 = 0$.

Locking materials with hysteresis

The natural generalization of (2.2) goes as follows

$$\varepsilon_{ij}(u) = \varepsilon^e_{ij} + \varepsilon^p_{ij} , \qquad \sigma_{ij} = \sigma^p_{ij} + \sigma^\ell_{ij}$$

$$\varepsilon^e_{ij} = \Lambda_{ijkh} \, \sigma_{kh}$$

There exists a convex set B in the strain space ($\underset{\sim}{E}$) which the plastic strain tensor is constrained to stay in

$$\varepsilon^p \in B \qquad \sigma^\ell \in \partial I_B(\varepsilon^p) \qquad B \text{ closed convex set in } E$$

There exists a convex set P in the stress space ($\underset{\sim}{E}$) which the first part of the stress tensor is constrained to stay in

$$\sigma^p \in P \quad , \quad \dot\varepsilon^p \in \partial I_P(\sigma^p) \qquad P \text{ closed convex set in } E$$

Therefore the constitutive law, written in a condensed form, amounts to :

$$\left\{ \begin{array}{lll} \varepsilon = \varepsilon^e + \varepsilon^p & , \ \sigma = \sigma^p + \sigma^\ell & , \ \varepsilon^e = \Lambda\sigma \\ \varepsilon^p \in B & , \ \sigma^\ell \in \partial I_B(\varepsilon^p) & \\ \sigma^p \in P & , \ \dot\varepsilon^p \in \partial I_P(\sigma^p) . & \end{array} \right. \qquad (2.4)$$

We claim that the two constitutive laws have the same structure : they both are *generalized standard materials*.

2.3. Generalized standard materials.

The theory of generalized standard materials, due to HALPHEN, NGUYEN QUOC SON takes its roots into ZIEGLER's and MOREAU's works. It proposes a general framework for the establishment of constitutive laws, accounting for the two laws of Thermodynamics (detailed exposures can be found in HALPHEN, NGUYEN QUOC SON [5], NGUYEN QUOC SON [6], GERMAIN [7], GERMAIN, NGUYEN QUOC SON and SUQUET [8], SUQUET [9]).

We admit the existence of a density of free energy depending on the state variables (ε,α) (†) :

$$\rho w = \rho w(\varepsilon,\alpha) \qquad (\rho \text{ density of the body})$$

ρw is supposed to be convex with respect to (ε,α).

The *state laws* define the thermodynamical forces

$$\sigma^R = \rho \frac{\partial w}{\partial \varepsilon}(\varepsilon,\alpha) \quad , \qquad A = - \rho \frac{\partial w}{\partial \alpha}(\varepsilon,\alpha)$$

σ^R is the reversible part of the stress tensor.

In case of a nondifferentiable free energy w the preceeding relations are to be understood in the sense of subdifferentials [10], [11]

(†) for the sake of simplicity we omit thermal effects. The temperature T will not be listed among the state variables.

$$(\sigma^R, -A) \in \rho\partial w(\varepsilon,\alpha) \tag{2.5}$$

We admit the existence of a potential of dissipation $\mathcal{D}$ convex function of its arguments $(\dot\varepsilon,\dot\alpha)$, which yields the *complementary laws*

$$\sigma^{IR} = \sigma - \sigma^R = \frac{\partial\mathcal{D}}{\partial\dot\varepsilon}(\dot\varepsilon,\dot\alpha) \quad , \quad A = \frac{\partial\mathcal{D}}{\partial\dot\alpha}(\dot\varepsilon,\dot\alpha)$$

or in a generalized sense :

$$(\sigma^{IR}, A) \in \partial\mathcal{D}(\dot\varepsilon,\dot\alpha) \tag{2.6}$$

σ^{IR} is the irreversible part of the stress tensor.

In the framework of generalized standard materials *a constitutive law is specified by the data of the two thermodynamical potentials* ρw *and* $\mathcal{D}$.

It can be proved [6, 7, 9] that the *mechanical dissipation* amounts to

$$d = \sigma^{IR}\dot\varepsilon + A\dot\alpha$$

Application to the specific situation of locking materials.

Ideal locking materials

The only state variable is the strain ε. The two thermodynamical potentials amount to

$$\begin{cases} \rho w(\varepsilon) = \frac{1}{2} a_{ijkh}\,\varepsilon_{kh}\,\varepsilon_{ij} + I_B(\varepsilon) \\ \mathcal{D}(\dot\varepsilon) = 0 \end{cases}$$

The state laws (2.5) and the complementary laws (2.6) yield

$$\begin{cases} \sigma^R \in a\varepsilon + \partial I_B(\varepsilon) \\ \sigma^{IR} = \sigma - \sigma^R = 0 \end{cases}$$

which is exactly (2.3) .

Remark. The ideal locking material is *not dissipative* ($\mathcal{D} = 0$) : it is an *hyperelastic* material.

Locking materials with hysteresis

The state variables are ε and $\alpha = \varepsilon^P$. The following choice of potentials is made

$$\begin{cases} \rho w(\varepsilon, \varepsilon^P) = \frac{1}{2} a(\varepsilon - \varepsilon^P)(\varepsilon - \varepsilon^P) + I_B(\varepsilon^P) \\ \mathcal{D}(\dot{\varepsilon}, \dot{\varepsilon}^P) = I_P^{\star}(\dot{\varepsilon}^P) \quad (\dagger) \end{cases}$$

The state laws (2.5) yield

$$\sigma^R = a(\varepsilon - \varepsilon^P) \quad , \quad A \in a(\varepsilon - \varepsilon^P) - \partial I_B(\varepsilon^P) \tag{2.7}$$

The complementary laws (2.6) yield

$$\sigma^{IR} = 0 \quad , \quad A \in \partial(I_P^{\star})(\dot{\varepsilon}^P) \quad \text{i.e.} \quad \dot{\varepsilon}^P \in \partial I_P(A) \tag{2.8}$$

Therefore the total stress reduces to the reversible stress

(†) $I_P^{\star}$ denotes the Legendre Fenchel transform of I_P [10,11] .

$$\sigma = \sigma^R = a(\varepsilon - \varepsilon^p)$$

Setting

$$\varepsilon^e = \varepsilon - \varepsilon^p \ , \quad \Lambda = a^{-1} \ , \quad \sigma^p = A \ , \quad \sigma^\ell = a(\varepsilon - \varepsilon^p) - \sigma^p \in \partial I_B(\varepsilon^p)$$

we see that the law defined by (2.7)(2.8) takes the form (2.4). The locking material under consideration here is *dissipative*, and the mechanical dissipation amounts to

$$d_1 = \sigma^{IR} \dot{\varepsilon} + A\dot{\alpha} = \sigma^p \dot{\varepsilon}^p \ .$$

Remark. ZIEGLER and PRAGER [2] also considered a non newtonian fluid for which the locking constraints acts on the *strain rate* $\dot{\varepsilon}$. This type of fluid is also a generalized standard material. The choice of state variables and potentials goes as follows

state variables : ε

potentials : $\begin{cases} \rho w(\varepsilon) = \Phi(\mathrm{Tr}\varepsilon) \\ \mathcal{D}(\dot{\varepsilon}) = I_B(\dot{\varepsilon}) \end{cases}$

the constitutive law amounts to

$$\sigma^R = -\, p \ \mathrm{Id} \qquad \text{where} \qquad p = -\frac{\partial \Phi}{\partial(\mathrm{Tr}\varepsilon)}$$

$$\sigma^{IR} \in \partial I_B(\dot{\varepsilon})$$

i.e. $\sigma \in -p \ \mathrm{Id} + \partial I_B(\dot{\varepsilon})$.

Remark. The interest of recognizing a generalized standard form in a constitutive law is that general theorem on variational principles, behavior at infinity... have been derived in this general setting.

2.4. Evolution problem for a locking material with hysteresis.

We now turn to the evolutive boundary problem posed by a locking material occupying a bounded domain Ω, submitted to body and boundary forces, to imposed displacements, and undergoing a quasi-static evolution.

In addition to the constitutive law (2.4) the stress and strain fields must obey further requirements

$$\left\{\begin{array}{llll} \varepsilon_{ij} = \varepsilon_{ij}(u) & \text{in } \Omega & \text{compatibility relations} \\ \dfrac{\partial \sigma_{ij}}{\partial x_j} + \rho f_i = 0 & \text{in } \Omega & \text{equilibrium equations} \\ u_i = U_i^d & \text{on } \partial\Omega_U & \text{imposed displacements on a part } \partial\Omega_U \text{ of } \partial\Omega \\ \sigma_{ij} n_j = F_i^d & \text{on } \partial\Omega_F & \text{imposed forces on a part } \partial\Omega_F \text{ of } \partial\Omega^{(\dagger)} \end{array}\right.$$

The loading $f(x,t)$, $F^d(x,t)$, $U^d(x,t)$ is given on $[0,T]$.

<u>Splitting of the problem.</u>

We first solve a purely elastic problem :

$$\left\{\begin{array}{l} \varepsilon_{ij}(u^{e\ell}) = \Lambda_{ijkh}\, \sigma^{e\ell}_{kh} \;,\; \dfrac{\partial \sigma^{e\ell}_{ij}}{\partial x_j} + \rho f_i = 0 \\ \sigma^{e\ell}_{ij}\, n_j = F_i^d \text{ on } \partial\Omega_F \;,\; u_i^{e\ell} = U_i^d \text{ on } \partial\Omega_U \end{array}\right.$$

Provided that :

. Ω is a bounded domain with a Lipschitz boundary

(†) $\partial\Omega_U$, $\partial\Omega_F$ are open and disjoint subsets of $\partial\Omega$ $\overline{\partial\Omega_U} \cup \overline{\partial\Omega_F} = \partial\Omega$

. $f \in W^{1,2}(0,T\ ;\ L^2(\Omega)^3)$, $F^d \in W^{1,2}(0,T\ ;\ L^2(\partial\Omega)^3)$,
$U^d \in W^{1,2}(0,T\ ;\ H^{1/2}(\partial\Omega)^3)$

. the elasticity matrix is symmetric, bounded and coercive,

the above elastic problem admits a solution $(\sigma^{e\ell}, u^{e\ell})$

$\sigma^{e\ell} \in W^{1,2}(0,T\ ;\ L^2(\Omega,E))$ (†)

$u^{e\ell} \in W^{1,2}(0,T)\ ;\ \mathbb{H}^1(\Omega))$ (††)

Setting

$$\bar{\sigma} = \sigma - \sigma^{e\ell} \quad \text{and} \quad \bar{u} = u - u^{e\ell}$$

we see that $\bar{\sigma}$, $\bar{u}$ satisfies the following set of equations

$$\begin{cases} \varepsilon(\bar{u}) = A\bar{\sigma} + \varepsilon^p \ , & \operatorname{div} \bar{\sigma} = 0 \ \text{ in } \ \Omega \\ \bar{\sigma}.n = 0 \ \text{ on } \ \partial\Omega_F \ , & \bar{u} = 0 \ \text{ on } \ \partial\Omega_U \end{cases}$$

For a given ε^p in $L^2(\Omega,E)$ we can associate the unique solution $\bar{\sigma}$ in $L^2(\Omega,E)$ of the preceding elastic problem :

$$\bar{\sigma} = - R\ \varepsilon^p$$

This equality defines a *linear, continuous self adjoint and maximal monotone operator* R *from* $L^2(\Omega,E)$ *into itself.*

From the definition of $\bar{\sigma}$ we derive

(†) $L^2(\Omega,E)$ = symmetric 3×3 tensors of order 2 with components in $L^2(\Omega)$

(††) $\mathbb{H}^1(\Omega) = H^1(\Omega)^3$

$$\sigma(t) = - R\,\varepsilon^p(t) + \sigma^{e\ell}(t) \quad \text{in} \quad L^2(\Omega,E)$$

Therefore *the determination of the stress field reduces to the determination of the field of plastic strains.*(†)

Evolution equation for the plastic strains

We note that

$$\sigma(t) = \sigma^p(t) + \sigma^\ell(t) \qquad \text{in } L^2(\Omega,E)$$

where $\quad \sigma^\ell(t) \in \partial I_{\mathbb{B}}\,(\varepsilon^p(t)) \qquad$ "

and $\quad \sigma^p(t) \in \partial(I^\star_{\mathbb{P}})\,(\dot\varepsilon^p(t)) \qquad$ "

$\mathbb{B}$ and $\mathbb{P}$ are respectively defined as

$$\mathbb{B} = \{e \in L^2(\Omega,E)\ ,\ e(x) \in B \quad \text{a.e.}\, x \in \Omega\}$$

$$\mathbb{P} = \{\tau \in L^2(\Omega,E)\ ,\ \tau(x) \in P \quad \text{a.e.}\, x \in \Omega\}$$

Therefore the field of plastic strains satisfies the following *non-linear evolution equation* in $L^2(\Omega,E)$

$$\begin{cases} \partial(I^\star_{\mathbb{P}})\,(\dot\varepsilon^p(t)) + \partial I_{\mathbb{B}}\,(\varepsilon^p(t) + R\,\varepsilon^p(t) \ni \sigma^{e\ell}(t) \\ \varepsilon^p(0) = \varepsilon^p_o \end{cases} \qquad (2.9)$$

(†) $\sigma^{e\ell}(t)$ is a given quantity.

The resolution of this evolution equation seems to be an open problem. It bears some ressemblance with an equation discussed by VISINTIN [12] in a problem issued from phase transition, but where compactness methods applied (this is not the case here).

This problem being too difficult, we turn to an easier one.

Rate principles for the plastic strains

Let us assume that the present state of plastic strain $\varepsilon^P(t)$ is known. We want to determine the *rate* $\dot{\varepsilon}^P(t)$ of plastic strain.

Since ε^P must belong to $\mathbb{B}$, the rate $\dot{\varepsilon}^P$ is constrained to stay in the projecting cone of $\mathbb{B}$ at $\varepsilon^P(t)$

$$\dot{\mathbb{B}}(\varepsilon^P) = \{\varepsilon^\star \mid (\varepsilon^\star, z) \leqslant 0 \quad \forall\, z \in \partial I_{\mathbb{B}}(\varepsilon^P)\}$$

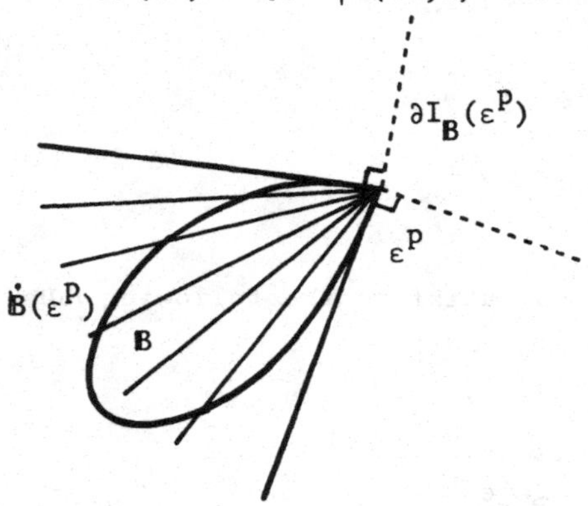

A field $\varepsilon^\star$ belonging to $\dot{\mathbb{B}}(\varepsilon^P)$ is said to be a *locking rate of plastic strain.*

- Figure 5 -

Variational principle for the strain rate $\dot{\varepsilon}^P$

$\dot{\varepsilon}^P$ *minimizes among all locking admissible rates of plastic strains the functional*

$$\inf_{\varepsilon^\star \in \dot{\mathbb{B}}(\varepsilon^P(t))} \quad I^\star_{\mathbb{P}}(\varepsilon^\star) + (R\varepsilon^P(t), \varepsilon^\star)_{L^2} - (\sigma^{e\ell}(t), \varepsilon^\star)_{L^2} \qquad (2.10)$$

Proof. We notice that

$$(\varepsilon^{\star},z)_{L^2} \leqslant 0 \qquad \forall\, z \in \partial I_{\mathbb{B}}\,(\varepsilon^p(t))$$

we shall use this inequality with $z = \sigma^{\ell}$

On the other hand

$$(\dot{\varepsilon}^p\,,\,\sigma^{\ell})_{L^2} = \frac{d}{dt}\,(I_{\mathbb{B}}\,(\varepsilon^p)) = \frac{d}{dt}\,0 = 0$$

Therefore

$$(\sigma,\varepsilon^{\star} - \dot{\varepsilon}^p)_{L^2} = (\sigma^p,\varepsilon^{\star} - \dot{\varepsilon}^p)_{L^2} + (\sigma^{\ell},\ \varepsilon^{\star} - \dot{\varepsilon}^p)_{L^2}$$

$$\leqslant (\sigma^p,\varepsilon^{\star} - \dot{\varepsilon}^p)_{L^2}$$

But $(\sigma,\varepsilon^{\star} - \dot{\varepsilon}^p)_{L^2} = -(R\varepsilon^p(t),\varepsilon^{\star} - \dot{\varepsilon}^p)_{L^2} + (\sigma^{e\ell}(t),\varepsilon^{\star} - \dot{\varepsilon}^p)_{L^2}$

Thus

$$I^{\star}_{\mathbb{P}}\,(\varepsilon^{\star}) - I^{\star}_{\mathbb{P}}\,(\dot{\varepsilon}^p) + (R\varepsilon^p(t),\varepsilon^{\star} - \dot{\varepsilon}^p)_{L^2} - (\sigma^{e\ell}(t),\varepsilon^{\star} - \dot{\varepsilon}^p)_{L^2}$$

$$\geqslant (\sigma^p,\varepsilon^{\star} - \dot{\varepsilon}^p)_{L^2} + (R\varepsilon^p(t),\varepsilon^{\star} - \dot{\varepsilon}^p)_{L^2} - (\sigma^{e\ell}(t),\varepsilon^{\star} - \dot{\varepsilon}^p)_{L^2}$$

$$\geqslant 0$$

Remark. Due to the high nonlinearity of the thermodynamical potentials, the variational principles established in NGUYEN QUOC SON [13] or SUQUET [9] do not apply.

2.5. Open problems.

. Under which set of assumptions on the loading is it possible to prove the existence of a solution of the variational principle (2.10) ?

. Same question for the evolution equation (2.9) .

. Consider a proportional loading

$$\lambda f \ , \ \lambda F^d \ , \ \lambda U^d$$

We can easily conceive that for small λ equation (2.9) admits a solution. Does there exist a limit value for λ (similar to the plastic limit load, or the locking limit load) ?

. Is it possible to discuss under general assumptions on the functionals φ and ψ , the following evolution equation :

$$\begin{cases} \partial\varphi(\dot{\alpha}) + \partial\psi(\alpha) + R\alpha \ni f(t) & \text{in a Hilbert space } H \\ \alpha(0) = \alpha_o & \end{cases}$$

where φ and ψ are convex, l.s.c., proper functionals on H , R is a linear, positive continuous operator from H into H .

3. IDEAL LOCKING MATERIALS

We consider the hyperelastic materials introduced originally by PRAGER, namely ideal locking materials. Let us recall the constitutive law (2.3) :

$$\left\{\begin{array}{ll} \sigma_{ij} = \sigma^{e}_{ij} + \sigma^{\ell}_{ij} & \\ \sigma^{e}_{ij} = a_{ijkh}\ \varepsilon_{kh}(u) & \\ \varepsilon(u) \in B & \\ (\sigma^{\ell}_{ij}, \varepsilon^{\star}_{ij} - \varepsilon_{ij}(u)) \leqslant 0 & \forall\ \varepsilon^{\star} \in B \end{array}\right. \tag{3.1}$$

The free energy of the material reduces to its elastic energy

$$\rho w(\varepsilon) = \left\{\begin{array}{ll} \frac{1}{2}\, a_{ijkh}\ \varepsilon_{ij}\ \varepsilon_{kh} & \text{if } \ \varepsilon \in B \\ + \infty & \text{otherwise} \end{array}\right.$$

A few assumptions on material datas (†) ensure that ρw is a convex

---

(†) B is bounded and contains 0 as an interior point

function, continuous on the interior of its domain ; moreover its conjugate function $\rho w^{\star}$ satisfies

$$\exists c_o, c_1 > 0 \ , \ \forall \ \xi \in E \ , \ c_o(|\varepsilon|_E - 1) \leq \rho w^{\star}(\varepsilon) \leq c_1(|\varepsilon|_E + 1) \tag{3.2}$$

The constitutive law (3.1) now reads

$$\begin{cases} \varepsilon(u(x)) \in B \\ \sigma(x) \in \partial \rho w^{\star}(\varepsilon(u(x))) \ , \qquad \text{a.e.x} \quad \text{in} \quad \Omega \end{cases}$$

W and $W^{\star}$ denote the strain energy and the complementary energy of the body respectively defined on $L^2(\Omega,E)$ and $L^1(\Omega,E)$ as

$$W^{\star}(\sigma) = \int_{\Omega} \rho w^{\star}(\sigma(x))dx$$

$$W(\varepsilon) = \int_{\Omega} \rho w(\varepsilon(x))dx$$

In addition to the constitutive law, the stress and strain must satisfy the following requirements :

$$\begin{cases} \dfrac{\partial \sigma_{ij}}{\partial x_j} + \rho f_i = 0 \quad \text{in } \Omega \text{ equilibrium equations} \\ \sigma_{ij} n_j = F_i^d \quad \text{on } \partial\Omega_F \\ u_i = U_i^d \quad \text{on } \partial\Omega_U \end{cases}$$

A few assumptions on the loadings (†) enable us to define the spaces

(†) $\rho f \in L^2(\Omega)^2$, $F^d \in L^2(\partial\Omega)^3$, $U^d \in H^{1/2}(\partial\Omega)^3$

of kinematically admissible fields of statically admissible fields, and the work of given external forces as :

$$\mathcal{U}_{ad} = \{u \in \mathbb{H}^1(\Omega) \ , \ u = U^d \quad \text{on} \quad \partial\Omega_U\}$$

$$\mathcal{B} \ = \{u \in \mathbb{H}^1(\Omega) \ , \ \varepsilon(u(x)) \in B \quad \text{a.e.} x \in \Omega\}$$

$$\mathcal{S}_{ad} = \{\sigma \in L^2(\Omega,E) \ , \ \text{div}\,\sigma + \rho f \ = 0 \quad \text{in} \quad \Omega \ , \ \sigma.n = F^d \quad \text{on} \quad \partial\Omega_F\}$$

$$L(u) = \int_\Omega \rho f u \, dx + \int_{\partial\Omega_F} F^d u \, ds \qquad \forall \ u \in \mathbb{H}^1(\Omega) \ .$$

3.1. Variational properties of the stress and displacement fields.

Provided that they have a minimum regularity, the solutions (σ,u) of the above problem satisfy the following variational principles (PRAGER [2], DUVAUT-LIONS [14]) in terms of the *strain energy* :

$$\inf_{u \in \mathcal{U}_{ad}} \ [W(\varepsilon(u)) - L(u)] \quad , \tag{3.3}$$

in terms of the *complementary energy* :

$$\sup_{\sigma \in \mathcal{S}_{ad}} \ [- W^\star(\sigma) + \int_{\partial\Omega_U} \sigma.n \ U^d \ ds] \tag{3.4}$$

Problem (3.3) *admits a solution*, provided that the following condition is satisfied :

$$\mathcal{B} \cap \mathcal{U}_{ad} \neq \phi \tag{3.5}$$

Indeed, W turns out to be a strictly convex lower semi continuous and coercive functional :

$$W(\varepsilon(u)) \geqslant \frac{\alpha}{2} \quad |\varepsilon(u)|_2^2$$

Korn's inequality proves the coercivity on the non empty closed convex subset $\mathcal{B} \cap \mathcal{U}_{ad}$ of $\mathbb{H}^1(\Omega)$ of the functional involved in (3.3) . Then, existence and uniqueness of a solution for the variational problem (3.3) is easily derived.

The variational problem (3.4) for the stress field σ is a more difficult one. Indeed, as a consequence of (3.2) , the functional $W^\star$ is only coercive on $L^1(\Omega,E)$ which is a non reflexive space. Therefore, proving the existence of a solution of (3.4) by means of classical arguments of coercivity, requires the introduction of a new functional space, accounting for *stress concentrations* :

$$\Sigma(\Omega) = \{\sigma \in M^1(\Omega,E) \text{ , div } \sigma \in L^2(\Omega,\mathbb{R}^N)\} \qquad (M^1 = \text{bounded measures})$$

A detailed study of $\Sigma(\Omega)$ can be found in [15], and the proof of the existence of a solution σ in $\Sigma(\Omega)$ is completed in [16]. In order that the solutions u and σ of (3.3) and (3.4) satisfy the extremality relations, i.e. the constitutive law (3.1) it is necessary [10] that

$$\text{Inf}(3.3) = \text{Sup}(3.4) \tag{3.6}$$

The proof of (3.6) can be found in DEMENGEL-SUQUET [4].

3.2. Locking limit analysis.

In the previous works concerned with locking materials the assumption

$$\mathcal{B} \cap \mathcal{U}_{ad} \neq \phi \tag{3.7}$$

was not discussed. By analogy to the plastic limit analysis we call such a discussion the *locking limit analysis* : it determines for which set of imposed displacements U^d , the condition (3.7) is fullfilled.

For sake of simplicity, U^d is assumed to be proportional to a load parameter

$$U^d = \lambda u_o$$

The space of kinematically admissible fields now depends on the load parameter

$$\mathcal{U}_{ad}(\lambda) = \{u \in \mathbb{H}^1(\Omega) \ , \ u = \lambda u_o \quad \text{on} \quad \partial\Omega_U\}$$

The locking limit analysis amounts to determine the admissible values of λ . Let us define

$$\bar{\lambda}_\ell = \text{Sup}\{\lambda \in \mathbb{R} \mid \mathcal{B} \cap \mathcal{U}_{ad}(\lambda) \neq \phi\} \qquad \textit{Problem } Q \tag{3.8}$$

Since a sup is computed in (3.8) $\bar{\lambda}_\ell$ is approximated by *lower values*. An approximation by *upper values* can be proposed which consists of the dual problem of (3.8) :

$$\inf_{\sigma \in S_o} \int_\Omega \pi_B(\sigma)\,dx \qquad \textit{Problem } Q^\star \tag{3.9}$$

$$\int_{\partial\Omega_U} \sigma.n\, u_o\, ds = 1$$

where

$$S_o = \{\sigma \in L^2(\Omega,E)\ ,\ \mathrm{div}\,\sigma = 0 \quad \text{a.e.} \quad \text{in } \Omega \quad ,\ \sigma.n = 0 \quad \text{on} \quad \partial\Omega_F\}$$

$$\pi_B(\sigma) = \sup_{e \in B} (\sigma,e)\ .$$

We shall prove the following theorem :

THEOREM 2. *Under the assumptions* H1,H2 *listed hereafter :* $Q^\star$(3.9) *is the dual problem of* Q (3.8) *and the primal dual relations hold :*

$$\text{Sup } Q = \text{Inf } Q^\star$$

If moreover H3 *is satisfied :*

$$\bar\lambda_\ell = \text{Inf } Q^\star = \text{Sup } Q < +\infty \tag{3.10}$$

$\bar\lambda_\ell$ *is the locking limit load .*

<u>Remark.</u> It is worth noting the analogy with the plastic limit analysis. Problem (3.8) is a kinematical approach of $\bar\lambda_\ell$ while (3.9) is a statical approach of $\bar\lambda_\ell$.

<u>H1</u> Ω is a bounded domain with a Lipschitz boundary. B is a bounded, closed convex set of E containing 0 as an interior point.

<u>H2</u> . $\rho f \in L^2(\Omega)^3$, $F^d \in L^2(\partial\Omega)^3$

$u_o \in W^{1,\infty}(\partial\Omega)^N$ and $u_o = \chi_o u_o$

where χ_o denotes the characteristic function of $\partial\Omega_U$.

H3 There exists $\sigma^o \in S_o$ such that

$$\int_{\partial\Omega_U} \sigma^o.n\, u_o\, ds \neq 0 .$$

Proof. The proof requires several steps. We shall complete the first one ; the other ones are treated in full details in [4] .

We set

$$V = \mathbb{H}^1(\Omega) \; , \quad Y = L^2(\Omega,E)$$

We define a linear operator $\Lambda \in \mathcal{L}(V,Y)$ by

$$\Lambda\, v = \varepsilon(v) .$$

We define on V and Y two functionals F and G by :

$$F(v) = \begin{cases} -\lambda & \text{if } u = \lambda u_o \text{ on } \partial\Omega_U \quad (u \in \mathcal{U}_{ad}(\lambda)) \\ +\infty & \text{otherwise} \end{cases}$$

$G(p) = I_{\mathbb{B}}(p)$ where $I_{\mathbb{B}}$ denotes the indicator function of $\mathbb{B}$.

Problem (3.8) now reads :

$$\operatorname*{Inf}_{v \in V} \{F(v) + G(\Lambda v)\}$$

and its dual problem is [5] :

$$\operatorname*{Sup}_{p^\star \in Y^\star} \{- G^\star(-p^\star) - F^\star(\Lambda^\star p^\star)\}$$

A theorem of KRASNOSELSKII ensures that :

$$G^\star(p^\star) = \int_\Omega \pi_B(p^\star)dx .$$

The computation of $F^\star(\Lambda^\star p^\star)$ is performed in the following proposition.

Proposition.

$$F^\star(\Lambda^\star p^\star) = \begin{cases} 0 \ \textit{if} \ p^\star \in S_o \quad \textit{and} \ \int_{\partial\Omega_U} p^\star.n\, u_o\, ds = 1 \\ + \infty \quad \textit{otherwise.} \end{cases}$$

Proof.

$$F^\star(\Lambda^\star p^\star) = \operatorname*{Sup}_{u \in \mathcal{U}_{ad}(\lambda)} [(\Lambda^\star p^\star, u) + \lambda] \qquad (3.11)$$

From assumption H2 we deduce that there exists $U_o \in W^{1,\infty}(\Omega)^N$ such that

$$U_o = u_o \text{ on } \partial\Omega \quad , \quad \text{i.e.} \quad \lambda U_o \in \mathcal{U}_{ad}(\lambda)$$

Then :

$$F^\star(\Lambda^\star p^\star) \geqslant \operatorname*{Sup}_{\lambda \in \mathbb{R}} \lambda [(\Lambda^\star p^\star, U_o) + 1]$$

This supremum equals $+\infty$ except when

$$(\Lambda^\star p^\star , U_o) + 1 = 0 \qquad (3.12)$$

We perform the computation of $F^\star(\Lambda^\star p^\star)$ in this specific situation. Using the fact that $1 = - (\Lambda^\star p^\star , U_o)$ we get from (3.12) :

$$F^\star(\Lambda^\star p^\star) = \sup_{u \in \mathcal{U}_{ad}(\lambda)} [(\Lambda^\star p^\star , u - \lambda U_o)]$$

$$= \sup_{\substack{v \in V \\ v = 0 \text{ on } \partial\Omega_U}} (\Lambda^\star p^\star , v)$$

The computation of this last supremum is classical (TEMAM [17]) :
$F^\star(\Lambda^\star p^\star) = +\infty$ except for the $p^\star$ satisfying :

$$\text{div } p^\star = 0 \text{ in } \Omega , \quad p^\star.n = 0 \quad \text{on } \partial\Omega_F \qquad (3.13)$$

With the help of Green's formula the condition (3.12) is equivalent to :

$$\int_{\partial\Omega_U} p^\star.n \, u_o \, ds + 1 = 0$$

Finally we have shown that a necessary condition to be satisfied by $p^\star$ in order to give a finite value to $F^\star(\Lambda^\star p^\star)$ is :

$$p^\star \in S_o \quad , \quad \int_{\partial\Omega_U} p^\star.n \, u_o \, ds + 1 = 0 \qquad (3.14)$$

Under these conditions it can be proved from (3.11)(†) that

$$F^{\star}(\Lambda^{\star} p^{\star}) = 0 \quad \text{for } p^{\star} \text{ satisfying (3.14)}$$

which completes the proof of the proposition.

It enables us to perform the computation of $Q^{\star}$ which amounts to :

$$- \sup_{p^{\star} \in S_o} \left[- \int_{\Omega} \pi_B(-p^{\star}) dx \right]$$

$$\int_{\partial\Omega_U} p^{\star}.n\, u_o\, ds + 1 = 0$$

or equivalently

$$\inf_{\sigma \in S_o} \left[\int_{\Omega} \pi_B(\sigma) dx \right]$$

$$\int_{\partial\Omega_U} \sigma.n\, u_o\, ds = 1$$

which is exactly (3.9) . The primal-dual relations yield :

$$\sup Q \leqslant \inf Q^{\star}$$

The proof of the reverse inequality is a technical one. It is due to F. DEMENGEL and uses a penalty method. It can be found in DEMENGEL-SUQUET [4] .

Let us emphasize once more that the stress problem (3.9) is *not*

(†) The complete justification of this point follows the appendix of section 2 in TEMAM [17] .

coercive in the classical space $L^2(\Omega,E)$ but only in $L^1(\Omega,E)$. Therefore we expect to find a stress field σ in $M^1(\Omega,E)$. The following example illustrates this point.

3.3. An example of locking limit analysis : torsion of cylindrical bars.

Let us consider a cylindrical bar, with a simply connected cross-section, made from a locking material, and submitted to a torsion experiment, with angle λ. The stress tensor and the displacement field exhibit the following classical form :

$$\sigma = \begin{pmatrix} 0 & 0 & \sigma_{13} \\ 0 & 0 & \sigma_{23} \\ \sigma_{13} & \sigma_{23} & 0 \end{pmatrix} \quad , \quad u = \begin{cases} u_1 = -\lambda x_2 x_3 \\ u_2 = \lambda x_1 x_3 \\ u_3 = u_3(x_1, x_2, \lambda) \end{cases}$$

where $\sigma_{ij} = \sigma_{ij}(x_1, x_2)$.

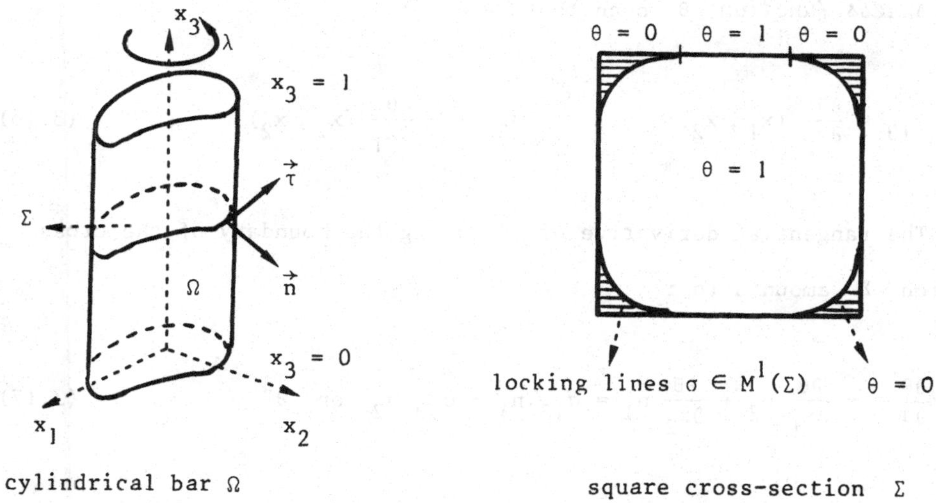

- Figure 6 -

Equilibrium equations reduce to

$$\frac{\partial \sigma_{13}}{\partial x_1} + \frac{\partial \sigma_{23}}{\partial x_2} = 0$$

The boundary conditions amount to

$$\sigma_{ij}\, n_j = 0 \quad \text{on lateral sides of the cylinder} \tag{3.15}$$

$$\begin{cases} u_1 = -\lambda\, x_2 & \\ u_2 = \lambda\, x_1 & \text{on the upper section of the cylinder, } x_3 = 1 . \\ \sigma_{33} = 0 & \\ & \\ u_1 = u_2 = 0 & \text{on the lower section of the cylinder, } x_3 = 0 . \\ \sigma_{33} = 0 & \end{cases}$$

A classical analysis of the equilibrium equations shows the existence of a *stress function* θ such that

$$\sigma_{13} = \frac{\partial \theta}{\partial x_2}(x_1 , x_2) \quad , \quad \sigma_{23} = - \frac{\partial \theta}{\partial x_1}(x_1 , x_2) \tag{3.16}$$

The tangential derivative of θ along the boundary of the cross section Σ amounts to :

$$\frac{\partial \theta}{\partial \tau} = - \frac{\partial \theta}{\partial x_1} n_2 + \frac{\partial \theta}{\partial x_2} n_1 = \sigma_{13}\, n_1 + \sigma_{23}\, n_2 \quad \text{on } \partial\Sigma . \tag{3.17}$$

According to (3.15) the boundary $\partial\Sigma$, is free of stress.

Therefore, recalling that the cross section is simply connected, we get

$$\theta = \text{constant on } \partial\Sigma$$

θ being defined up to an additive constant, we can choose the constant on $\partial\Sigma$ to be 0. Therefore if σ is in $L^2(\Omega,E)$, θ belongs to $H^1_o(\Sigma)$.

B is the ball of center 0 and radius k_o. The locking limit analysis problem can be explicitely solved :

$$\int_\Omega \pi_B(\sigma)\,dx = \int_\Omega k_o\,|\sigma|\,dx = \int_\Sigma k_o\,|\nabla\theta|\,dx_1\,dx_2\,,$$

$$\int_{\partial\Omega} \sigma.n\,u_o\,ds = \int_{x_3=0} 0 + \int_{x_3=h} (-x_2\,\sigma_{13} + x_1\,\sigma_{23})ds$$

$$= -\int_\Sigma \left(\frac{\partial\theta}{\partial x_2}\,x_2 + \frac{\partial\theta}{\partial x_1}\,x_1\right)dx_1\,dx_2 = \int_\Sigma \theta\,dx_1\,dx_2\,.$$

The locking limit analysis problem amounts to

$$\overline{\lambda}_\ell = \underset{\substack{\theta \in H^1_o(\Sigma) \\ \int_\Sigma \theta\,dx = 1}}{\text{Min}} \int_\Sigma k_o\,|\nabla\theta|\,dx \qquad (3.18)$$

This minimization problem has been already encountered by STRANG[18] in the determination of the limit load of a vertical column submitted to body forces (anti-shear problem). The following conclusions of STRANG are especially meaningful in the example here considered :

a) The problem (3.18) *does not admit* a solution θ in $H^1_o(\Sigma)$, nor in $W^{1,1}_o(\Sigma)$. The proper space to work with, is $BV(\Sigma)$; the stress tensor σ is therefore in $M^1(\Sigma)$ by virtue of (3.16). In $BV(\Sigma)$ *the solutions* θ *of* (3.18) *are characteristic functions of sets* with rather smooth boundaries. According to (3.16), the stress tensor is a *Dirac distribution* on the boundaries of these sets : these lines will be called *locking lines*.

b) locking lines necessarily intersect the boundary $\partial\Sigma$ of the cross-section. Therefore the stress function θ is no more constant on $\partial\Sigma$ since it jumps from 0 to 1. As a consequence its tangential derivative given by (3.17) contains Dirac distributions on $\partial\Sigma$, and in particular does not vanish in $\mathcal{D}'(\Sigma)$. *The condition of free edge* (3.15) *is not satisfied* in a distributional sense. In a more general context the boundary conditions of imposed forces have to be relaxed (cf. [16]).

c) in case of a square cross section, STRANG found out the explicit solution of (3.18). Locking lines are plotted on figure 6. Jumps of θ on $\partial\Sigma$ are noticeable. The curved parts of locking lines are circle portions.

3.4. Possible extensions.

A theory of locking of *structures* can be proposed. Strains are taken in a generalized sense : rotations, deflections, angles... In the example of a robot, the limit strain ε_0 can model the free motion of the joints. The problem is to determine which joints will be locked under a specified loading and which imposed displacements are admissible. We expect the stresses to concentrate on the locked joints which therefore require an adequate reinforcement.

4. CONCLUSIONS

A possible modelling of hysteresis phenomena, accounting for locking effects has been proposed. The case of ideal locking materials has been considered and a special emphasis has been set on the locking limit analysis.

REFERENCES

1. Prager, W., On ideal locking materials, *Transactions of the Society of Rheology*, 1, 1957, p. 169-175.

2. Prager, W., Unilateral constraints in Mechanics of Continua, *Estratto dagli atti del Simposio Lagrangiano*, Academia delle Scienze di Torino, 1964, p. 1-11.

3. Prager, W., Elastic solids of limited compressibility, *Proceedings of the 9th Int. Congress of Appl. Mech.*, Brussels, 1958, 5, p. 205-211.

4. Demengel, F., Suquet, P., On locking materials, to be published.

5. Halphen, B., Nguyen Quoc Son, Sur les matériaux standard généralisés, *J. de Mécanique*, 14, 1975, p. 39-63.

6. Nguyen Quoc Son, *Problèmes de Plasticité et de Rupture*, Cours D.E.A. Orsay, Polycopié 1982.

7. Germain, P., *Cours de Mécanique des Milieux Continus*, Masson, Paris 1974.

8. Germain, P., Nguyen Quoc Son, Suquet, P., Continuum Thermodynamics, *J. Appl. Mech* ,,, 50, 1983, p. 1010-1020.

9. Suquet, P., Local and global aspects in the mathematical theory of Plasticity, *Symposium "Plasticity Today"*, Udine 1983. Ed. A. Bianchi, A. Sawczuk, To be published.

10. Ekeland, I., Temam, R., *Analyse Convexe et Problèmes variationnels*, Dunod, Paris 1974.

11. Moreau, J.J., *Fonctionnelles convexes*, Cours au Collège de France, Polycopié Paris 1966.

12. Visintin, A., A Phase transition problem with delay, *Control and Cybernetics*, 11, 1982, p. 5-18.

13. Nguyen Quoc Son, Bifurcation et stabilité des systèmes irréversibles obéissant au principe de dissipation maximale, to be published in *J. Méca. Théo. Appl.*

14. Duvaut, G., Lions, J.L., *Les inéquations en Mécanique et en Physique*, Dunod, Paris 1972.

15. Demengel, F., Théorèmes de trace et de densité pour des espaces fonctionnels de la mécanique non linéaire, To be published in *Journal de Maths pures et appliquées*.

16. Demangel, F., Relaxation et existence pour le problème des matériaux à blocage, To be published.

17. Temam, R., *Problèmes mathématiques en Plasticité*, Dunod 1983.

18. Strang, G., A minimax problem in Plasticity Theory, in *Functional analysis Methods in Numerical Analysis*, Lecture Notes in Math., n° 701, 1979, p. 319-333.

9. [illegible], P., [illegible]cal and [illegible]obal a[illegible] of [illegible] mathematical theory of plasticity, [illegible] Symposium [illegible], Dunod, Paris, 1983, [illegible]. A. Sawczuk, [illegible], to be published.

10. [illegible], G., [illegible], R., [illegible] convexes [illegible] Dunod, Paris [illegible].

11. Moreau, J.J., Fonctionnelles convexes, Séminaire Collège de France, [illegible] Paris 1966.

12. [illegible], A phase transition problem [illegible], Control and Cybernetics, [illegible], 1982, [illegible].

13. [illegible], Bifurcation et stabilité des systèmes irréversibles [illegible] principe de dissipation maximale, [illegible] de Méca. Théo. Appl., [illegible].

14. Duvaut, G., Lions, J.L., Les inéquations en mécanique et en physique, Dunod, Paris, 1972.

15. Demengel, F., Théorèmes de traces [illegible] de densité pour des [illegible] fonctionnelles de mécanique non linéaire. To be published in Journal de Math. pures et appliquées.

16. Demengel, F., Relaxation [illegible] existence pour le problème des matériaux plastiques. To be published.

17. Temam, R., Problèmes mathématiques en plasticité, Dunod 198[illegible].

18. [illegible]ang, [illegible] minimax problem in plasticity theory, in Functional Analysis Methods in Numerical Analysis, Lecture Notes in Math. 701, [illegible].

UN PROBLEMA DINAMICO PER LA PIASTRA SU SUOLO ELASTICO UNILATERALE*

R. Toscano
Istituto di Matematica
University of Naples

In questo lavoro si studia il problema dinamico della piastra elastica poggiata su suolo unilateralmente elastico, modellato secondo Winkler.

Alla piastra, che occupa una regione limitata Ω del piano x_1, x_2 (il riferimento cartesiano ortogonale 0, x_1, x_2, x_3 è destrorso), libera al bordo ed inizialmente soggetta a spostamenti u_0 e velocità u_1 trasversali assegnati, è applicato un carico trasversale distribuito f dipendente dal tempo t.

Sono pure assegnate, e indipendenti da t, la densità di massa μ (x) e la reazione specifica del sottosuolo k (x) al variare del punto $x = (x_1, x_2)$ in Ω.

In tale situazione il problema di determinare l'evoluzione dello spostamento trasversale, supposto piccolo, nell'intervallo di tempo [0, T], consiste nel ricercare una funzione u (t, x) soddisfacente all'equazione:

$$\mu \frac{\partial^2 u}{\partial t^2} + \Delta \Delta u + k u^+ = f \qquad \text{su }]0, T[\times \Omega, \tag{1}$$

sotto le condizioni:

(*) Ricerca effettuata con fondi erogati dal M.P.I.

$$\left.\begin{aligned} &\boldsymbol{M}\,u = 0 \\ &\boldsymbol{T}\,u = 0 \end{aligned}\right\} \quad \text{su }]\,0, T\,[\times \partial\Omega, \tag{2}$$

$$\left.\begin{aligned} &u\,(0, x) = u_0\,(x) \\ &\frac{\partial u}{\partial t}\,(0, x) = u_1\,(x) \end{aligned}\right\} \quad \text{su } \Omega, \tag{3}$$

dove:

Δ è l'operatore di Laplace,

$$\boldsymbol{M}\,u = \nu\,\Delta u + (1 - \nu)\left[\frac{\partial^2 u}{\partial x_1^2}\,n_1^2 + 2\,\frac{\partial^2 u}{\partial x_1\,\partial x_2}\,n_1 n_2 + \frac{\partial^2 u}{\partial x_2^2}\,n_2^2\right],$$

$$\boldsymbol{T}\,u = -\frac{\partial}{\partial n}\,\Delta u + (1 - \nu)\,\frac{\partial}{\partial s}\left[\frac{\partial^2 u}{\partial x_1^2}\,n_1 n_2 - \frac{\partial^2 u}{\partial x_1\,\partial x_2}\,(n_1^2 - n_2^2) + \right.$$
$$\left. - \frac{\partial^2 u}{\partial x_2^2}\,n_1\,n_2\right],$$

essendo $n = (n_1, n_2)$ il versore normale esterno a $\partial\Omega$ e $\nu \in\,]-1, 1[$ il modulo di Poisson.

In primo luogo ci occuperemo della formulazione del problema (1), (2), (3) in senso debole, che consentirà di dimostrare l'esistenza, l'unicità, nonché proprietà qualitative e quantitative della soluzione, quest'ultime dedotte in base a noti risultati relativi ai problemi lineari. Successivamente stabiliremo un teorema di dipendenza continua dai dati. Infine dimostreremo un teorema di regolarità da cui si deduce che la soluzione debole è anche soluzione in senso classico, cioè che verifica puntualmente la (1) e le (2).

I procedimenti dimostrativi rientrano nelle tecniche che abitualmente si seguono per lo studio di questioni di questo tipo.

1. Prima di formulare il problema in senso debole, precisiamo anzitutto le ipotesi sui dati e le notazioni che utilizzeremo.

Supponiamo che Ω sia un aperto in R^2, limitato e connesso, di classe C^0. Supponiamo ancora che μ e k siano elementi non nulli di $L^\infty(\Omega)$ con:

$$\mu \geqslant \mu_0 \ (\mu_0 = \text{cost.} > 0) \text{ e } k \geqslant 0 \qquad \text{q.o. su } \Omega.$$

Indichiamo con:

$(\cdot, \cdot)$ il prodotto scalare in $L^2(\Omega)$,

$<\cdot, \cdot>$ la dualità tra $H^2(\Omega)$ e $(H^2(\Omega))'$,

$|\cdot|$ la norma in $L^2(\Omega)$,

$\|\cdot\|$ la norma in $H^2(\Omega)$,

e poniamo, per ogni $v, z \in H^2(\Omega)$:

$$a(v, z) = \int_\Omega \left[\frac{\partial^2 v}{\partial x_1^2} \frac{\partial^2 z}{\partial x_1^2} + 2(1-\nu) \frac{\partial^2 v}{\partial x_1 \partial x_2} \frac{\partial^2 z}{\partial x_1 \partial x_2} + \right.$$

$$\left. + \nu \frac{\partial^2 v}{\partial x_1^2} \frac{\partial^2 z}{\partial x_2^2} + \nu \frac{\partial^2 v}{\partial x_2^2} \frac{\partial^2 z}{\partial x_1^2} + \frac{\partial^2 v}{\partial x_2^2} \frac{\partial^2 z}{\partial x_2^2} \right] dx.$$

Rileviamo che:

$$a(v, v) \geqslant c_0 \sum_{|r|=2} |D^r v|^2 \qquad \forall\, v \in H^2(\Omega) \qquad (c_0 = \text{cost.} > 0), \tag{4}$$

e che quindi:

$$a(v, v) + |v|^2 \geqslant c_1 \|v\|^2 \quad \forall v \in H^2(\Omega) \qquad (c_1 = \text{cost.} > 0). \tag{5}$$

Assegnati $f \in L^2(0, T; L^2(\Omega))$, $u_0 \in H^2(\Omega)$ e $u_1 \in L^2(\Omega)$, consideriamo il seguente

PROBLEMA (P). *Trovare* $u \in L^2(0, T; H^2(\Omega)) \cap H^1(0, T; L^2(\Omega))$ *tale che* $u(0) = u_0$ *e*

$$\int_0^T [a(u(t), v(t)) - (\mu u'(t), v'(t)) + (ku^+(t), v(t)) - (f(t), v(t))]\, dt =$$

$$= (\mu u_1, v(0)) \quad \forall v \in L^2(0, T; H^2(\Omega)) \cap H^1(0, T; L^2(\Omega)) \textit{ con } v(T) = 0.$$

OSSERVAZIONE 1. Da notare che l'appartenenza di u ad $H^1(0, T; L^2(\Omega))$ implica che u è uguale q.o. su $]0, T[$ ad una funzione assolutamente continua da $[0, T]$ a $L^2(\Omega)$ [1] (Appendice), e quindi si può parlare di valore della u nel punto 0.

E' utile per il seguito osservare che il problema (P) può porsi equivalentemente come segue:

Trovare $u \in L^2(0, T; H^2(\Omega)) \cap H^1(0, T; L^2(\Omega))$ *soddisfacente alle condizioni*:

$$\mu u' \in H^1(0, T; (H^2(\Omega))'), \tag{7}$$

$$<(\mu u')'(t), v> + a(u(t), v) + (ku^+(t), v) =$$

$$= (f(t), v) \quad \forall v \in H^2(\Omega) \qquad \textit{q.o. su }]0, T[, \tag{8}$$

$$u(0) = u_0, \tag{9}$$

$$(\mu u')(0) = \mu u_1 \quad \textit{nel senso di } (H^2(\Omega))'. \tag{10}$$

OSSERVAZIONE 2. La (7) comporta l'esistenza di una funzione L assolutamente continua da [0, T] allo spazio $(H^2(\Omega))'$ tale che $(\mu u'(t), v) = \langle L(t), v \rangle \ \forall v \in H^2(\Omega)$ q.o. su $]0, T[$ [1] (Appendice). Pertanto la (10) va intesa nel senso che $\langle L(0), v \rangle = (\mu u_1, v) \ \forall v \in H^2(\Omega)$.

L'equivalenza del problema (P) col (7) - (10) è pressoché immediata. Invero, se u è soluzione del problema (P) e $v \in H^2(\Omega)$, dalla (6) si trae che, per ogni $\phi \in C_0^\infty(]0, T[)$,

$$\int_0^T (\mu u'(t), v)\, \phi'(t)\, dt = \int_0^T [a(u(t), v) + (ku^+(t), v) - (f(t), v)]\, \phi(t)\, dt.$$

Ciò significa che la funzione:

$$t \to (\mu u'(t), v)$$

appartiene ad $H^1(]0, T[)$ e risulta:

$$\frac{d}{dt}(\mu u'(t), v) = -[a(u(t), v) + (ku^+(t), v) - (f(t), v)] \text{ q.o. su }]0, T[.$$

Quanto ora detto equivale alle (7), (8) [1] (Appendice). Scelto poi ϕ in $C^1([0, T])$ con $\phi(0) = 1$ e $\phi(T) = 0$, utilizzando la (6) con $v(t) = \phi(t)\, v$ e tenendo conto delle (7), (8) si ottiene:

$$\langle (\mu u')(0), v \rangle = (\mu u_1, v),$$

cioè la (10). Il viceversa è ovvio.

TEOREMA 1. *Il problema (P) ammette una e una sola soluzione* u. *Inoltre per la* u *si ha:*

$$u \in C^0([0, T], H^2(\Omega)), \quad u' \in C^0([0, T], L^2(\Omega)), \quad u'(0) = u_1,$$

$$|\sqrt{\mu}\, u'(t)|^2 + a(u(t), u(t)) + |\sqrt{k}u^+(t)|^2 = |\mu u_1|^2 +$$

$$+ a(u_0, u_0) + |\sqrt{k}u_0^+|^2 + 2\int_0^t (f(s), u'(s))\, ds \quad \forall t \in [0, T]$$ (uguaglianza dell'energia).

Dim. Circa l'unicità, siano u e $\bar{u}$ soluzioni del problema (P). Poniamo $w = u - \bar{u}$ e, fissato s in $]0, T]$, $w_1(t) = \int_s^t w(\tau)\, d\tau$ per ogni $t \in [0, T]$. Tenendo conto della (8), intanto si ha:

$$\int_0^s <(\mu w')'(t), w_1(t)> dt + \int_0^s a(w(t), w_1(t))\, dt =$$
$$= \int_0^s (k(\bar{u}^+(t) - u^+(t)), w_1(t))\, dt.$$

D'altra parte, poiché $w_1 \in H^1(0, T; H^2(\Omega))$, si ha anche:

$$\int_0^s <(\mu w')'(t), w_1(t)> dt = - \int_0^s (\mu w'(t), w(t))\, dt = -\frac{1}{2} |\sqrt{\mu}\, w(s)|^2,$$
$$\int_0^s a(w(t), w_1(t))\, dt = -\frac{1}{2} a(w_1(0), w_1(0)).$$

Ne segue:

$$|\sqrt{\mu}\, w(s)|^2 \leqslant 2 \int_0^s (k(u^+(t) - \bar{u}^+(t)), w_1(t))\, dt \leqslant$$
$$\leqslant 2 \|k\|_{L^\infty(\Omega)} \int_0^s |w(t)| \left(\int_t^s |w(\tau)|\, d\tau\right) dt \leqslant \|k\|_{L^\infty(\Omega)} \left(\int_0^s |w(t)|\, dt\right)^2.$$

Dunque:

$$|w(s)| \leqslant \left(\frac{1}{\mu_0} \|k\|_{L^\infty(\Omega)}\right)^{1/2} \int_0^s |w(t)|\, dt \qquad \forall s \in [0, T],$$

e ciò, per il lemma di Gronwall, implica che $w(s) = 0 \quad \forall s \in [0, T]$.

Acquisita l'unicità della soluzione, ne dimostriamo l'esistenza utilizzando il metodo di Faedo-Galerkin. Applicazioni di questo metodo si trovano, ad esempio, in [2] (Cap. 3) nel caso di equazioni lineari, ed in [3] (Cap. 1), [4] (Cap. 1), [5] con riferimento ad alcuni problemi non lineari.

Sia dunque $\{v_j\}$ una successione di elementi di $H^2(\Omega)$ linearmente indipendenti tale che, indicato per ogni $m \in N$ con V_m il sottospazio generato da $\{v_1, \ldots, v_m\}$, si abbia:

$$\overline{\bigcup_{m \in N} V_m} = H^2(\Omega). \tag{11}$$

Denotiamo con u_{0m} e u_{1m} le proiezioni ortogonali su V_m rispettivamente di u_0 in $H^2(\Omega)$ e di u_1 in $L^2(\Omega)$, sicché:

$$u_{0m} = \sum_{1}^{m}{}_j\, \alpha_{jm} v_j, \qquad u_{1m} = \sum_{1}^{m}{}_j\, \beta_{jm} v_j \qquad (\alpha_{jm}, \beta_{jm} \in R),$$

e, per la (11):

$$u_{0m} \to u_0 \text{ in } H^2(\Omega), \tag{12}$$

$$u_{1m} \to u_1 \text{ in } L^2(\Omega). \tag{13}$$

Evidentemente, per ogni $m \in N$, esiste un'unica m-pla $(g_{1m}, \ldots, g_{mm}) \in H^2(0, T; R^m)$ soluzione del problema di Cauchy:

$$\sum_{1}^{m} {}_i (\mu v_i, v_j)\, g''_{im}(t) + \sum_{1}^{m} {}_i\, a(v_i, v_j)\, g_{im}(t) + (k[\sum_{1}^{m} {}_i\, g_{im}(t)\, v_i]^+, v_j) =$$

$$= (f(t), v_j) \quad \text{q.o. su }]0, T[,$$

$$g_{jm}(0) = \alpha_{jm},\ g'_{jm}(0) = \beta_{jm} \quad \forall j \in \{1, \ldots, m\}.$$

Posto allora $u_m(t) = \sum_{1}^{m} {}_i\, g_{im}(t)\, v_i$ per ogni $t \in [0, T]$, si ha:

$$u_m \in H^2(0, T; H^2(\Omega)),\ u_m(0) = u_{0m},\ u'_m(0) = u_{1m},$$

$$(\mu u''_m(t), v_j) + a(u_m(t), v_j) + (ku^+_m(t), v_j) =$$

$$= (f(t), v_j) \quad \forall j \in \{1, \ldots, m\} \quad \text{q.o. su }]0, T[. \tag{14}$$

Moltiplicando ambo i membri della (14) per $g'_{jm}(t)$ e sommando per $j = 1, \ldots, m$, si ottiene:

$$(\mu u''_m(t), u'_m(t)) + a(u_m(t), u'_m(t)) + (ku^+_m(t), u'_m(t)) =$$

$$= (f(t), u'_m(t)) \quad \text{q.o. su }]0, T[,$$

da cui:

$$|\sqrt{\mu}\, u'_m(t)|^2 + a(u_m(t), u_m(t)) + |\sqrt{k}u^+_m(t)|^2 = |\sqrt{k}u^+_{0m}|^2 +$$

$$+ |\sqrt{\mu}\, u_{1m}|^2 + a(u_{0m}, u_{0m}) + 2\int_0^t (f(s), u'_m(s))ds \ \forall t \in [0, T]. \tag{15}$$

La (15) ed il lemma di Gronwall, sussistendo le (12), (13), assicurano che:

$$\sup_{t \in [0,T]} |u'_m(t)| \leqslant c_2 \qquad \text{con } c_2 = c_2(f, u_0, u_1, \mu, k), \tag{16}$$

e che:

$$\sup_{t \in [0,T]} \| u_m(t) \| \leqslant c_3 \quad \text{con } c_3 = c_3(f, u_0, u_1, \mu, k) \tag{17}$$

non appena si tiene conto delle (4), (5) e della relazione:

$$u_m(t) = u_{0m} + \int_0^t u'_m(s)\, ds \quad \forall t \in [0, T]. \tag{18}$$

Le (16), (17) implicano l'esistenza di un $u \in L^2(0, T; H^2(\Omega)) \cap H^1(0, T; L^2(\Omega))$ e di una successione $\{m_h\}$ strettamente crescente di interi positivi tali che:

$$u_{m_h} \to u \quad \text{in } L^2(0, T; H^2(\Omega)) \text{ debolmente,} \tag{19}$$

$$u'_{m_h} \to u' \quad \text{in } L^2(0, T; L^2(\Omega)) \text{ debolmente} \tag{20}$$

ed inoltre:

$$u_{m_h} \to u \quad \text{in } C^0([0, T], L^2(\Omega)), \tag{21}$$

stante la (18) e la compattezza dell'immersione di $H^2(\Omega)$ in $L^2(\Omega)$.

Mostriamo che u è la soluzione del problema (P). A partire dalla (14) e utilizzando le (19), (20), (21), si perviene alla relazione:

$$- \int_0^T (\mu u'(t), v_j)\, \phi'(t)\, dt + \int_0^T a(u(t), v_j)\, \phi(t)\, dt + \int_0^T (ku^+(t), v_j) \phi(t)\, dt =$$
$$= \int_0^T (f(t), v_j)\, \phi(t)\, dt \quad \forall j \in N \quad \text{e} \quad \forall \phi \in C_0^\infty(]0, T[),$$

la quale, tenendo presente la (11), vale anche se si sostituisce v_j con un qualsiasi $v \in H^2(\Omega)$. Dunque u soddisfa alle (7), (8). La (9) è conseguenza immediata delle (12), (21). Circa la (10), scelto ϕ in $C^1([0, T])$ con $\phi(0) = 1$ e $\phi(T) = 0$, e osservato che:

$$(\mu u_{1m_h}, v_j) = (\mu u'_{m_h}(0), \phi(0) v_j) =$$
$$= - \int_0^T (\mu u''_{m_h}(t), \phi(t) v_j)\, dt - \int_0^T (\mu u'_{m_h}(t), \phi'(t) v_j)\, dt,$$

si ha, in virtù delle (7), (8), (14), (19), (20), (21):

$$(\mu u_{1m_h}, v_j) \to \langle (\mu u')(0), v_j \rangle \quad \text{per } h \to +\infty .$$

Pertanto dev'essere, per la (13):

$$\langle (\mu u')(0), v_j \rangle = (\mu u_1, v_j) \ \forall j \in N,$$

e di qui la (10) a causa della (11).

Per quanto attiene alle proprietà della soluzione u precisate nell'enunciato, introdotto in $L^2(\Omega)$ il prodotto scalare:

$$(v, w)_\mu = \int_\Omega v\, w\, \mu\, d x \qquad \forall v, w \in L^2(\Omega),$$

che genera una norma equivalente a quella usuale, si ha che u è la soluzione del problema:

$$z \in L^2(0, T; H^2(\Omega)) \cap H^1(0, T; L^2(\Omega)),\ z(0) = u_0, \tag{22}$$

$$\int_0^T [a(z(t), v(t)) - (z'(t), v'(t))_\mu + (\mu^{-1}(ku^+(t) - f(t)), v(t))_\mu]\, dt =$$

$$= (u_1, v(0))_\mu \quad \forall v \in L^2(0, T; H^2(\Omega)) \cap H^1(0, T; L^2(\Omega)) \text{ con } v(T) = 0. \tag{23}$$

Le suddette proprietà seguono allora da un teorema di Torelli [6], [2] (Cap. 3, teorema 8.2).

Denotiamo con W lo spazio di Banach delle $v \in C^0([0, T], H^2(\Omega)) \cap C^1([0, T], L^2(\Omega))$, la norma di $v \in W$ essendo definita la relazione:

$$\| v \|_W = \| v \|_{C^0([0,T], H^2(\Omega))} + \| v' \|_{C^0([0,T], L^2(\Omega))};$$

denotiamo ancora, per ogni $(f, u_0, u_1) \in L^2(0, T; L^2(\Omega)) \times H^2(\Omega) \times L^2(\Omega)$, con $F(f, u_0, u_1)$ la corrispondente soluzione del problema (P).

TEOREMA 2. *L'operatore* F *è continuo dallo spazio* $L^2(0, T; L^2(\Omega)) \times H^2(\Omega) \times L^2(\Omega)$ *allo spazio* W.

Dim. Il procedimento dimostrativo si basa sostanzialmente sulla uguaglianza dell'energia, cui soddisfa la soluzione di un'equazione lineare del secondo ordine in t [6], [2] (Cap. 3, lemma 8.3).

Siano, infatti, $(f, u_0, u_1) \in L^2(0, T; L^2(\Omega)) \times H^2(\Omega) \times L^2(\Omega)$ e $\{(f_n, u_{0n}, u_{1n})\}$ una successione di elementi dello stesso spazio tali che:

$$\begin{aligned} & f_n \to f \quad \text{in } L^2(0, T; L^2(\Omega)), \\ & u_{0n} \to u_0 \quad \text{in } H^2(\Omega), \\ & u_{1n} \to u_1 \quad \text{in } L^2(\Omega). \end{aligned} \tag{24}$$

Posto $u = F(f, u_0, u_1)$ e $u_n = F(f_n, u_{0n}, u_{1n})$, poiché $u - u_n$ è la soluzione del problema (22), (23) con $u_0 - u_{0n}$, $u_1 - u_{1n}$, $k(u^+ - u_n^+)$ e $f - f_n$ in luogo rispettivamente di u_0, u_1, ku^+ e f, vale per essa l'uguaglianza dell'energia:

$$|\sqrt{\mu}(u'(t) - u_n'(t))|^2 + a(u(t) - u_n(t), u(t) - u_n(t)) =$$

$$= |\sqrt{\mu}(u_1 - u_{1n})|^2 + a(u_0 - u_{0n}, u_0 - u_{0n}) +$$

$$+ 2\int_0^t (f(s) - f_n(s), u'(s) - u'_n(s))\, ds +$$

$$+ 2\int_0^t (k(u_n^+(s) - u^+(s)), u'(s) - u'_n(s))\, ds \qquad \forall t \in [0, T].$$

Pertanto, tenendo conto della (5), si ha:

$$\mu_0 |u'(t) - u'_n(t)|^2 + c_1 \| u(t) - u_n(t) \|^2 \leqslant$$

$$\leqslant |\sqrt{\mu}(u_1 - u_{1n})|^2 + a(u_0 - u_{0n}, u_0 - u_{0n}) + 2|u_0 - u_{0n}|^2 +$$

$$+ \| f - f_n \|^2_{L^2(0,T;L^2(\Omega))} + (1 + \| k \|_{L^\infty(\Omega)} + 2T) \int_0^t (\| u(s) - u_n(s) \|^2 +$$

$$+ |u'(s) - u'_n(s)|^2)\, ds \quad \forall t \in [0, T],$$

e di qui, sfruttando il lemma di Gronwall e le (24):

$$u_n \to u \qquad \text{in } W.$$

2. Al fine di stabilire che la soluzione del problema (P) è anche soluzione del problema (1), (2), (3), basta ovviamente dimostrare il seguente teorema di regolarità:

TEOREMA 3. *Si supponga* Ω *di classe* $C^{3,1}$, $u_0 \in H^4(\Omega)$, $u_1 \in H^2(\Omega)$. *Se*

i) $f \in H^1(0, T; L^2(\Omega))$,

ii) $\mathcal{M} u_0 = 0 = \mathcal{T} u_0$,

allora per la soluzione u *del problema (P) si ha:*

$$u \in L^\infty(0, T; H^4(\Omega)) \cap H^{1,\infty}(0, T; H^2(\Omega)) \cap H^{2,\infty}(0, T; L^2(\Omega)).$$

Dim. La dimostrazione si ispira al procedimento introdotto per le equazioni lineari [7] (Cap. 5). Essa infatti si fonda sul metododi Faedo-Galerkin e su un noto risultato di regolarità relativo alle equazioni ellittiche. Procederemo come è stato fatto per il teorema 1, scegliendo $\{ v_j \}$ in modo che u_0 appartenga a V_1 con la precisazione che ora u_{1m} denota la proiezione ortogonale di u_1 su V_m in $H^2(\Omega)$, sicché:

$$u_{0m} = u_0 \ \forall m \in N, \tag{25}$$

$$u_{1m} \to u_1 \qquad \text{in } H^2(\Omega). \tag{26}$$

Ciò premesso, rileviamo dapprima che:

$$u_m^+ \in H^1(0, T; L^2(\Omega)), \tag{27}$$

$$\left| \frac{d}{dt} u_m^+ (t) \right| \leqslant | u'_m (t) | \qquad \text{q.o. su }]\,0, T\,[, \tag{28}$$

e che per la (27) e la i):

$$u_m \in H^3 (0, T; H^2 (\Omega)).$$

Dunque ambo i membri della uguaglianza:

$$(\mu\, u''_m (t), v_j) + a (u_m (t), v_j) + (k u_m^+ (t), v_j) =$$
$$= (f (t), v_j) \qquad \forall j \in \{ 1, \ldots, m \}, \tag{29}$$

sono assolutamente continui in [0, T] e l'uguaglianza stessa è valida per ogni $t \in [0, T]$. A partire dalla (29) si ottengono la (16) e le relazioni:

$$| \sqrt{\mu}\, u''_m (t) |^2 + a (u_m (t), u''_m (t)) + (k u_m^+ (t), u''_m (t)) =$$
$$= (f (t), u''_m (t)) \qquad \forall t \in [0, T], \tag{30}$$

$$(\mu\, u'''_m (t), u''_m (t)) + a (u'_m (t), u''_m (t)) + (k \frac{d}{dt} u_m^+ (t), u''_m (t)) =$$
$$= (f' (t), u''_m (t)) \qquad \text{q.o. su }]\,0, T\,[. \tag{31}$$

La (30), unitamente alla ii) e alla (25), dà luogo alla maggiorazione:

$$| u''_m (0) | \leqslant 1/\mu_0 \, [\, | f (0) | + | \Delta\Delta\, u_0 | + \| k \|_{L^\infty(\Omega)} \cdot | u_0^+ | \,];$$

dalla (31), tenendo conto delle (5), (16), (26), (28) e della disuguaglianza precedente, si deduce che:

$$| u''_m (t) |^2 + \| u'_m (t) \|^2 \leqslant c_4 + (1 + \| k \|_{L^\infty(\Omega)}) \int_0^t [\, | u''_m (s) |^2 +$$
$$+ \| u'_m (s) \|^2 \,]\, ds \qquad \forall t \in [0, T], \text{ con } c_4 = c_4 (f, u_0, u_1, \mu, k),$$

e di qui, in virtù del lemma di Gronwall:

$$\sup_{t \in [0,T]} [\, | u''_m (t) | + \| u'_m (t) \| \,] \leqslant c_5, \text{ con } c_5 = c_5 (f, u_0, u_1, \mu, k). \tag{32}$$

E' facile allora constatare l'esistenza di $\tilde{u} \in H^1(0, T; H^2(\Omega)) \cap H^2(0, T; L^2(\Omega))$ e di una successione $\{m_h\}$ strettamente crescente di interi positivi tali che:

$$u_{m_h} \to \tilde{u} \quad \text{in } L^2(0, T; H^2(\Omega)) \quad \text{debolmente},$$

$$u'_{m_h} \to \tilde{u}' \quad \text{in } L^2(0, T; H^2(\Omega)) \quad \text{debolmente},$$

$$u''_{m_h} \to \tilde{u}'' \quad \text{in } L^2(0, T; L^2(\Omega)) \quad \text{debolmente},$$

$$u_{m_h} \to \tilde{u} \quad \text{in } C^1([0, T], L^2(\Omega)).$$

Per quanto si è visto nel corso della dimostrazione del teorema 1, possiamo quindi concludere che $\tilde{u} = u$ e che:

$$u \in H^{1,\infty}(0, T; H^2(\Omega)) \cap H^{2,\infty}(0,T; L^2(\Omega)),$$

non appena si tiene conto della (32). Il fatto poi che $u''(t) \in L^2(\Omega)$ consente di scrivere la (8) nella forma:

$$(\mu u''(t), v) + a(u(t), v) + (ku^+(t), v) =$$
$$= (f(t), v) \qquad \forall v \in H^2(\Omega) \qquad \text{q.o. su }]0, T[. \tag{33}$$

Resta da mostrare che:

$$u \in L^\infty(0, T; H^4(\Omega)). \tag{34}$$

A tale scopo, detto P_1 lo spazio dei polinomi su Ω al più di primo grado, e indicati con $P_1^\perp$ e P_1^* gli ortogonali di P_1 rispettivamente in $L^2(\Omega)$ e in $H^4(\Omega)$, sussistendo la (4) e tenendo conto che [8] (Cap. 2):

$$c' [\sum_{|r|=2} |D^r v|^2]^{1/2} \leq \| [v] \|_{H^2(\Omega)/P_1} \leq c'' [\sum_{|r|=2} |D^r v|^2]^{1/2} \qquad \forall [v] \in H^2(\Omega)/P_1,$$

con c' e c'' costanti positive indipendenti da v, per ogni $g \in P_1^\perp$ l'equazione variazionale:

$$z \in H^2(\Omega) \; : \; a(z, v) = (g, v) \qquad \forall v \in H^2(\Omega),$$

ammette infinite soluzioni il cui insieme costituisce un elemento $[z]$ di $H^4(\Omega)/P_1$ e che l'applicazione Ψ:

$$g \to [z],$$

è un isomorfismo continuo di $P_1^{\perp}$ in $H^4(\Omega)/P_1$ [2] (Cap. 2). Quanto ora asserito, stante la (33), assicura che:

$$u(t) \in H^4(\Omega) \qquad \text{q.o. su }]0, T[,$$

e che, posto $u(t) = \bar{u}(t) + \bar{p}(t)$, con $\bar{u}(t) \in P_1^{\circ}$ e $\bar{p}(t) \in P_1$, risulta:

$$\bar{u}(t) = J \circ \Psi(-\mu u''(t) - ku^{+}(t) + f(t)) \qquad \text{q.o. su }]0, T[, \tag{35}$$

essendo J l'isomorfismo canonico di $H^4(\Omega)/P_1$ su P_1°. La (35) implica che:

$$\bar{u} \in L^{\infty}(0, T; H^4(\Omega)).$$

Da ciò e dall'appartenenza di u a $L^{\infty}(0, T; H^2(\Omega))$ si trae che:

$$\bar{p} \in L^{\infty}(0, T; H^2(\Omega)),$$

nonché, conseguentemente:

$$\bar{p} \in L^{\infty}(0, T; H^4(\Omega)).$$

La (34) è così acquisita, e con essa il teorema.

OSSERVAZIONE 3. Il metodo seguito per dimostrare i teoremi 1, 2, 3 consente ovvie generalizzazioni. Ad esempio i risultati ottenuti sussistono ancora se Ω è un aperto di R^n e se si sostituisce all'operatore biarmonico l'operatore:

$$A = \sum_{\substack{|r|=2 \\ |s|=2}} D^s (a_{rs} D^r) \qquad (a_{rs} \in L^{\infty}(\Omega) \text{ e } a_{rs} = a_{sr}),$$

soddisfacente alla condizione (4) con:

$$a(v, z) = \sum_{\substack{|r|=2 \\ |s|=2}} \int_{\Omega} a_{rs} D^r v D^s z \, dx \qquad \forall v, z \in H^2(\Omega).$$

$\boldsymbol{M}$ e $\boldsymbol{T}$ denoteranno allora gli operatori frontiera definiti nell'insieme:

$$\{ v \in H^2(\Omega), | \, A v \in L^2(\Omega) \ \text{ in senso debole} \}, \tag{36}$$

a valori rispettivamente in $H^{-1/2}(\partial \Omega)$ e $H^{-3/2}(\partial \Omega)$, tali da verificare la formula di Green:

$$a(v, z) = \int_{\Omega} z \, Av dx + \langle \boldsymbol{M} v, \gamma_1 z \rangle + \langle \boldsymbol{T} v, \gamma_0 z \rangle,$$

qualunque siano $z \in H^2(\Omega)$ e v nell'insieme (36), dove γ_i è l'operatore traccia di ordine i. Evidentemente per la validità del teorema 3 basta supporre $a_{rs} \in C^{1\,1}(\bar{\Omega})$.

RIFERIMENTI

1. Brézis, H., *Opérateurs Maximaux Monotones et Semigroupes de Contractions dans les Espace de Hilbert,* Math. Studies 5, North-Holland, 1973.

2. Lions, J.L., et Magenes, E., *Problèmes aux Limites non Homogènes et Applications,* Vol. 1, Dunod, 1968.

3. Duvaut, G., et Lions, J.L., *Les Inéquations en Mécanique et en Physique,* Dunod, Paris, 1972.

4. Lions, J.L., *Quelques Méthodes de Résolution des Problèmes aux Limites non Linéaires,* Dunod, Gauthier-Villars, Paris, 1969.

5. Lions, J.L., and Strauss, W.A., Some non-linear evolution equations, *Bull. Soc. Math. Fr.,* t. 93, p. 43-96, 1965.

6. Torelli, G., Un complemento ad un teorema di J.L. Lions sulle equazioni differenziali astratte del secondo ordine, *Rend. Sem. Mat. Univ. Padova,* t. 34, 1964, p. 224-241.

7. Lions, J.L., et Magenes. E., *Problèmes aux Limites non Homogènes et Applications,* Vol. 2., Dunod, 1968.

8. Nečas, J., *Les Méthodes directes en théorie des équations elliptiques,* Masson, 1967.

Printed in the United States
By Bookmasters